SELENIUM IN THE ENVIRONMENT AND HUMAN HEALTH

PROCEEDINGS OF THE 3RD INTERNATIONAL CONFERENCE ON SELENIUM IN THE ENVIRONMENT AND HUMAN HEALTH, HEFEI, CHINA, 10–14 NOVEMBER 2013

Selenium in the Environment and Human Health

Editors

G.S. Bañuelos
US Department of Agriculture-Aricultural Research Service, Parlier, CA, USA

Z.-Q. Lin
Department of Biological Sciences & Environmental Sciences Program, Southern Illinois University, Edwardsville, IL, USA

Xuebin Yin
Suzhou Institute for Advanced Study, University of Science and Technology of China, Suzhou, China

CRC Press
Taylor & Francis Group
Boca Raton London New York

CRC Press is an imprint of the
Taylor & Francis Group, an **informa** business
A BALKEMA BOOK

Published by:
CRC Press/Balkema
P.O. Box 447, 2300 AK Leiden, The Netherlands
e-mail: Pub.NL@taylorandfrancis.com
www.crcpress.com – www.taylorandfrancis.com

Typeset by V Publishing Solutions Pvt Ltd., Chennai, India

CRC Press/Balkema is an imprint of the Taylor & Francis Group, an informa business

ISBN 13: 978-1-138-00017-9 (hbk)
ISBN 13: 978-0-203-77141-9 (ebk)

Typeset by V Publishing Solutions Pvt Ltd., Chennai, India

Cover photo:
Full moon and rape board © Klagyivik Viktor
Courtesy of Shutterstock

Selenium in the Environment and Human Health – Bañuelos, Lin & Yin (Eds)
© 2014 Taylor & Francis Group, London, ISBN 978-1-138-00017-9

Table of contents

5 *Cellular and molecular functions of selenium*

6 *Selenium pollution and control*

Selenium in the Environment and Human Health – Bañuelos, Lin & Yin (Eds)
© 2014 Taylor & Francis Group, London, ISBN 978-1-138-00017-9

Preface

Although selenium was shown to affect biological systems over 80 years ago and was recognized as an essential micronutrient in the mid-20th century, the "flowering" of the selenium field has taken place in the 21st century. The biological function of selenium through selenoproteins has been firmly established and selenoproteomes have been characterized in many species. Scientists throughout the world are at work determining the function of selenoproteins and, thereby, advancing knowledge of physiology as well as uncovering disease mechanisms.

The importance of selenium in the environment has been recognized. Many areas of the world are selenium deficient, requiring selenium supplementation programs for animals and humans. Other areas have high selenium bioavailability and produce agricultural products that are rich in the element. Selenium mobility and excessive accumulation in the environment, often anthropogenic, can lead to illness in animals and humans.Thus, managing its impact on the biological, physical and ecological environment is also of great importance for the selenium research community.

To evaluate the many roles of selenium in today's complex world, the 3rd International Conference on Selenium in the Environment and Human Health will be held on 10-14 November 2013 in Hefei, China, and will provide an interactive platform for individual and multi-disciplinary selenium researchers to discuss many important topics in the selenium field. Some primary objectives of the conference are to share and evaluate the most recent research findings obtained worldwide and to relate various research areas of selenium to each other. A total of 90 extended abstracts have been peer reviewed and are included in these proceedings entitled "Selenium in the Environment and Human Health". These scientific papers cover a variety of selenium research fields—geology, biogeochemistry, selenium deficiency and toxicity, selenium pollution control and remediation, selenium interaction with pollutants, selenoproteins, selenium-related human and animal health, and development of selenium-enriched agricultural products—just to name a few. I am also happy to see there are many research projects that were conducted by graduate students and many from developing countries. The cooperation among selenium researchers from around the world is also evident from the research papers published in this book. For this, I would like to acknowledge the effort made by the council of the International Society for Selenium Research and the conference organizers.

Raymond F. Burk, M.D.
Professor of Medicine
Vanderbilt University
Nashville, TN, USA

Selenium in the Environment and Human Health – Bañuelos, Lin & Yin (Eds)
© 2014 Taylor & Francis Group, London, ISBN 978-1-138-00017-9

Acknowledgements

The editors would like to acknowledge the contribution from the council members of the International Society for Selenium Research: Joel Caton (North Dakota State University, USA), Lutz Schomburg (Charité Medical Hospital Berlin, Germany), Martin Broadley (University of Nottingham, UK), Allan Chilimba (Ministry of Agriculture and Food Security, Malawi), Karaj S. Dhillon (Punjab Agriculture University, India), Gijs Du Laing (Ghent University, Belgium), Graham Lyons (University of Adelaide, Australia) and Milton Ferreira de Moraes (Ferderaz University of Parana, Brazil). We are thankful for the encouragement and guidance from the honorary members of the International Society for Selenium Research, Raymond Burk (Vanderbilt University, USA) and Steve McGrath (Rothamsted Research, UK). We also would like to thank Ning Duan (Chinese Research Academy of Environmental Science) and Linxi Yuan (University of Science and Technology of China) for their assistance with the conference and the proceedings.

We deeply appreciate the gracious financial support provided by the Chinese Academy of Science, the Chinese Research Academy of Environmental Science, the National Science Foundation of China, and the University of Science and Technology of China. We hope you enjoy the academic atmosphere and social friendships at this conference. We also look forward to meeting you again at the 4th International Conference on Selenium in the Environment and Human Health that will be held in Cuiaba, Brazil in 2015.

G.S. Bañuelos
Z.-Q. Lin
Xuebin Yin

 ISBN 978-1-138-00017-9

List of contributors

Abou Faycal, C. Ibrahim: Radiobiology Group, UMR1052, Cancer Research Centre, Lyon, France
Albrecht, J.C.: Embrapa CPAC, Planaltina, DF, Brazil
Alfthan, G.: Department of Chronic Disease Prevention, National Institute for Health and Welfare, Helsinki, Finland; georg.alfthan@thl.fi
Alifar, N.: Land Management Department, Faculty of Agriculture, University Putra Malaysia; alifar_n@yahoo.com
Ander, E.L.: British Geological Survey, Keyworth, Nottingham, NG12 5GG, UK
Ardelli, F.: ISTerre, University Joseph Fourier, CNRS, BP 53, F-38041 Grenoble Cedex 1, France
Arnér, Elias S.J.: Division of Biochemistry, Department of Medical Biochemistry and Biophysics, Karolinska Institutet, SE-171 77 Stockholm, Sweden; elias.arner@ki.se
Awang, Y.: Crop Science Department, Faculty of Agriculture, University Putra Malaysia, Malaysia

Bai, Guihua: USDA-ARS, Hard Winter Wheat Genetics Research Unit, Manhattan, KS 66506, USA
Banning, H.M.: Karlsruhe Institute of Technology, Institute of Mineralogy & Geochemistry, Karlsruhe, Germany; helena.banning@kit.edu
Bañuelos, G.S.: USDA-ARS, Parlier, California, USA; gbanuelos@fresno.ars.usda.gov
Bao, Zhengyu: Faculty of Earth Science, China University of Geosciences, Wuhan 430074, China; zybao@cug.edu.cn
Becker, Niels-Peter: Institute for Experimental Endocrinology, Charité-Medical University Berlin, Germany; niels-peter.becker@charite.de
Berg, Michael: Eawag, Swiss Federal Institute of Aquatic Science and Technology, 8600 Dübendorf, Switzerland
Bian, Jianchao: Shandong Institute for Endemic Disease Control and Research, Jinan, Shandong, China
Black, C.R.: University of Nottingham, School of Biosciences, Sutton Bonington Campus, Loughborough, LE12 5RD, UK
Blazina, Tim: Eawag, Swiss Federal Institute of Aquatic Science and Technology 8600 Dübendorf, Switzerland; tim.blazina@eawag.ch
Bochot, C.: ISTerre, University Joseph Fourier, CNRS, BP 53, F-38041 Grenoble Cedex 1, France
Bodgi, L.: INSERM, Radiobiology Group, UMR1052, Cancer Research Centre, Lyon, France
Brendel, R.: Molecular Cell Biology, Botanical Institute, and Center for Functional Nanostructures, Karlsruhe Institute of Technology, Kaiserstraße 2, D-76131 Karlsruhe, Germany
Broadley, M.R.: University of Nottingham, School of Biosciences, Sutton Bonington Campus, Loughborough, LE12 5RD, UK
Bueno, M.: Laboratory of Bioinorganic Analytical and Environmental Chemistry (LCABIE), University of Pau et des Pays de l'Adour/CNRS, UMR 5254, IPREM, Hélioparc, Pau Cedex 9, France

Cao, Junling: Institute of Endemic Diseases, College of Medicine, Xi'an Jiaotong University, Xi'an 710061, Shaanxi, China; Key Laboratory of Environment and Genes Related to Diseases (Xi'an Jiaotong University), Ministry of Education, China; Key Laboratory of Microelement and Endemic Disease (Xi'an Jiaotong University), Ministry of Health, China; caojl@mail.xjtu.edu.cn
Cao, X.M.: Research Center on Life Sciences and Environmental Sciences, Harbin University of Commerce, Harbin, Heilongjiang, China
Carvalho, J.G.: Federal University of Lavras, Lavras, MG, Brazil
Castrec-Rouelle, M.: UPMC, UMR 7618 BioEMCo, PO Box 120, 4 place Jussieu, 75252 Paris cedex 05, France; maryse.rouelle@upmc.fr

Caterson, Bruce: Laboratory of Connective Tissue Biology, School of Biosciences, Cardiff University, Cardiff, CF10 3US, UK
Caton, J.S.: Center for Nutrition and Pregnancy, Animal Sciences Department, North Dakota State University, ND 58108, USA; joel.caton@ndsu.edu
Charlet, L.: ISTerre, University Joseph Fourier, CNRS, BP 53, F-38041 Grenoble Cedex 1, France; charlet38@gmail.com
Chen, Jinghong: Institute of Endemic Diseases, College of Medicine, Xi'an Jiaotong University, Xi'an 710061, Shaanxi, China; Key Laboratory of Environment and Genes Related to Diseases (Xi'an Jiaotong University), Ministry of Education, China; Key Laboratory of Microelement and Endemic Disease (Xi'an Jiaotong University), Ministry of Health, China
Chen, P.: College of Life Sciences, Shenzhen Key Laboratory of Microbial Genetic Engineering, Shenzhen University, Shenzhen 518060, China
Chen, X.L.: College of Material Science and Chemical Engineering, China University of Geosciences & State Key Laboratory of Geological and Environmental Geological, Wuhan 430074, China
Cheng, N.: Research Center on Life Sciences and Environmental Sciences, Harbin University of Commerce, Harbin, Heilongjiang, China
Chilimba, A.D.C.: Ministry of Agriculture and Food Security, Department of Agricultural Research Services, Lunyangwa Research Station, P.O. Box 59, Mzuzu, Malawi; achilimba@gmail.com
Christophersen, O.A.: Department of Animal and Aquacultural Sciences, The Norwegian University of Life Sciences, 1432 Aas, Norway
Chu, J.Z.: College of Life Sciences, Hebei University, Baoding 071002, China

Dan, Zhigang: Technology Center for Heavy Metal Cleaner Production Engineering, Chinese Research Academy of Environmental Sciences, Beijing 100012, China
Deng, Yongjin: Sericulture Research Institute, Anhui Academy of Agricultural Sciences, Hefei 230031, Anhui, China; plantprotection2006@126.com
Devic, C.: INSERM, Radiobiology Group, UMR1052, Cancer Research Centre, Lyon, France
Domingos, C.R.S.: Federal University of Parana, Curitiba-PR, Brazil
Dong, Fang: Research Center on Life Sciences and Environmental Sciences, Harbin University of Commerce, Harbin 150076, China; Engineering Research Center of Natural Anticancer Drugs, Ministry of Education, Harbin 150076, China
Du Laing, G.: Ghent University, Ghent, Belgium; gijs.dulaing@ugent.be
Du, Xiubo: College of Life Sciences, Shenzhen Key Laboratory of Microbial Genetic Engineering, Shenzhen University, Shenzhen 518060, China; xiubo.du@gmail.com
Duan, Ning: Technology Center for Heavy Metal Cleaner Production Engineering, Chinese Research Academy of Environmental Sciences, Beijing 100012, China; ningduan@craes.org.cn

Eiche, E.: Institute of Mineralogy and Geochemistry, Karlsruhe Institute of Technology, Adenauerring 20b, D-76131 Karlsruhe, Germany
Esaki, N.: Institute for Chemical Research, Kyoto University, Uji, Kyoto 611-0011, Japan
Evangelista, A.: Center for Agricultural Research, 85813-450, Cascavel-PR, Brazil

Faisal, M.: Department of Microbiology and Molecular Genetics, University of the Punjab, Quaid-e-Azam Campus, Lahore-54590, Pakistan
Feng, Xin-Bin: Institute of Geochemistry, Chinese Academy of Sciences, Guiyang, 550002, China; fengxinbin@vip.skleg.cn
Ferlazzo, M.: INSERM, Radiobiology Group, UMR1052, Cancer Research Centre, Lyon, France
Foray, N.: INSERM, Radiobiology Group, UMR1052, Cancer Research Centre, Lyon, France
Franco, F.A.: Center for Agricultural Research, 85813-450, Cascavel-PR, Brazil
Fu, Qiang: Institute of Endemic Diseases, College of Medicine, Xi'an Jiaotong University, Xi'an 710061, Shaanxi, China; Key Laboratory of Environment and Genes Related to Diseases (Xi'an Jiaotong University), Ministry of Education, China; Key Laboratory of Microelements and Endemic Disease (Xi'an Jiaotong University), Ministry of Health, China

Gailer, J.: University of Calgary, Calgary, Alberta, Canada
Gao, Xiaojuan: Technology Center for Heavy Metal Cleaner Production Engineering, Chinese Research Academy of Environmental Sciences, Beijing 100012, China
George, G.N.: University of Saskatchewan, Saskatoon, Saskatchewan, Canada

Gondwe, C.: School of Agriculture & Wine, University of Adelaide, Waite Campus, Glen Osmond, SA 5064, Australia
Gong, Pan: Jiangsu Provincial Key Laboratory of Crop Genetics and Physiology; Key Laboratory of Plant Functional Genomics of Ministry of Education, Yangzhou University, Yangzhou, China
Granzotto, A. INSERM, Radiobiology Group, UMR1052, Cancer Research Centre, Lyon, France
Gu, Shiliang: Jiangsu Provincial Key Laboratory of Crop Genetics and Physiology; Key Laboratory of Plant Functional Genomics of Ministry of Education, Yangzhou University, Yangzhou 225009, China
Guan, Wenwen: Jiangsu Bio-Engineering Research Centre on Selenium, Suzhou, Jiangsu 215123, China
Guilherme, L.R.G. Federal University of Lavras, Postal Code 37200-000, Lavras-MG, Brazil; guilherm@dcs.ufla.br
Guo, L.: College of Resources and Environment, Northwest A & F University, Yangling, Shaanxi, 712100 China
Guo, Wenlin: Technology Center for Heavy Metal Cleaner Production Engineering, Chinese Research Academy of Environmental Sciences, Beijing 100012, China
Guo, Y.: Guangzhou Institute of Geological Survey, Guangzhou 510440, China

Haddad, S.: Department of Biological Sciences, Southern Illinois University, Edwardsville, Illinois 62026, USA
Hammer, C.J.: Center for Nutrition and Pregnancy, Animal Sciences Department, North Dakota State University, ND 58108, USA
Haug, A.: Department of Animal and Aquacultural Sciences, The Norwegian University of Life Sciences, 1432 Aas, Norway
Hong, J.: Environmental Sciences Program, Southern Illinois University, Edwardsville, Illinois 62026, USA
Hong, Xin: Anhui Normal University, Wuhu, Anhui 241000, China
Hou, Q.Y.: Chinese University of Geosciences, Beijing, China
Hu, Y.: Flinders Centre for Innovation in Cancer, Flinders University of South Australia, Bedford Park South Australia 5042, Australia
Huang, Zhen: Department of Chemistry & Department of Biology, Georgia State University, Atlanta, GA 30303, USA; huang@gsu.edu

Ishak, C.F.: Land Management Department, Faculty of Agriculture, University Putra Malaysia, Malaysia

Jain, V.K.: Chemistry Division, Bhabha Atomic Research Centre, Trombay, Mumbai-400085, India
Ji, B.Y.: The 5th Institute of Geology and Mineral Exploration of Qinghai, Xining, Qinghai, China
Ji, Yu-Bin: Research Center on Life Sciences and Environmental Sciences, Harbin University of Commerce, Harbin 150076, China; Engineering Research Center of Natural Anticancer Drugs, Ministry of Education, Harbin 150076, China; jyb@hrbcu.edu.cn
Jiang, Lin-Hua: Technology Center for Heavy Metal Cleaner Production Engineering, Chinese Research Academy of Environmental Sciences, Beijing 100012, China
Jiang, Xia: Department of Nutrition and Food Hygiene, School of Public Health, Soochow University, China
Jiang, Yanyan: College of Public Health, Shandong University, Jinan, Shandong, China
Jin, L.X.: Sichuan Institute of Geological Survey, Chengdu, Sichuan, China
Johnson, Annette C.: Eawag, Swiss Federal Institute of Aquatic Science and Technology 8600 Dübendorf, Switzerland
Johnson, T.M.: Department of Geology, University of Illinois at Urbana-Champaign, Urbana, Illinois 61801, USA
Jones, Leighann: Department of Biological Sciences, Southern Illinois University, Edwardsville, Illinois 62026, USA
Joy, E.J.M.: School of Biosciences, University of Nottingham, Sutton Bonington, Leicestershire LE12 5RD, UK

Kabiru, E.: School of Public Health, Kenyatta University, Kenya
Kong, Juanjuan: Anhui Normal University, Wuhu, Anhui 241000, China
Konrad, G.: Institute of Mineralogy and Geochemistry, Karlsruhe Institute of Technology, Adenauerring 20b, D-76131 Karlsruhe, Germany

Kosmala, R.: Institute of Mineralogy and Geochemistry, Karlsruhe Institute of Technology, Adenauerring 20b, D-76131 Karlsruhe, Germany
Kunwar, A.: Radiation & Photochemistry Division, Bhabha Atomic Research Centre, Trombay, Mumbai-400085, India
Kurihara, T.: Institute for Chemical Research, Kyoto University, Uji, Kyoto 611-0011, Japan

La Porte, P.F.: Emory University, Atlanta, Georgia, USA
Lammel, J.: Yara International, Research Centre, Hanninghof 48249, Duelmen, Germany
Lan, Guofang: Jiangsu Provincial Key Laboratory of Crop Genetics and Physiology; Key Laboratory of Plant Functional Genomics of Ministry of Education, Yangzhou University, Yangzhou 225009, China
Larsen, R.: Intermountain Farmers Association, Salt Lake City, Utah 84104, USA
Lassen, Thorjørn: Norwegian Institute for Water Research, Gaustadalléen 21, 0349 Oslo, Norway
Lavu, R.V.S.: Ghent University, Ghent, Belgium
Le Hécho, I.: Laboratory of Bioinorganic Analytical and Environmental Chemistry (LCABIE), University of Pau et des Pays de l'Adour/CNRS, UMR 5254, IPREM, Hélioparc, Pau Cedex 9, France
Lei, Hongling: Key Laboratory of Biological Resources Protection and Utilization of Hubei Province, China; College of Biological Sciences and Technology, Hubei University for Nationalities, Enshi, Hubei, China; leihl123366@163.com
Lenz, Markus: University of Applied Sciences and Arts Northwestern Switzerland, Institute for Ecopreneurship, CH-4132 Muttenz, Switzerland; Sub-Department of Environmental Technology, Wageningen University, 6700 EV Wageningen, The Netherlands
Leslie, E.M.: University of Alberta, Edmonton, Alberta, Canada
Li, Dan: Nano Science and Technology Institute & Advanced Lab for Selenium and Human Health, University of Science and Technology of China, Suzhou, Jiangsu 215123, China
Li, Fei: Jiangsu Bio-Engineering Research Centre on Selenium, Suzhou, Jiangsu 215123, China
Li, J.: College of Life Science, Shenzhen Key Laboratory of Microbial Genetic Engineering, Shenzhen University, Shenzhen 518060, China
Li, Jin: Key Laboratory of Chemical Synthesis and Pollution Control, Sichuan Province, China; College of Chemistry and Chemical Engineering, China West Normal University, Nanchong, Sichuan, China
Li, L.: USDA-ARS & Cornell University, Ithaca, NY, USA
Li, Miao: Key Laboratory of Agri-Food Safety of Anhui Province, School of Plant Protection, Anhui Agriculture University, Hefei, Anhui 230036, China; miaoli@ustc.edu.cn
Li, Qin: School of Biological Science and Technology, Hubei Institute for Nationalities, Hubei 445000, China; Key Laboratory of Biologic Resources Protection and Utilization of Hubei Province, Enshi, Hubei, China;
Li, Tao: Jiangsu Provincial Key Laboratory of Crop Genetics and Physiology; Key Laboratory of Plant Functional Genomics of Ministry of Education, Yangzhou University, Yangzhou 225009, China; taoli@yzu.edu.cn
Liang, Dong-Li: Northwest A & F University, Yangling, Shaanxi 712100, China; dlliang@nwsuaf.edu.cn
Lin, Xinying: College of Public Health, Shandong University, Jinan, Shandong, China; xinyingll@sdu.edu.cn
Lin, Z.-Q.: Environmental Sciences Program, Southern Illinois University, Edwardsville, Illinois 62026, USA; zhlin@siue.edu
Ling, N.: Research Center on Life Sciences and Environmental Sciences, Harbin University of Commerce, Harbin, Heilongjiang, China; lingnaqd@163.com
Liu, Dan: Key Laboratory of Chemical Synthesis and Pollution Control, Sichuan Province, China; College of Chemistry and Chemical Engineering, China West Normal University, Nanchong, Sichuan, China
Liu, Dan: Research Center on Life Sciences and Environmental Sciences, Harbin University of Commerce, Harbin 150076, China
Liu, J.: College of Life Science, Shenzhen Key Laboratory of Marine Bioresources and Ecology, Shenzhen 518060, China
Liu, J.J.: Northwest A & F University, Yangling, Shannxi 712100, China
Liu, Qiong: Department of Marine Biology, Shenzhen Key Laboratory of Marine Biotechnology and Ecology, Shenzhen 518060, China; liuqiong@szu.edu.cn
Liu, Shoufeng: Jiangsu Bio-Engineering Research Centre on Selenium, Suzhou 215123, Jiangsu, China

Liu, Yan: Key Laboratory of Chemical Synthesis and Pollution Control, Sichuan Province, China; College of Chemistry and Chemical Engineering, China West Normal University, Nanchong, Sichuan, China
Liu, Ying: Advanced Lab for Selenium and Human Health, Suzhou Institute for Advanced Study, University of Science and Technology of China, Suzhou, Jiangsu 215123, China; louie@ustc.edu.cn
Longchamp, M.: LGCgE, Lille1 University, 59655 Villeneuve d'Ascq Cedex, France
Lu, Xiaoqi: School of Earth and Space Sciences & Advanced Lab for Selenium and Human Health, University of Science and Technology of China, Hefei, Anhui 230026, China
Luo, Lei: Jiangsu Bio-Engineering Research Centre on Selenium, Suzhou, Jiangsu 215123, China
Luo, Mingxiu: Institute of Endemic Diseases, College of Medicine, Xi'an Jiaotong University, Xi'an, Shannxi 710061, China; Key Laboratory of Environment and Genes Related to Diseases (Xi'an Jiaotong University), Ministry of Education, China; Key Laboratory of Microelement and Endemic Disease (Xi'an Jiaotong University), Ministry of Health, China
Lv, Y.Y.: China University of Geosciences, Beijing, China
Lyons, G.H.: School of Agriculture & Wine, University of Adelaide, Waite Campus, Glen Osmond, SA 5064, Australia; graham.lyons@adelaide.edu.au

Ma, Zhenzhen: Faculty of Earth Science, China University of Geosciences, Wuhan 430074, China
Man, N.: College of Resources and Environment, Northwest A & F University, Yangling, Shaanxi 712100, China
Marshall, James R.: Cancer Prevention and Control, Roswell Park Cancer Institute, Buffalo, NY 14263, USA; james.marshall@roswellpark.org
Martitz, Anine: Institute for Experimental Endocrinology, Charité-Medical University Berlin, Germany
McGrath, S.P.: Rothamsted Research, Harpenden, Hertfordshire AL5 2JQ, UK; steve.mcgrath@rothamsted.ac.uk
McIntosh, G.H.: Flinders Centre for Innovation in Cancer, Flinders University of South Australia, Bedford Park South Australia 5042, Australia; mcintoshs@iprimus.com.au
Meye, A.M.: Division of Animal Sciences, University of Missouri, Columbia, MO 65211, USA
Mihara, H.: College of Life Sciences, Ritsumeikan University, Kusatsu, Shiga 525-8577, Japan; mihara@fc.ritsumei.ac.jp
Moraes, M.F.: Federal University of Mato Grosso, Postal Code 78600-000, Barra do Garças-MT, Brazil
Morrison, P.D.: RMIT University, Melbourne, Victoria, Australia
Morrissy, Jennifer: Environmental Sciences Program, Southern Illinois University, Edwardsville, Illinois 62026, USA

Neumann, T.: Institute of Mineralogy and Geochemistry, Karlsruhe Institute of Technology, Adenauerring 20b, D-76131 Karlsruhe, Germany
Neville, T.L.: Center for Nutrition and Pregnancy, Animal Sciences Department, North Dakota State University, ND 58108, USA
Ni, J.Z.: Changchun Institute of Applied Chemistry, Chinese Academy of Sciences, Changchun, China; College of Life Sciences, Shenzhen University, Shenzhen, China; University of Chinese Academy of Sciences, Chinese Academy of Sciences, Beijing, China; jzni@szu.edu.cn
Nick, P.: Molecular Cell Biology, Botanical Institute, and Center for Functional Nanostructures, Karlsruhe Institute of Technology, Kaiserstraße 2, D-76131 Karlsruhe, Germany
Nothstein, A.K.: Institute of Mineralogy and Geochemistry, Karlsruhe Institute of Technology, Adenauerring 20b, D-76131 Karlsruhe, Germany; alexandra.nothstein@kit.edu

Orso, G.: Federal University of Parana, Curitiba-PR, Brazil
Otieno, S.B.: Department of Community Health, Kenyatta University, Kenya; samwelbotieno@yahoo.com

Pannier, F.: Laboratory of Bioinorganic Analytical and Environmental Chemistry (LCABIE), University of Pau et des Pays de l'Adour/CNRS, UMR 5254, IPREM, Hélioparc, Pau Cedex 9, France
Pascoalino, J.A.L.: Federal University of Parana, Curitiba-PR, Brazil
Pickering, I.J.: University of Saskatchewan, Saskatoon, Saskatchewan, Canada; ingrid.pickering@usask.ca
Pilon-Smits, E.A.H.: Colorado State University, Biology Department, Fort Collins, CO 80523, USA; epsmits@lamar.colostate.edu

Poblaciones, M.J.: Department of Agronomy and Forest Environment Engineering, University of Extremadura, Badajoz, Spain
Ponomarenko, O.: University of Saskatchewan, Saskatoon, Saskatchewan, Canada
Prabhu, P.: Chemistry Division, Bhabha Atomic Research Centre, Trombay, Mumbai-400085, India
Pratti, T.V.L.: Ghent University, Ghent, Belgium
Priyadarsini, K. Indira: Radiation & Photochemistry Division, Bhabha Atomic Research Centre, Trombay, Mumbai-400085, India; kindira@barc.gov.in

Qin, Hai-Bo: State Key Laboratory of Environmental Geochemistry, Institute of Geochemistry Chinese Academy of Sciences, Guiyang 550002, China
Qin, Li-Qiang: Department of Nutrition and Food Hygiene, School of Public Health, Soochow University, China; qinliqiang@suda.edu.cn
Qin, Long: Research Center on Life Sciences and Environmental Sciences, Harbin University of Commerce, Harbin 150076, China
Qiu, Shi: Department of Marine Biology, Shenzhen Key Laboratory of Marine Biotechnology and Ecology, Shenzhen 518060, China

Rahman, Z.A.: Land Management Department, Faculty of Agriculture, University Putra Malaysia, Malaysia
Ralston, N.V.C.: Energy & Environmental Research Center, University of North Dakota, Grand Forks, ND 58202, USA; nralston@undeerc.org
Ramos, S.J.: Vale Institute of Technology, Postal Code 31140-130, Belo Horizonte-MG, Brazil
Rayman, M.P.: Faculty of Health and Medical Sciences, University of Surrey, Guildford, GU2 7XH, UK; m.rayman@surrey.ac.uk
Raymond, L.J.: Energy & Environmental Research Center, University of North Dakota, Grand Forks, ND 58202, USA
Reis, A.R.: Federal University of Lavras, Postal Code 37200-000, Lavras-MG, Brazil; andrekun@gmail.com
Reynolds, L.P.: Center for Nutrition and Pregnancy, Animal Sciences Dept., North Dakota State University, ND 58108, USA
Riemann, M.: Molecular Cell Biology, Botanical Institute, and Center for Functional Nanostructures, Karlsruhe Institute of Technology, Kaiserstraße 2, D-76131 Karlsruhe, Germany
Rodrigo, S.M.: Department of Agronomy and Forest Environment Engineering, University of Extremadura. Badajoz, Spain
Rutzke, M.: USDA-ARS & Cornell University, Ithaca, NY, USA

Scheeren, P.L.: Embrapa Wheat, Passo Fundo-RS, Brazil
Schoenberg, R.: University of Tübingen, Geochemistry Group, Tübingen, Germany
Schomburg, Lutz: Institute for Experimental Endocrinology, Charité- Medical University Berlin, Germany; lutz.schomburg@charite.de
Schubert, R.: Eawag, Swiss Federal Institute of Aquatic Science and Technology 8600 Dübendorf, Switzerland
Sever, Victoria: Department of Biological Sciences, Southern Illinois University, Edwardsville, Illinois 62026, USA
Sharifianpour, G.: Land Management Department, Faculty of Agriculture, University Putra Malaysia, Malaysia
Si, C.: College of Life Sciences, Hebei University, Baoding 071002, China
Singh, B.G.: Radiation & Photochemistry Division, Bhabha Atomic Research Centre, Trombay, Mumbai-400085, India
Singh, Swechhya: Environmental Sciences Program, Southern Illinois University, Edwardsville, IL 62026, USA
Small, D.M.: RMIT University, Melbourne, Victoria, Australia; darryl.small@rmit.edu.au
Souza, G.A.: Federal University of Lavras, Lavras, MG, Brazil
Spallholz, J.: Texas Tech University, Lubbock, Texas, USA
Stelling, M.E.: Institute of Mineralogy and Geochemistry, Karlsruhe Institute of Technology, Adenauerring 20b, D-76131 Karlsruhe, Germany
Strait, K.: Emory University, Atlanta, Georgia, USA
Stranges, S.: Division of Health Sciences, University of Warwick Medical School, Medical School Building, Gibbet Hill Campus Coventry, CV4 7AL, UK

Sun, Fayu: Jiangsu Provincial Key Laboratory of Crop Genetics and Physiology; Key Laboratory of Plant Functional Genomics of Ministry of Education, Yangzhou University, Yangzhou 225009, China
Sun, Q.Y.: Research Center on Life Sciences and Environmental Sciences, Harbin University of Commerce, Harbin, Heilongjiang, China
Sun, X.F.: Soil and Fertilizer Institute, Qinghai Academy of Agricultural and Forestry Sciences, Xining, Qinghai 810016, China

Tack, F.M.G.: Ghent University, Ghent, Belgium
Takahashi, Y.: Department of Earth and Planetary Systems Science, Graduate School of Science, Hiroshima University, 1-3-1 Kagamiyama, Higashi-Hiroshima, Hiroshima 739-8526, Japan
Tang, Yin: College of Public Health, Nanchang University, Nanchang 330006, China; 352011328@qq.com
Taylor, J.B.: United States Department of Agriculture, Agricultural Research Service, US Sheep Experiment Station, Dubois, Idaho 83423, USA; bret.taylor@ars.usda.gov
Terry, Norman: Department of Plant and Microbial Biology, University of California, Berkeley, California 94720, USA; nterry@berkeley.edu
Thiry, Y.: Andra, Research and Development Division, Châtenay-Malabry Cedex, France
Tian, H.: College of Material Science and Chemical Engineering, China University of Geosciences & The State Key Laboratory of Geological and Environmental Geological, Wuhan 430074, China
Tian, J.: College of Life Science, Shenzhen Key Laboratory of Microbial Genetic Engineering, Shenzhen University, Shenzhen 518060, China; jing.tianjingtj@gmail.com
Tjong, W.: RMIT University, Melbourne, Victoria, Australia
Tong, Xinzhao: School of Nanoscience & Advanced Lab for Selenium and Human Health, University of Science and Technology of China, Suzhou, Jiangsu 215123, China
Tullo, P.Di.: Laboratory of Bioinorganic Analytical and Environmental Chemistry (LCABIE), University of Pau et des Pays de l'Adour/CNRS, UMR 5254, IPREM, Hélioparc, Pau Cedex 9, France; pamela.ditullo@univ-pau.fr

Van de Wiele, T.: Ghent University, Ghent, Belgium
Villetti, H.L.: Federal University of Parana, Curitiba-PR, Brazil
von Brasch, M.: Institute of Mineralogy and Geochemistry, Karlsruhe Institute of Technology, Adenauerring 20b, D-76131 Karlsruhe, Germany
Vonnahme, K.A.: Center for Nutrition and Pregnancy, Animal Sciences Department, North Dakota State University, ND 58108, USA
Vriens, Bas: Eawag, Swiss Federal Institute of Aquatic Science and Technology 8600 Dübendorf, Switzerland; bas.vriens@eawag.ch

Wang, An: Jiangsu Provincial Key Laboratory of Crop Genetics and Physiology; Key Laboratory of Plant Functional Genomics of Ministry of Education, Yangzhou University, Yangzhou 225009, China
Wang, Bo: Suzhou Center for Disease Prevention and Control, Suzhou, China
Wang, Chao: Department of Marine Biology, Shenzhen Key Laboratory of Marine Biotechnology and Ecology, Shenzhen 518060, China
Wang, Dacheng: Key Laboratory of Chemical Synthesis and Pollution Control, Sichuan Province, China; College of Chemistry and Chemical Engineering, China West Normal University, Nanchong, Sichuan, China; dachengwang@hotmail.com
Wang, Jian: School of Earth and Space Sciences & Advanced Lab for Selenium and Human Health, University of Science and Technology of China, Hefei, Anhui 230026, China
Wang, L.X.: Department of Marine Biology, Shenzhen Key Laboratory of Marine Biotechnology and Ecology, Shenzhen 518060, China
Wang, Qian: Key Laboratory of Agri-Food Safety of Anhui Province, School of Plant Protection, Anhui Agriculture University, Hefei, Anhui 230036, China
Wang, Zhi: College of Life Sciences, Shenzhen Key Laboratory of Microbial Genetic Engineering, Shenzhen University, Shenzhen 518060, China
Watts, M.J.: British Geological Survey, Keyworth, Nottingham, NG12 5GG, UK
Waza, K.: Kenneth Kaunda Foundation, Zambia
Wei, C.H.: College of Material Science and Chemical Engineering, China University of Geosciences & State Key Laboratory of Geological and Environmental Geological, Wuhan 430074, China; chwei1000@163.com

Wen, Yu-Cheng: Technology Center for Heavy Metal Cleaner Production Engineering, Chinese Research Academy of Environmental Sciences, Beijing 100012, China
Were, F.: School of Medicine, University of Nairobi, Kenya
Winkel, Lenny H.E.: Eawag, Swiss Federal Institute of Aquatic Science and Technology, 8600 Dübendorf, Switzerland & Swiss Federal Institute of Technology (ETH) Zurich, 8092 Zurich, Switzerland; lenny.winkel@eawag.ch, lwinkel@ethz.ch
Wu, Jie: Key Laboratory of Agri-Food Safety of Anhui Province, School of Resources and Environment & Plant Protection, Anhui Agriculture University, Hefei, Anhui 230036, China
Wu, Zhilin: Key Laboratory of Agri-Food Safety of Anhui Province, School of Plant Protection, Anhui Agriculture University, Hefei, Anhui 230036, China

Xi, Gongfang: Anhui Normal University, Wuhu, Anhui 241000, China; Institute for Agro-Product Quality Standards and Testing Technologies, Shanghai Academy of Agricultural Sciences, Shanghai 201106, China
Xie, J.Y.: College of Resources and Environment, Northwest A & F University, Yangling, Shaanxi 712100, China
Xu, Fuyuan: Technology Center for Heavy Metal Cleaner Production Engineering, Chinese Research Academy of Environmental Sciences, Beijing 100012, China
Xu, Yunhui: Key Laboratory of Agri-Food Safety of Anhui Province, School of Plant Protection, Anhui Agriculture University, Hefei, Anhui 230036, China

Yang, Limin: Key Laboratory of Agri-Food Safety of Anhui Province, School of Plant Protection, Anhui Agriculture University, Hefei, Anhui 230036, China
Yang, Z.F.: China University of Geosciences, Beijing, China
Yao, C.-X.: Institute for Agro-Product Quality Standards and Testing Technologies, Shanghai Academy of Agricultural Sciences, Shanghai 201106, China; chunxia.yao@163.com
Yao, X.Q.: College of Life Sciences, Hebei University, Baoding 071002, China; yaoxiao301@126.com
Yasin, M.: Department of Microbiology and Molecular Genetics, University of the Punjab, Quaid-e-Azam Campus, Lahore-54590, Pakistan; yasin_mmg@yahoo.com
Yin, Xue-Bin: Suzhou Key Laboratory for Selenium and Human Health, University of Science and Technology of China, Suzhou, China; xbyin@ustc.edu.cn
You, Z.: College of Computer Science and Software Engineering, Shenzhen University, Shenzhen 518060, China
Young, G.P.: Flinders Centre for Innovation in Cancer, Flinders University of South Australia, Bedford Park South Australia 5042, Australia
Young, S.D.: School of Biosciences, University of Nottingham, Sutton Bonington, Leicestershire LE12 5RD, UK
Yu, D.S.: Northwest A & F University, Yangling, Shannxi 712100, China
Yu, Miao: Research Center on Life Sciences and Environmental Sciences, Harbin University of Commerce, Harbin 150076, China; Engineering Research Center of Natural Anticancer Drugs, Ministry of Education, Harbin 150076, China
Yu, T.: China University of Geosciences, Beijing, China; yutao@cugb.edu.cn
Yuan, Linxi: Jiangsu Bio-Engineering Research Centre on Selenium, Suzhou, Jiangsu 215123, China; yuanli@ustc.edu.cn

Zambrano, M.C.: USDA-ARS, Water Management Unit, Parlier, California, USA
Zeng, Z.Y.: College of Life Sciences, Shenzhen University, Shenzhen, China
Zhang, B.J.: College of Public Health, Nanchang University, Nanchang, Jiangxi 330006, China; Jiangxi Institute of Soil and Water Conservation, Nanchang, Jiangxi 330029, China; zhangbaoj04@163.com
Zhang, Chi: School of Biological Science and Technology, Hubei Institute for Nationalities, Hubei 445000, China; Key Laboratory of Biologic Resources Protection and Utilization of Hubei Province, Enshi, Hubei, China; zhtzu@163.com
Zhang, Ge: Technology Center for Heavy Metal Cleaner Production Engineering, Chinese Research Academy of Environmental Sciences, Beijing 100012, China
Zhang, Heyu: Key Laboratory of Agri-Food Safety of Anhui Province, School of Plant Protection, Anhui Agriculture University, Hefei, Anhui 230036, China
Zhang, Hua: Institute of Geochemistry, Chinese Academy of Sciences, Guiyang 550002, China; Norwegian Institute for Water Research, Gaustadalléen 21, 0349 Oslo, Norway

Zhang, L.Y.: Institute of Medicinal Fungi Research, Nanjing Xiaozhuang University, Nanjing 211171, China
Zhang, Man: College of Public Health, Shandong University, Jinan, Shandong, China
Zhang, R.: Soil and Fertilizer Institute, Qinghai Academy of Agricultural and Forestry Sciences, Xining, Qinghai 810016, China
Zhang, R.L.: Shenzhen Center for Disease Control and Prevention, Shenzhen, China
Zhang, T.Q.: Agricultural and Agri-Food Canada, Greenhouse and Processing Crops Research Centre, Harrow, Ontario, N0R 1G0, Canada; tiequan.zhang@agr.gc.ca
Zhang, Y.F.: The 5th Institute of Geology and Mineral Exploration of Qinghai, Xining, Qinghai, China
Zhang, Yu: Key Laboratory of Agri-Food Safety of Anhui Province, School of Resources and Environment & Plant Protection, Anhui Agriculture University, Hefei, Anhui 230036, China
Zhang, Zengtie: Institute of Endemic Diseases, College of Medicine, Xi'an Jiaotong University, Xi'an 710061, Shaanxi, China; Key Laboratory of Environment and Genes Related to Diseases (Xi'an Jiaotong University), Ministry of Education, China; Key Laboratory of Microelement and Endemic Disease (Xi'an Jiaotong University), Ministry of Health, China
Zhang, Zhi: Technology Center for Heavy Metal Cleaner Production Engineering, Chinese Research Academy of Environmental Sciences, Beijing 100012, China
Zhao, Qiguo: State Key Laboratory of Soil and Sustainable Agriculture, Institute of Soil Science, Chinese Academy of Sciences, Nanjing 210008, China; qgzhao@issas.ac.cn
Zheng, J.: College of Life Science, Shenzhen Key Laboratory of Marine Bioresources and Ecology, Shenzhen 518060, China
Zheng, Li-Rong: Beijing Synchrotron Radiation Facility, Institute of High Energy Physics, Chinese Academy of Sciences, Beijing 10039, China
Zheng, Qing: Jiangsu Bio-Engineering Research Centre on Selenium, Suzhou, Jiangsu 215123, China
Zhong, X.: Research Center on Life Sciences and Environmental Sciences, Harbin University of Commerce, Harbin, Heilongjiang, China
Zhou, Chao: Technology Center for Heavy Metal Cleaner Production Engineering, Chinese Research Academy of Environmental Sciences, Beijing 100012, China
Zhou, Dazhai: Key Laboratory of Biological Resources Protection and Utilization of Hubei Province, China; College of Biological Sciences and Technology, Hubei University for Nationalities, Enshi, Hubei, China
Zhou, Qiankun: Anhui Normal University, Wuhu, Anhui 241000, China
Zhou, Shoubiao: Anhui Normal University, Wuhu, Anhui 241000, China; zhoushoubiao@vip.163.com
Zhu, Chun-Lei: Technology Center for Heavy Metal Cleaner Production Engineering, Chinese Research Academy of Environmental Sciences, Beijing 100012, China
Zhu, Fa-Hua: State Power Environmental Protection Research Institute, Nanjing, Jiangsu 210031, China
Zhu, Jian-ming: State Key Laboratory of Environmental Geochemistry, Institute of Geochemistry, Chinese Academy of Sciences, Guiyang 550002, China; zhujianming@vip.gyig.ac.cn
Zhu, Yuanyuan: Jiangsu Bio-Engineering Research Centre on Selenium, Suzhou, Jiangsu 215123, China
Zou, C.: College of Life Science, Shenzhen Key Laboratory of Marine Bioresources and Ecology, Shenzhen 518060, China
Zou, Xiaoyan: College of Public Health, Shandong University, Jinan, Shandong, China

1 *Biogeochemistry of selenium*

Selenium in the Environment and Human Health – Bañuelos, Lin & Yin (Eds)
© 2014 Taylor & Francis Group, London, ISBN 978-1-138-00017-9

Role of climatic factors on the terrestrial distribution of selenium

L.H.E. Winkel
Eawag, Swiss Federal Institute of Aquatic Science and Technology, Dübendorf, Switzerland
Swiss Federal Institute of Technology (ETH) Zurich, Switzerland

T. Blazina, B. Vriens, R. Schubert & C. Annette Johnson
Eawag, Swiss Federal Institute of Aquatic Science and Technology, Dübendorf, Switzerland

1 INTRODUCTION

Selenium (Se) is an element that is of key importance for human health. However, it has a narrow range between the doses that are essential to humans and those that are associated with adverse effects.

Globally, the geographic distribution of Se is highly uneven. Areas of selenium excess are known but regions of Se deficiency are more widespread and larger in size. Globally, it is estimated that up to one billion people are affected by Se deficiency (Haug *et al.*, 2007). Therefore, Se supplementation and fortification of foods, as well as Se fertilization of agricultural soils, are increasingly discussed as possible strategies of improving human Se status (Combs, 2001).

Despite the growing awareness of the importance of this element, the sources and behavior of Se in the natural environment are still poorly understood. Nevertheless, in order to prevent health hazards due to distribution and redistribution of Se in the environment it is very important to have a better understanding of the mechanisms that that control the environmental behavior of Se.

Areas in which natural Se accumulates in excess in soils, plants and groundwater can mostly be attributed to the nearby presence of Se-rich rocks. Nevertheless, the distribution of low and medium level Se soils remains a mystery that has puzzled scientists for decades. Since the distribution of this essential trace element in the natural environment is closely related to human health issues, it is of major importance to acquire a better understanding of the factors that control this distribution.

Hypothetically, atmospheric input can explain the large-scale low to medium Se distribution in the terrestrial environment (Wen & Carignan, 2007). In this presentation we will address the different sources and pathways that contribute to terrestrial Se distributions. Furthermore, we will focus on atmospheric Se as a source for terrestrial Se in surface environments.

2 THE ATMOSPHERE AS A SINK OF SELENIUM

The atmosphere is an important transient reservoir for Se, which is largely fed by volatile Se compounds formed via biomethylation in marine environments (Challenger & North, 1934; Amouroux *et al.*, 1997, 2000 & 2001). In addition, biomethylation from terrestrial environments can potentially exist as a significant process in the Se biogeochemical cycle, and the subsequent volatilization of Se can be an additional source to the atmospheric budget. Previously, biomethylation of Se in terrestrial environments had been studied with a focus on Se bioremediation processes (*e.g.*, Terry *et al.*, 1992; Frankenberger & Arshad, 2001; Banuelos & Lin, 2007; Pilon-Smits & LeDuc, 2009) in naturally high Se environments, such as the Kesterson Reservoir in California. However, on a global scale, high Se environments are small in number and size. Little information is yet available on the volatilization of Se from terrestrial environments with ordinary to low Se content, and its relevance to the natural global atmospheric Se budget remains unknown. We will present some first results of a Se biomethylation study in a pristine wetland and address the implications of obtained results.

3 THE ATMOPSHERE AS A SOURCE OF SELENIUM

The atmosphere can function as a source of terrestrial Se when it is returned to the Earth's surface via wet and dry deposition (Cutter & Church, 1986; Mosher & Duce, 1987; Haygarth *et al.*, 1993). We present new evidence on how climatic conditions play a major role in the large-scale terrestrial Se distribution, both directly as a source of Se (via atmospheric deposition) and indirectly by controlling pedoclimatic regimes and thus Se bioavailability.

We expect that projected changes in climatic conditions will have a significant impact on the environmental Se distribution. We will finally show how such climatic factors can be used to predict large-scale terrestrial Se distributions. The ability to predict dynamic terrestrial Se distributions will be crucial in helping to prevent future health hazards resulting from the uneven distributions of Se in the environment.

REFERENCES

Amouroux, D. & Donard, O.F.X. 1997. Evasion of selenium to the atmosphere via biomethylation processes in the Gironde estuary, France. *Marine Chemistry* 58: 173–188.

Amouroux, D., Pecheyran, C. & Donard, O.F.X. 2000. Formation of volatile selenium species in synthetic seawater under light and dark experimental conditions. *Applied Organometallic Chemistry* 14: 236–244.

Amouroux, D., Liss, P.S., Tessier, E., Hamren-Larsson, M. & Donard, O.F.X. 2001. Role of oceans as biogenic sources of selenium. *Earth and Planetary Science Letters* 189: 277–283.

Banuelos, G.S. & Lin, Z.-Q. 2007. Acceleration of selenium volatilization in seleniferous agricultural drainage sediments amended with methionine and casein. *Environ. Pollut.* 150: 306–312.

Challenger, F. & North, H.E. 1934. The production of organometalloidal compounds by micro-organisms-II: dimethyl selenide. *Journal of the Chemical Society*, 68–71.

Combs, J.F. Jr. 2001. Selenium in global food system. *Br. J. Nutr.* 85: 517–547.

Cutter, G.A. & Church, T.M. 1986. Selenium in western Atlantic precipitation. *Nature* 322: 720–724.

Frankenberger, W.T., Jr. & Arshad, M. 2001. Bioremediation of selenium-contaminated sediments and water. *Biofactors* 14(1–4): 241–254.

Haug, A., Graham, R.D., Christophersen, O.A. & Lyons, G.H. 2007. How to use the world's scarce selenium resources efficiently to increase the selenium concentration in food. *Microb. Ecol. Health Dis.* 19 (4): 209–228.

Haygarth, P.M., Harrison, A.F. & Jones, K.C. 1993. Geographical and seasonal variation in deposition of selenium to vegetation. *Environ. Sci. Technol.* 27: 2878–2884.

Mosher, B.W. & Duce, R.A. 1987. A global atmospheric selenium budget. *Journal of Geophysical Research* 92 (D11): 13289–13298.

Pilon-Smits, E.A.H. & LeDuc, D.L. 2009. Phytoremediation of selenium using transgenic plants. *Current Opinion in Biotechnology* 20(2): 207–212.

Terry, N., Carlsen, C., Raab, T.K. & Zayed, A.M. 1992. Rates of selenium volatilization among crop species. *J. Environ. Qual.* 21(3): 341–344.

Wen, H. & Carignan, J. 2007. Reviews on atmospheric selenium: Emissions, speciation and fate. *Atmos. Environ.* 41: 7151–7165.

Selenium in the Environment and Human Health – Bañuelos, Lin & Yin (Eds)
© 2014 Taylor & Francis Group, London, ISBN 978-1-138-00017-9

Assessing global cycling of selenium: Analytical tools to quantify volatile species and their fluxes

Bas Vriens & Lenny H.E. Winkel
Eawag, Swiss Federal Institute of Aquatic Science and Technology, Dübendorf, Switzerland
Institute of Biogeochemistry and Pollutant Dynamics, ETH Zurich, Zurich, Switzerland

Markus Lenz
University of Applied Sciences and Arts Northwestern Switzerland, Institute for Ecopreneurship, Muttenz, Switzerland
Sub-Department of Environmental Technology, Wageningen University, EV Wageningen, The Netherlands

Michael Berg
Eawag, Swiss Federal Institute of Aquatic Science and Technology, Dübendorf, Switzerland

1 INTRODUCTION

Selenium (Se) is an essential trace element for life. However, it can negatively affect human health due to the narrow range between its toxicity and beneficence and its unequal distribution in the surface environment. To date, sources, sinks and fluxes in the global Se cycle are poorly understood and inadequately quantified (Winkel *et al.*, 2011). One potentially significant cycling process is Se biomethylation and subsequent volatilization of Se to the atmosphere. This process has been studied for bioremediation purposes in (naturally) high Se environments such as the Kesterson Reservoir in California (Terry *et al.*, 1992; Frankenberger & Arshad, 2001; Pilon-Smits & LeDuc, 2009). On a global scale, however, high Se environments are small in number and size. Little information is available on the volatilization of Se from terrestrial environments with average to low Se content, and its relevance to the global atmospheric Se budget remains unknown.

Biomethylation and volatilization is also known for arsenic (As) and sulfur (S) (Bentley & Chasteen, 2004; Mestrot *et al.*, 2011). In the natural environment, Se and As often occur simultaneously with S (Fernandez-Martinez & Charlet, 2009; Langner *et al.*, 2012), and it is therefore important to study the biogeochemical cycles of Se and As in relation with that of major elements like S. However, to quantify volatilization of Se, S and As in the field, highly sensitive and robust pre-concentration methods that can resolve speciation at very low concentrations are required.

We developed two complementary analytical speciation techniques for the analysis of organic and inorganic Se, S and As in the environment. Both methods are able to resolve trace-concentrations of species of these elements in the aqueous and gaseous phase.

2 SAMPLING AND ANALYTICAL METHODS

The first method is a chemotrapping method using concentrated nitric acid (Winkel *et al.*, 2010). Preservation of information on the original speciation in the gas phase is maintained through the formation of oxidized, non-volatile derivative compounds. Species trapped in the trapping liquid can subsequently be resolved and quantified using a state-of-the-art laboratory speciation technique (HLPC-HR-ICP-MS). In combination with a flow-through chamber system (Fig. 1), the chemotrapping set-up can be deployed for field studies in various environments. With minor adaptations, the HPLC-HR-ICP-MS method is also suitable for

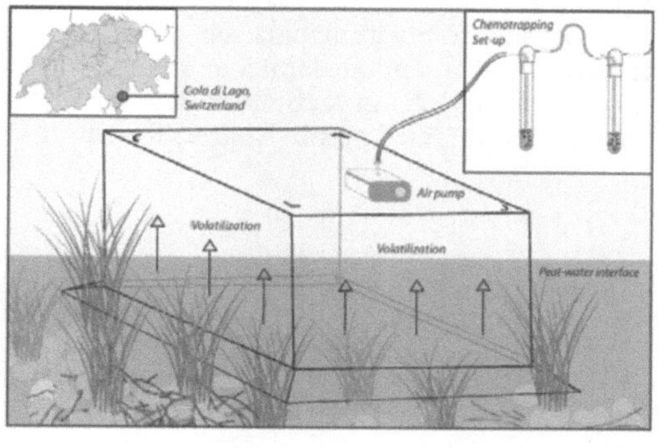

Figure 1. Location of *Gola di Lago* in Southern Switzerland (top left), and schematic representation of the flow-through chamber set-up coupled to chemotraps.

the analysis of dissolved, methylated species of Se, S and As in surface waters.

The second method is a solid-phase micro-extraction (SPME) method coupled to GC-MS, which allows for simultaneous detection and quantification of trace concentrations of volatile, methylated Se and S species in both the gaseous (headspace-SPME) and aqueous (direct immersion-SPME) phases. This presentation will introduce these novel techniques, their capabilities and limitations, and how these methods complement each other in the analysis of Se, S and As species in complex environmental samples.

3 FIELD APPLICATION

Both speciation methods were applied in a minerotrophic (surface and ground water fed) peatland in southern Switzerland (*Gola di Lago*, Fig. 1). This alpine peat bog has previously been studied for Se and As concentrations in the peat material (González *et al.*, 2006; Langner *et al.*, 2012), but trace element volatilization has never been quantified.

Using aforementioned speciation techniques, we were able to detect and quantify emissions of Se, S and As from the peatland in the form of non-, mono—and di-methylated species, and determined gas fluxes of Se (0.11 μg Se/m^2/d), S (37 μg S/m^2/d) and As (0.16 μg As/m^2/d). In addition, the methods were applied to study the distribution and speciation of Se, S and As in the surface water of the peat bog.

4 CONCLUSIONS

The presented analytical methods are straightforward to use, quantitative, reproducible, and robust, and can be applied to qualitatively and quantitatively study Se, S and As volatilization in a wide range of natural environments, demonstrated by the reproducible and accurate analysis of species of Se, S and As in a complex environmental system. The obtained results enable, for the first time, a comparison of volatile fluxes and speciation with total elemental concentrations and speciation of Se, S and As in the peat and surface water. Our results suggest that trace element volatilization from wetlands, and possibly other terrestrial environments, may be significantly contributing to their global cycles, especially in the case of Se.

REFERENCES

Bentley, R. & Chasteen, T.G. 2004. Environmental VOSCs—formation and degradation of dimethyl sulfide, methanethiol and related materials. *Chemosphere* 55: 291–317.

Fernandez-Martinez, A. & Charlet, L. 2009. Selenium environmental cycling and bioavailability: a structural chemist point of view. *Reviews in Environmental Science and Biotechnology* 8: 81–110.

Frankenberger, W.T., Jr. & Arshad, M. 2001. Bioremediation of selenium-contaminated sediments and water. *Biofactors* 14: 241–254.

González, A.Z.I., Krachler, M., Cheburkin, A.K. & Shotyk, W. 2006. Spatial distribution of natural enrichments of arsenic, selenium, and uranium in a minerotrophic peatland, Gola di Lago, Canton Ticino, Switzerland. *Environmental Science & Technology* 40: 6568–6574.

Langner, P., Mikutta, C. & Kretzschmar, R. 2012. Arsenic sequestration by organic sulphur in peat. *Nature Geoscience* 5: 66–73.

Mestrot, A., Feldmann, J., Krupp, E.M., Hossain, M.S., Roman-Ross, G. & Meharg, A.A. 2011. Field fluxes and speciation of arsines emanating from soils. *Environmental Science & Technology* 45: 1798–1804.

Pilon-Smits, E.A.H. & LeDuc, D.L. 2009. Phytoremediation of selenium using transgenic plants. *Current Opinion in Biotechnology* 20: 207–212.

Terry, N., Carlson, C., Raab, T.K. & Zayed, A.M. 1992. Rates of selenium volatilization among crop species. *Journal of Environmental Quality* 21: 341–344.

Winkel, L., Feldmann, J. & Meharg, A.A. 2010. Quantitative and qualitative trapping of volatile methylated selenium species entrained through nitric acid. *Environmental Science & Technology* 44: 382–387.

Winkel, L.H.E., Johnson, C.A., Lenz, M., Grundl, T., Leupin, O.X., Amini, M. *et al.* 2011. Environmental selenium research: from microscopic processes to global understanding. *Environmental Science & Technology* 46: 571–579.

Selenium in the Environment and Human Health – Bañuelos, Lin & Yin (Eds)

Selenium in soil profiles and the rhizosphere soils of typical Kaschin-Beck disease areas

Y.Y. Lv, T. Yu, Z.F. Yang & Q.Y. Hou
China University of Geosciences, Beijing, China

B.Y. Ji & Y.F. Zhang
The 5th Institute of Geology and Mineral Exploration of Qinghai, Xining, Qinghai Province, China

L.X. Jin
Sichuan Institute of Geological Survey, Chengdu, Sichuan Province, China

1 INTRODUCTION

Kaschin-Beck Disease (KBD) is regarded as an endemic osteoarthropathy characterized by deformity of affected joint cartilage and epiphyseal plate cartilage of the four limbs, which poses a serious threat to the quality of people's life in china (Tan & Huang, 1991; Wang & Xie, 2005). It is reported recently that KBD afflicts at least 2.5 million people in China and neighboring parts of Russia and North Korea (Stone, 2009). Obviously, the KBD distributes in the low selenium-belt from northeast to southwest of China, shaped like a large saddle (Zhang *et al.*, 2011). Up to now, only Aba in Sichuan Province, Tibet Autonomous Region, and south of Qinghai Province have occurred new KBD cases after preventing and controlling in China (Zhang *et al.*, 2010). Since 1957, important scientific research has helped to make possible a shift in thinking about the genesis of KBD and the effect of selenium (Se) on public health (Schwarz & Foltz, 1957). Selenium, a trace element, is an essential component of the biologically enzyme glutathione peroxidase (GPx), which acts as an antioxidant preventing tissue degeneration (Keshavarzi *et al.*, 2012).

2 MATERIALS AND METHODS

2.1 *Study Areas*

Tibetan Autonomous Prefecture of Aba, Sichuan Province (including Zamtang, Aba, Jinchuan, Barkam, Hongyuan county) and Tibetan Autonomous Prefecture of Hainan, Qinghai Province (Xinghai county), as typical KBD-affected areas, are located in the eastern edge of the Qinghai-Tibet Plateau in China. At some distance from above areas, Yanting county, Sichuan Province and Ping'an county, Qinghai Province were treated as control areas respectively.

The average altitude of these KBD-affected areas is 3300 m. These areas are underlain by different geomorphic features. The weather in hummocky plateau landscape is typically plateau of continental, which features faint difference in temperature of its four seasons and the mountain plateau landscape has a cool subhumid climate, clearly between dry and wet seasons.

2.2 *Methods*

In the field of medical geology, element Se is influenced by the large rock-soil-water-plant-human geochemical cycle (Dissanayake & Chandrajith, 2009). Besides the importance of speciation, ionic strength, pH or redox potential on the bioavailability of Se (Fairweather-Tait *et al.*, 2011), changes in topographical features and leaching/erosion processes control the migration of Se into soils (Fernandez-Martinez & Charlet, 2009) can never be ignored. We used a combined method of pedology and geochemistry for soil profiles and rhizosphere soils to discuss the integrated constraints on Se in respective KBD-affected/unaffected areas, including new and old stricken areas, serious, slight, and unaffected areas.

3 RESULTS AND DISCUSSION

3.1 *Selenium in soil profiles*

Concentrations of Se in soil profiles of typical KBD areas (*e.g.*, Aba area and Xinhai county) resembled a low-selenium area in Ping'an county (Fig. 1a). A comparison of the vertical distribution of Se concentrations between serious and slight/free KBD areas is shown in Figure 1b.

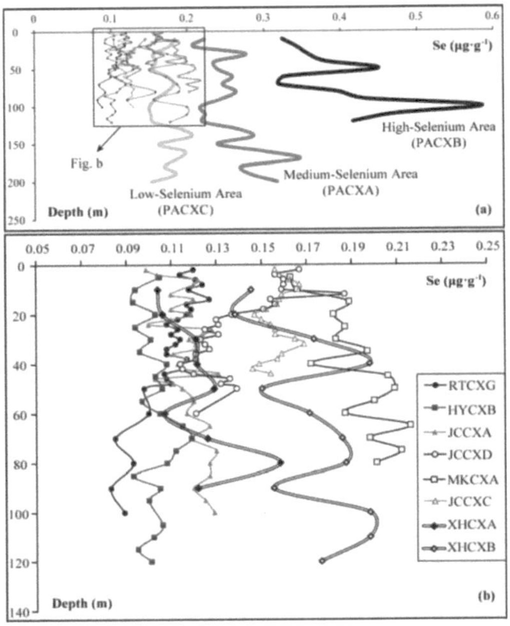

Figure 1. Selenium concentations in soil profiles. (RTCXG, HYCXB, JCCXA, and XHCXA were sampled at serious KBD areas; JCCXD, MKCXA, and XHCXB were sampled at slight KBD areas; JCCXC was sampled at KBD-unaffected area).

Howere, Selenium content in typical KBD areas varied from 0.08 to 0.215 $\mu g \cdot g^{-1}$, which coincided with lack (≤ 0.125 μg/g) and margin (0.125–0.175 μg/g) of selenium threshold (Tan, 1989). Generally, Se contents in serious areas were lower than slight/free areas, especially Se in topsoil.

Otherwise, Se concentration in the profile for new stricken area was slightly higher than the old one. Considering the majority of new stricken areas appeared on the mountainside, and old stricken ones are scattered on hummocky plateau, Se eluviation in topsoil maybe an important factor contributing to new KBD cases.

3.2 *Selenium in rhizosphere soils*

The gap of Se content in rhizosphere soils between Aba area and Yanting county was small (the average total Se content in topsoil is 0.12 and 0.11 μg/g, respectively). Selenium in crops in Yanting county was greater, due to the result that soil water-soluble Se (including selenite and selenate) is bound up with Se enrichment of crops. Furthermore, human daily average Se intake may be influented to some degree.

4 CONCLUSIONS

The recent data in soil profiles displayed that typical KBD areas underscored a feature of lower Se content, especially in serious KBD areas. Selenium eluviation on the mountainside might result in the occurrence of new KBD cases. Moreover, water-soluble Se in rhizosphere soils played a significant role in Se bioavailability, because it is easily absorbed by plants and subsequently taken in by humans.

REFERENCES

Dissanayake, C.B. & Chandrajith, R. 2009. *Introduction to Medical Geology*. Berlin Heidelberg: Springer-Verlag: 205–257.

Fairweather-Tait, S.J., Bao, Y., Broadley, M.R., Collings, R., Ford, D., Hesketh, J.E. *et al.* 2011. Selenium in human health and disease. *Antioxidants and Redox Signaling* 14(7): 1337–1383.

Fernandez-Martinez, A. & Charlet, L. 2009. Selenium environmental cycling and bioavailability: a structural chemist point of view. *Review of Environmental Science and Biotechnology* 8: 81–110.

Keshavarzi, B., Moore, F., Najmeddin, A. & Rahmani, F. 2012. The role of selenium and selected trace elements in the etiology of esophageal cancer in high risk Golestan province of Iran. *Science of the Total Environment* 433: 89–97.

Schwarz, K. & Foltz, C.M. 1957. Selenium as an integral part of Factor 3 against dietary necrotic liver degeneration. *Journal of the American Chemical Society* 79(12): 3292–3293.

Stone, R. 2009. A Medical Mystery in Middle China. *Science* 324: 1378–1381.

Tan, J.A. 1989. *The Atlas of Endemic Diseases and Their Environments in the People's Republic of China*. Beijing: Science Press. (in Chinese).

Tan, J.A. & Huang, Y.J. 1991. Selenium in Geo-ecosystem and its relation to endemic diseases in China. *Water, Air, and Soil Pollution* 57–58: 59–68.

Wang, J. & Xie, H. 2005. A Preliminary Assessment of Environmental Geology for Kaschin-Beck Disease in Zamgtang, Aba. *Acta Geologica Sichuan* 35(3): 169–172. (in Chinese).

Zhang, B., Yang, L., Wang, W. *et al.* 2011. Environmental selenium in the Kaschin-Beck disease area, Tibetan Plateau, China. *Environmental Geochemistry and Health* 33(5): 495–501.

Zhang, F.C., Wen, D.G., Guo, J.Q., Zhang, E.Y., Hao, A.B., AN, Y.h. *et al.* 2010. Research progress and prospect of geological environment in main endemic disease area. *Geology in China* 37(3): 551–562. (in Chinese).

Selenium in the Environment and Human Health – Bañuelos, Lin & Yin (Eds)
© 2014 Taylor & Francis Group, London, ISBN 978-1-138-00017-9

Selenium distribution linked to monsoon climate in the Chinese Loess Plateau

Tim Blazina & Lenny H.E. Winkel
Eawag, Swiss Federal Institute of Aquatic Science and Technology, Dübendorf, Switzerland
Institute of Biogeochemistry and Pollutant Dynamics, ETH Zurich, Zurich, Switzerland

Michael Berg
Eawag, Swiss Federal Institute of Aquatic Science and Technology, Dübendorf, Switzerland

1 INTRODUCTION

Selenium (Se) is a vital trace element for human health with a narrow range between deficiency and toxicity. Diets deficient in Se have been linked with various diseases, while those with an excess can lead to toxicity (Rayman, 2012). Both Se toxicity and deficiency are problems within China. However, Se deficiency is a much more prevalent problem. In Enshi, Hubei Province, high soil Se concentrations caused by the underlying bedrock have led to many cases of Se toxicity (Zhu & Zheng, 2001). Within central China, extremely low levels of soil Se has led to wide spread health problems. These include Kashin-Beck Disease (Stone, 2009), a chronic bone and cartilage disease affecting children as young as 2 or 3 years old and leads to deformed bones and joints, and Keshan Disease (Chen *et al.*, 1980), a chronic heart disease characterized by enlargements of the heart and congestive heart failure, which primarily affecting children 2–7 years of age and women of child-bearing age. Predicting where Se deficiency/toxicity occurs in China is very challenging due to selenium's heterogeneous distribution and our limited knowledge on what controls its heterogeneity.

It has been proposed that the largest natural flux of Se to the terrestrial environment is through wet deposition from the atmosphere (Ross, 1985; Wen & Carignan, 2007). These estimates show that the flux of atmospheric Se deposition via precipitation is 67–417% larger than the highest estimates of Se input through dry deposition. In China the precipitation gradient is controlled by the East Asian Summer Monsoon, which begins to bring moisture from the China Sea in late April-early May, and then progressively moves north and reaches its northern most extent in Northern China by the end of June and then subsequently retreats south (Yihui & Chan, 2005).

2 THE STUDY AREA

This study investigated the relationship between monsoonal climate and Se in the Chinese Loess Plateau, located in north-central China. The extensive, interbedded deposits of windblown mineral dust (loess) and paloesols that compose the Chinese Loess Plateau are an widely researched continental climate record, which has archived changes in the strength of the East Asian Monsoon system over the last 7 million years (Ma) (Liu & Ding, 1998).

Loess layers within these stratigraphic sections represent past glacial times where the East Asian Winter Monsoon was particular strong and large amounts of dust were transported from the deserts and inland basins lying to the north-northeast. The paleosol layers represent warmer interglacial times when the strength of the East Asian Summer Monsoon increased, and more weathering and pedogenesis of the loess occurred. The Chinese Loess Plateau, with its clear dichotomy between precipitation dominated times and dust dominated times offers a unique environmental archive, which has recorded Se deposition over a 6.8 Ma period.

In this paper we present the first long term (~6.8 Ma) record of atmospheric Se deposition in the Chinese Loess Plateau. The study site is at the Lingtai Loess Section, located in the wetter southern part of the Chinese Loess Plateau. The Lingtai loess section is composed of ~168 m of alternating Quaternary aged loess-paleosol layers underlain by approximately 120 m of the late Neogene Red Clay formation, which is a nearly continuous paleosol sequence with a basal age of ~6.8 Ma. This section has been extensively studied in the context of paleoclimate and much is known about changes in the paleomonsoon over the last ~6.8 Ma (Ding *et al.*, 1999; Sun *et al.*, 2010).

3 RESEARCH FINDINGS

We analyzed sediment samples from Lingtai for Se, and major and trace element chemistry (Fig. 1), using a combination of hydride generation atomic fluorescence spectroscopy (HG-AFS) and inductively coupled plasma mass spectrometry (ICP-MS). We have made comparisons between the HG-AFS and ICP-MS data and monsoon proxies that were determined for the same sample set (Sun *et al.*, 2010).

In paleosols between 2.6 and 1.3 Ma, we find very strong positive correlations between Se concentration and the monsoon proxies. In the paleosols between 1.26 and 0.16 Ma, the strong positive correlation with monsoon proxies is no longer present or there is a negative correlation.

Our results show that both precipitation and dust inputs associated with the East Asian Summer Monsoon and the East Asian Winter Monsoon, respectively, are important for Se distribution in the Chinese Loess Plateau. The distribution of this vital trace element (Se) is directly related to climatic factors. These findings will aid future efforts in predicting Se deficiency worldwide.

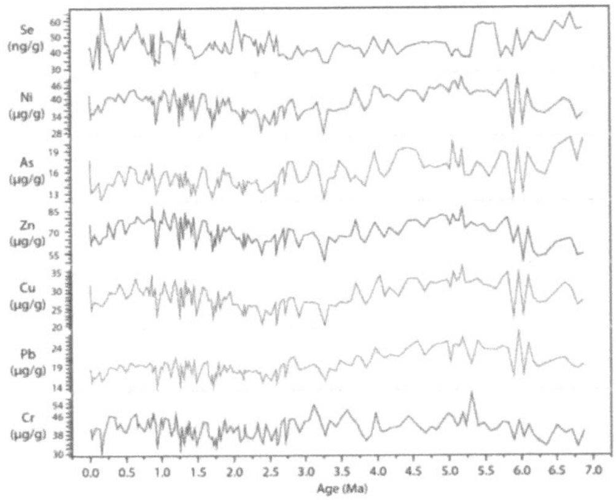

Figure 1. Elemental depth profiles for the Lingtai loess section for selected trace elements analyzed in this study.

REFERENCES

Chen, X., Yang, G., Chen, J., Chen, X., Wen, Z. & Ge, K. 1980. Studies on the relations of selenium and Keshan disease. *Biological Trace Element Research* 2: 91–107.

Ding, Z.L., Xiong, S.F., Sun, J.M., Yang, S.L., Gu, Z.Y. & Liu, T.S. 1999. Pedostratigraphy and paleomagnetism of a ~7.0 Ma eolian loess–red clay sequence at Lingtai, Loess Plateau, north-central China and the implications for paleomonsoon evolution. *Palaeogeography, Palaeoclimatology, Palaeoecology* 152: 49–66.

Liu, T. & Ding, Z. 1998. Chinese loess and the paleomonsoon. *Annual Review of Earth and Planetary Sciences* 26: 111–145.

Rayman, M.P. 2012. Selenium and human health. *The Lancet* 379: 1256–1268.

Ross, H.B. 1985. An atmospheric selenium budget for the region 30° N to 90° N. *Tellus* 37B: 78–90.

Stone, R. 2009. A Medical Mystery in Middle China. *Science* 324: 1378–1381.

Sun, Y., An, Z., Clemens, S.C., Bloemendal, J. & Vandenberghe, J. 2010. Seven million years of wind and precipitation variability on the Chinese Loess Plateau. *Earth and Planetary Science Letters* 297: 525–535.

Wen, H. & Carignan, J. 2007. Reviews on atmospheric selenium: Emissions, speciation and fate. *Atmospheric Environment* 41: 7151–7165.

Yihui, D. & Chan, J.C.L. 2005. The East Asian summer monsoon: an overview. *Meteorol. Atmos. Phys.* 89: 117–142.

Zhu, J. & Zheng, B. 2001. Distribution of selenium in a mini-landscape of Yutangba, Enshi, Hubei Province, China. *Applied Geochemistry* 16: 1333–1344.

Selenium in the Environment and Human Health – Bañuelos, Lin & Yin (Eds)
© 2014 Taylor & Francis Group, London, ISBN 978-1-138-00017-9

Fractionation and speciation of selenium in surface soil with deficient and adequate selenium

Dacheng Wang, Jin Li, Dan Liu & Yan Liu
Key Laboratory of Chemical Synthesis and Pollution Control, Sichuan Province, China
College of Chemistry and Chemical Engineering, China West Normal University, Nanchong, Sichuan, China

1 INTRODUCTION

Keshan disease and Kaschin-Beck Disease has been distributed in 309 and 303 counties, respectively, in 15 provinces in China due to the lack of Selenium (Se) in soil (Tan *et al.*, 1987). Though reduced dramatically in recent years, new cases of the two diseases are still reported in Sichuan, Qinghai, Tibet, Yunnan and other provinces (National Surveillance Team on Keshan Disease, 2008; Liu *et al.*, 2009). Until now the Se statues of a soil in China has been based on total Se concentrations, which do not always correlate with plant Se. Water soluble Se in soil is better correlated with plant Se (Nye and Peterson, 1975). Among main Se species in soil, only selenite and selenate are more available in plants, while humic acids bound Se (constitutes 30–50% of water soluble Se) cannot be used by plant (Li & Wang, 2002). The selenite and selenate fraction are better indicators for classifying the Se state of a soil. The present study investigated the ecological significance of total Se, water soluble Se, selenite, selenate and humic acid bound Se in soils that are in the Se deficiency and control areas in China.

2 MATERIALS AND METHODS

Soil samples were collected (0–20 cm) from Keshan and Kaschin-Beck Disease areas in Song-nen Plane in Heilongjiang, Aba in Sichuan, and Mouding in Yunnan. Samples from Pingan in Qinghai, Yanting and Nanchong in Sichuan were also collected as non-disease control. In Nanchong rapeseed plant and soil samples were collected to determine Se plant availability. The soil samples were air-dried, extracted with DDI water (w:w = 1:5) at the room temperature for 5 min. After the suspension settled overnight, it was vacuum filtered through 0.45 μm membrane filter. Selenite and selenate were separated with AG2-X8 anion exchange resin column, and humic acid bound Se was separated with XAD-8 column before Se analysis with 2,3-diaminonaphthalene (DAN) fluorometry (Wang *et al.*, 1994).

3 RESULTS AND DISCUSSION

According to the Ecological Threshold of Se in the Environment (Tan *et al.*, 1987), water soluble Se concentrations (<3 μg/kg) in Aba and Mouding indicate Se deficient soils and Song-nen Plane, Yanting and Nanchong levels (3–6 μg/kg) indicate sub-deficient, whereas Pingan in Qinghai is only considered adequate (6–8 μg/kg) (Table 1). Aba and Mouding are severely Se deficient areas with new cases of Kaschin-Beck Disease and Keshan disease reported recently. Yanting and Nanchong are, however, absolutely not Se deficient in soil and cereals.

The sum of the plant available selenite and selenate in Se deficient Aba and Mouding is 0.431 μg/kg and 0.168 μg/kg respectively, compared with Se adequate Yanting and Nanchong, 2.14 μg/kg and 0.584 μg/kg (Table 1). Interesting by the difference of the plant available Se in soil between severe deficient and absolute adequate is very small.

The average total Se in soil of Mouding is 87.6 μg/kg, and is considered deficient according to Tan's classification (<125 μg/kg). Selenite and selenate together accounted for only 0.19% of the total Se in soil. This portion of Se in soil predominantly controls the biogeochemical cycling

Table 1. The Se speciation in soil in different areas in China (μg/kg).

Sampling area	Water soluble	Selenite	Selenate	Humic bound
Soils from Se deficiency areas:				
Songnen plane (n = 42)	5.03	1.09	0.70	1.43
Aba (n = 38)	0.72	0.19	0.25	0.25
Mouding (n = 36)	0.58	0.08	0.09	0.133
Soils from high Se areas:				
Pingan (n = 13)	6.01	2.41	0.74	0.93
Yanting (n = 20)	3.50	1.80	0.34	0.83
Nanchong (31)	3.89	0.29	0.30	0.29

Table 2. The total and water soluble Se concentrations (μg/kg) in soil in different areas in China (Ecological Environment Research Team, 1991).

Location	Year	Total Se	Water soluble Se
Mouding (n = 57)	1984	64	0.795
Mouding (n = 36)	2012	87.6	0.583
Nanchong (n = 31)	2010	105	3.89
Songnen plane (n = 42)	2008	235	5.03

Table 3. The Se speciation in soil in Kashan disease areas (μg/kg, n = 3).

Location	Disease	Total Se	Wat. sol.	Selenite	Selenate
Luopingguan	non	85.0	0.545	0.080	0.165
Xindi	light	74.5	0.887	0.085	0.098
Hongdoushu	severe	62.7	0.595	0.101	0.089
Longwumiao	severe	52.6	0.797	0.149	0.079

of Se in food chain and is a more accurate measurement for the classification of the Se status of a soil.

The Se content of rapeseed plant sampled with surface soil in Nanchong exhibited no correlation with total Se in soil ($r = 0.0273$, $n = 27$), but it was significantly correlated with water soluble Se ($r = 0.361$, $P < 0.05$) and selenite in soil ($r = 0.529$, $P < 0.01$). Although the plant availability of selenate is 8–10 times higher than that of selenite (Ylaranta, 1983; Shand *et al.*, 1992), better correlation between rapeseed plant Se and selenate in soil was not found ($r = -0.0362$). To establish soil Se threshold with selenite and/or selenate more information is needed.

New incidences of both Keshan disease and Kaschin-Beck Disease has been reduced significantly in the last 30 years. In Mouding of Yunnan province, the total Se in soil showed a slight increase compared with that of the 1980's (Table 2), but the water soluble Se in soil decreased. The Se in soil, both total and water soluble, has remained deficient for last 28 years without sign of improvement. The reduced Keshan disease cases may due to the economical development and new food sources, even though it is often times insufficient to prevent locals from experiencing the Se deficiency disease. With a scattered occurrence of new Keshan disease new cases, sudden cardiac-related deaths in the area have been also related to the Se deficiency (Yang *et al.*, 2007). Effective measures are urgently needed to guarantee a sufficient Se intake in local population.

The Se speciation in cultivated soil of four villages in same catchment downstream in Chuxiong, Yunnan, which were affected by increase in Kashan disease in 1980's, showed a decrease in both total Se and selenate in soil (Table 3). In 1984, total Se concentration in soil, cereals and human hair in these villages were also decreased in same order (Tan *et al.*, 1991). The difference of the Se deficiency state in the area has not changed till today, and such small variation in total Se and Se speciation in soil caused local population with or without Keshan disease.

3 CONCLUSIONS

In Se deficiency areas in China, the humic acid bound Se is the largest portion (23–35%) in water soluble Se in soil, and selenite and selenate together is <0.431 μg/kg, while in areas with high and adequate Se in soil, selenite is the main species (up to 51%), and selenite and selenate is >0.584 μg/kg. The Se content of rapeseed plant has a better correlation with selenite in soil, and may be used as indicator for plant available Se. The Se deficient state in former Keshan disease area Mouding in Yunnan Province is still Se deficient since 1984. To prevent Se deficiency related diseases improvement of Se daily intake in local population is urgently needed.

REFERENCES

Ecological Environment Research Team, 1991. The research on ecological environment of Keshan disease area in Chuxiong. In W.H. Yu & F. Wang (eds.), *Collected Works of Comprehensive Survey of Keshan Disease in Chuxong, Yunnan, China,* 48–83. Beijing: People's Health Press. (in Chinese).

Li, Y.H. & Wang, W.Y. 2002. Progress on the research of soil environmental chemistry of selenium. *Chinese Journal of Soil Science* 33(2): 230–233. (in Chinese).

Liu, H., Zhou, L.W. & Liu, Y.Q. 2009. Analysis of national surveillance on Kaschin-Beck disease condition from 2000 to 2007. *Chinese Journal of Endemiology* 28(5): 554–556. (in Chinese).

National Surveillance Team on Keshan Disease, 2008. Analysis of national surveillance on Keshan disease in 2007. *Chinese Journal of Endemiology* 27(4): 412–415. (in Chinese).

Nye, S.M. & Peterson, P.J. 1975. The content and distribution of selenium in soils and plants from seleniferous areas in Eire and England. *Trace Substances in Environmental Health* 9: 113–121.

Shand, C., Coults, G., Duffs, E. & Atkinson, D. 1992. Soil selenium treatments to ameliorate selenium deficiency in herbage. *J. Sci. Food Agric.* 59: 27–35.

Tan, J.A., Li, R.B., Zheng, D.X., Zhu, Z.Y., Hou, S.F., Wang, W.Y. *et al.* 1987. Selenium ecological chemico-geography and endemic Keshan disease and Kaschin-Beck disease in China. In G.F. Combs, J.E. Spajjholz, O.A. Levander & J.E. Oldfield (eds.), *Selenium in Biology and Medicine; Proceedings of the Third International Symposium on Selenium in Biology and Medicine. Beijing, May 27-June1, 1984.* New York: Van Nostrand Reinhold Company.

Wang, D.C., Alfthan, G. & Aro, A. 1994. Determination of total and dissolved selenium species in natural water samples using fluorometry. *Environmental Science & Technology* 28: 383–387.

Yang, L., Huang, W.L., Wang, Y.B., Yang, L.P., Ma, L. & Li, Z.X. 2007. The correlation of unknown cause sudden cardiac death and selenium in cereals in Yunnan, China. *Journal of Preventive Medicine Information* 23(2): 73–74. (in Chinese).

Ylaranta, T. 1983. Effect of applied selenite and selenate on the selenium content of barley. *Ann. Agr. Finn.* 22: 166–179.

Selenium in the Environment and Human Health – Bañuelos, Lin & Yin (Eds)
© 2014 Taylor & Francis Group, London, ISBN 978-1-138-00017-9

Selenium speciation in rock from a high selenium region in Ziyang, China

Zhenzhen Ma & Zhengyu Bao
Faculty of Earth Science, China University of Geosciences, Wuhan, China

1 INTRODUCTION

Selenium (Se) is a potentially toxic trace element, however, its essentiality for humans has also been recognized. The toxicity, mobility and bioavailability of Se are controlled by chemical speciation in the environment. The fractionation of Se compounds in the rocks indicates that the weathering and rainfall causes the mobilization, conversion and transport of large amounts of originally immobile Se from the deposits into the surrounding environment (Bujdoš *et al.*, 2005). The determination of Se speciation is a key aspect to understanding its potential environmental consequences. Previous investigations have focused on studying the sources and distribution of total Se in the environment and its ecological effects. However, few reports are about the Se speciation trends in Se-rich rocks which are distributed in Ziyang, China. Therefore, the specific objectives of this study were mainly to investigate Se fractionation and transformation to analyze the environmental behavior of Se.

2 MATERIALS AND METHODS

Rock samples were collected from three local villages including Gaoqiao, Wamiao and Shuangan in a high Se region of Ziyang. Rock samples were air-dried at the room temperature (25°C), initially ground using a wood roller to pass through a No. 20-mesh sieve, and then further mechanically ground to less than 200 μm diameter powders for chemical analysis. All preparation and instrumental analyses were conducted at the Faculty of Material Science and Chemistry Engineering, China University of Geosci-ences. A five-step sequential extraction procedure was used according to Kulp and Pratt (2004) and Zhang *et al.* (1999) for the fractionation of Se in Se-rich rock samples as described in Table 1.

3 RESULTS AND DISCUSSIONS

3.1 *Distribution of total Se in the study area*

Mean concentrations of Se in the rock samples collected from Gaoqiao, Wamiao and Shuangan villages was 15.05 mg/kg, 13.30 mg/kg, and 26.83 mg/kg, respectively. The high Se samples in the Gaoqiao village are argillaceous limestone form Nan Muwan mine which include well crystal pyrite. In Wamiao village, there are high concentrations of Se that are thinly-bedded carbonaceous slate including high content of carbon and sludge. They are located in the horizon of *Lujiaping Formation* (lower Cambrian). Li *et al.* (1999) and Luo (2001) have collected a large number of samples on the geological strata in Ziyang, and their findings showed that, apart from *Yaolinghe Formation*, other formations have different levels of Se enrichment. The black shales of *Lujiaping Formation* and *Daguiping Formation* have the highest Se enrichment factor.

3.2 *Chemical fractionation of Se*

The Se concentrations in different fractions named as F1, F2, F3, F4 and F5 (see Table 1) were 1.42, 0.96, 5.05, 3.19 and 14.02 mg/kg, respectively. The proportions of Se in the fraction of F1, F2, F3, F4, and F5 were 5.7, 3.9, 20.5, 13.0, and 56.9%, respectively (Table 2). The average Se content in each fraction was in the order of F5 > F4 > F3 > F1 > F2. The sum of F1 and F2 fractions accounted for 10% of the total Se content. Sharmasarkar and Vance (1995) proposed the concepts of available, potentially available, and weatherable fractions. Available forms of Se includes water

Table 1. The selective sequential extraction procedure for Se fractionation analysis.

Fraction	Form of Se	Extractant	Method
F1	Water soluble Se	DDI water	2h, 25°C
F2	Ligand exchangeable Se	0.1MK_2HPO_4-KH_2PO_4	2h, 25°C
F3	Organic bound Se	0.1MNaOH	2h, 90°C
F4	Sulfide/selenide Se	KClO3+HCl	45 min
F5	Residue Se	HNO_3+H_2O_2+HF	digestion

Table 2. Selenium partitioning among different chemical fractions in rock collected from Shuangan village.

	Se (mg/kg)	
Fraction	Range (min-max)	Mean (n = 9)
F1	0.44–2.83	1.42
F2	0.11–2.19	0.96
F3	1.23–10.70	5.05
F4	0.62–12.13	3.19
F5	1.47–31.87	14.02

soluble fraction and ligand exchangeable fraction comprised approximately 20% and organic bound Se possessed major part. The available forms of Se are also prone to leaching in soil and easily taken up by plants. More research is needed on studying Se fraction and speciation in rock to soil transformation.

4 CONCLUSIONS

Available Se includes water soluble fraction and ligand exchangeable fraction comprised approximately 20% and organic-bound Se possessed major part. Bioavailable Se is subject to ground surface runoff, leaching into soil, and absorbed by plants. Future studies need to evaluate the rock-soil transformation, rock Se leaching experiments, and Se speciation analysis will also need to be carried out in the future.

REFERENCES

Bujdoš, M., Mulová, A., Kubová, J. & Medved, J. 2005. Selenium fractionation and speciation in rocks, soils, waters and plants in polluted surface mine environment. *Environmental Geology* 47: 353–360.

Kulp, T.R. & Pratt, L.M. 2004. Speciation and weathering of selenium in upper Cretaeeous chalk and shale from South Dakota and Wyoming, USA. *Geochimica et Cosmochimica Acta* 68(18): 3687–3701.

Li, D.P. 1999. The character of trace elements and advice for developing enrichment selenium crop in Ziyang. *Shanxi geology* 17(1): 67–73. (in Chinese).

Luo, K.L., Pan, Y.T., Wang, W.Y. & Tan, J.A. 2001. Selenium contents and distribution pattern in the Paleozoic strata in the southern Qinling Mountain. *Geological Review* 47(2): 211–217. (in Chinese).

Sharmasarkar, S. & Vance, G.F. 1995. Fractional partitioning for assessing solid phase speciation and geochemical transformations of soil selenium. *Soil Science* 160(1): 43–55.

Zhang, Zhong, Zhou, Liyi, & Zhang Qin. 1999. Speciation of selenium in geochemical samples by partial dissolution technique. *Rock and Mineral Analysis* 16(4): 255–261. (in Chinese).

Selenium in the Environment and Human Health – Bañuelos, Lin & Yin (Eds)
© 2014 Taylor & Francis Group, London, ISBN 978-1-138-00017-9

Selenium speciation in soils from selenosis area: Comparison between a Sequential Extraction Procedure and XAFS

Hai-Bo Qin & Jian-Ming Zhu
State Key Laboratory of Environmental Geochemistry, Institute of Geochemistry Chinese Academy of Sciences, Guiyang, China

Y. Takahashi
Department of Earth and Planetary Systems Science, Graduate School of Science, Hiroshima University, Kagamiyama, Higashi-Hiroshima, Hiroshima, Japan

Li-Rong Zheng
Beijing Synchrotron Radiation Facility, Institute of High Energy Physics, Chinese Academy of Sciences, Beijing, China

1 INSTRUCTION

Selenium (Se) has attracted considerable research attention due to its narrow range between dietary deficiency and toxicity for humans, and potentially paleoenvironmental proxy. Selenium naturally exists in multiple oxidation states (*e.g.*, -II, -I, 0, IV and VI) and as different organic Se compounds, with specific geochemistry properties (Kulp and Pratt, 2004; Lenz and Lens, 2009; Qin *et al.*, 2012). Therefore, knowledge of Se speciation is essential for understanding its potential bioavailability and impact on the environment.

Methods for determining Se speciation generally include Sequential Extraction Procedures (SEP) and X-ray Absorption Fine Structure (XAFS). The SEP methods have been developed to characterize the speciation, mobility, and bioavailability of Se in soils and sediments. However, there are some inherent limitations for SEP, including (i) the transformation and nonspecific extraction of Se species during multiple extraction steps; and (ii) the inconsistent comparisons in speciation results between different extraction protocols (Kulp and Pratt, 2004; Wright *et al.*, 2003). XAFS has been shown to provide direct, in situ information on Se speciation in natural samples with minimal preparation or treatment prior to analysis, but the limitation of XAFS is its relatively low sensitivity and it requires samples with high Se concentration (Pickering *et al.*, 1995; Ryser *et al.*, 2006). Because of these limitations in both methods, it is useful to combine SEP and XAFS to quantify Se speciation in soils and sediments.

Recently, this combined technique has been used to identify arsenic and lead speciation in natural samples (Funatsuki *et al.*, 2012; Huggins *et al.*, 2002). However, there are few comparable studies on Se speciation in natural samples by SEP and XAFS methods. In this study, SEP and XAFS methods were utilized to compare Se speciation in soils from Enshi, where previous studies have reported that Se-rich soils can be found (Qin *et al.*, 2012; Zhu *et al.*, 2008).

2 MATERIALS AND METHODS

Three soil (0–30 cm, S1-S3) and one weathered carbonaceous rock (S4) samples were collected from Enshi, where geology has been well described in a previous study by Zhu *et al.* (2008).

Selenium fractionations in soils were determined mainly by modifying a SEP summarized by Kulp and Pratt (2004): water soluble (MQ water) (F1), ligand exchangeable (0.1 M K2HPO4-KH_2PO4) (F2), base soluble (0.1 M NaOH) (F3), elemental (1 M Na_2SO_3) (F4), acetic acid soluble (15% CH_3COOH) (F5), sulfide/selenide (0.5 M $CrCl_2$) (F6), and residual (HNO_3+HF+H_2O_2) (F7).

Selenium K-edge XAFS spectra was obtained at beamline 1 W1B at the Beijing Synchrotron Radiation Facility (BSRF), Institute of High Energy Physics (IHEP), BL01B1 at SPring-8 (Hyogo, Japan). The XAFS data were analyzed using REX2000 software (Rigaku Co. Ltd.) and FEFF 7.02.

3 RESULTS AND DISCUSSIONS

3.1 *SEP*

Results of SEP for soil samples are shown in Figure 1. The organic matter associated Se (F3) is the predominant form in sample S1 and S3, which is in agreement with previous studies (Kulp and Pratt, 2004; Qin *et al.*, 2012), while Se mainly

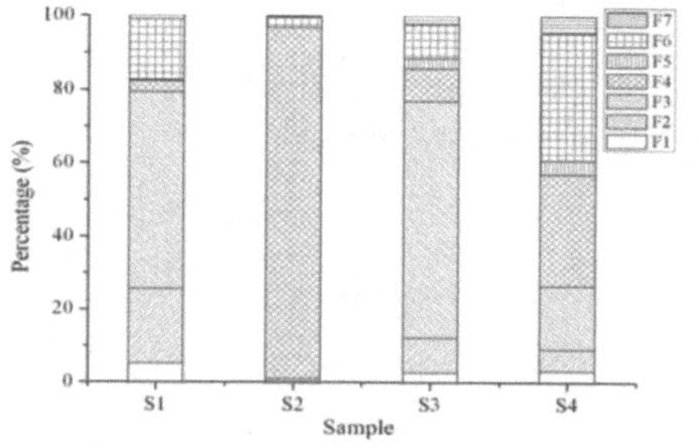

Figure 1. Selenium fractionation in soils in Enshi. S1–4: different soil samples; F1–7: different fractions of Se.

occurs in elemental Se (F4) fraction in sample S2 with extreme high Se content. However, for weathered carbonaceous rock sample, elemental and sulfide/selenide Se accounted for 65% of total Se in sample S4. This finding is consistent with similar studies on weathered shales (Kulp and Pratt, 2004; Matamoros-Veloza *et al.*, 2011).

3.2 *XAFS*

Numerous studies have demonstrated that the white line will shift as valency increases, and different XANES spectral reflects different Se oxidation state (Pickering *et al.*, 1995; Ryser *et al.*, 2006). The XANES spectral (Fig. 2) fitting results showed that Se(IV) is the main species in sample S1(72.6%) and S3(40.4%), Se(0) accounted to 93.7% of total Se in sample S2, and the predominant Se speciation in sample S4 is Se(0, -II) (80.4%).

3.3 *Assessment of the SEP/XAFS comparison*

Previous studies suggested that Se speciation in NaOH extract mainly exists in Se (IV) (Kulp and Pratt, 2004; Qin *et al.*, 2012). Hence, the dominant F3 fraction means Se (IV) is the main Se species in sample S1 and S3, which agrees with their XAFS results. For sample S2, elemental Se was the predominant Se form that was observed using both SEP and XAFS method. However, the SEP result is distinctly different with XAFS for sample S4. A previous study suggested that some selenide can be preferentially extracted by Na_2SO_3 that is used to extract elemental Se (Wright *et al.*, 2003). Hence, for samples with high selenide/sulfide Se, SEP method may overestimate elemental Se and underestimate selenide/sulfide Se in sample. Hence, caution must be taken when interpreting SEP data. However, SEP result can give some bioavailability information base on F1, F2, and F3 fractionations (Kulp and Pratt, 2004; Qin *et al.*, 2012), which is not reflected by XAFS technique. Therefore, combination of SEP and XAFS can identify and

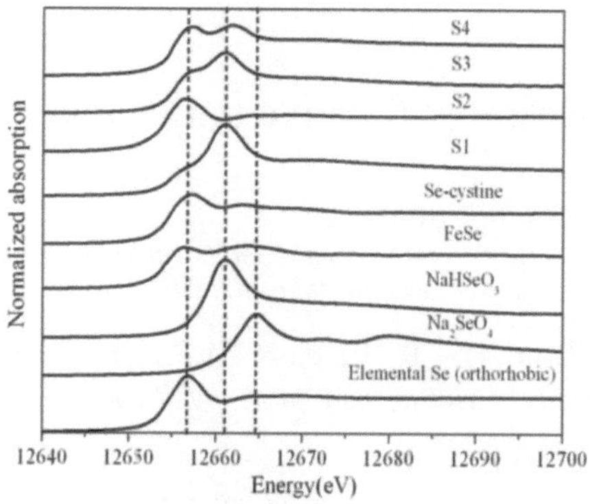

Figure 2. XAFS spectra of Se in soils collected from a high Se area in Enshi.

quantify Se speciation with high reliability and precision.

REFERENCES

Funatsuki, A., Takaoka, M., Oshita, K. & Takeda, N. 2012. Methods of determining lead speciation in fly ash by X-ray Absorption Fine-Structure Spectroscopy and a sequential extraction p\Procedure. *Analytical Sciences* 28: 481–490.

Huggins, F.E., Huffman, G.P., Kolker, A., Mroczkowski, S.J., Plamer, C.A. & Finkelman, R.B. 2002. Combined application of XAFS spectroscopy and sequential leaching for determination of arsenic speciation in coal. *Energy Fuels* 16: 1167–1172.

Kulp, T.R. & Pratt, L.M. 2004. Speciation and weathering of selenium in Upper Cretaceous chalk and shale from South Dakota and Wyoming, USA. *Geochim. Cosmochim. Acta* 68: 3687–3701.

Lenz, M. & Lens, P.N.L. 2009. The essential toxin: The changing perception of selenium in environmental sciences. *Sci. Total. Environ.* 407: 3620–3633.

Matamoros-Veloza, A., Newton, R.J. & Benning, L.G. 2011. What controls selenium release during shale weathering. *Appl. Geochem.* 26: S222–S226

Pickering, I.J., Brown, G.E. & Tokunaga, T.K. 1995. Quantitative speciation of selenium in soil using x-ray absorption spectroscopy. *Environ. Sci. Technol.* 29: 2456–2459.

Qin, H.B., Zhu, J.M. & Su, H. 2012. Selenium fractions in organic matter from Se-rich soils and weathered stone coal in selenosis areas of China. *Chemosphere* 86: 626–633.

Wright, M.T., Parker, D.R. & Amrhein, C. 2003. Critical evaluation of the ability of sequential extraction procedures to quantify discrete forms of selenium in sediments and soils. *Environ. Sci. Technol.* 37: 4709–4716.

Zhu, J.M., Wang, N., Li, S.H., Li, L. & Su, H.C. 2008. Distribution and transport of selenium in Yutangba, China: Impact of human activities. *Sci. Total. Environ.* 392: 252–261.

Selenium in the Environment and Human Health – Bañuelos, Lin & Yin (Eds)
© 2014 Taylor & Francis Group, London, ISBN 978-1-138-00017-9

Behaviour of native and applied selenium in Malawi soils

A.D.C. Chilimba
Ministry of Agriculture and Food Security, Department of Agricultural Research Services, Lunyangwa Research Station, Mzuzu, Malawi

S.D. Young, C.R. Black & M.R. Broadley
University of Nottingham, School of Biosciences, Sutton Bonington Campus, Loughborough, UK

J. Lammel
Yara International, Research Centre, Dülmen, Germany

1 INTRODUCTION

Selenium is an essential element for humans and livestock (Fairweather-Tait *et al.*, 2011). Suboptimal dietary Se intake is common in Malawi due to the low levels of available Se in most soils (Chilimba *et al.*, 2011). Chilimba *et al.* (2012a) reported a linear response to Se application in grain ($R^2 > 0.90$) and concluded that application of 10 g Se ha^{-1} to maize crops in Malawi would achieve adequate dietary intake. Thus, agronomic biofortification can increase the concentration of Se in maize grain. Although application of Se-enriched inorganic fertiliser increases grain Se in maize, it is vital to elucidate factors that affect Se dynamics in the soil. Different soil types adsorb selenite and selenate to differing extents due to variation in their physical, chemical and biological characteristics. The present study therefore examined the influence of soil properties on the behaviour of applied Se and native Se in different soil types in Malawi.

2 MATERIALS AND METHODS

Field experiments involved applications of 0 and 100 g Se/ha at Mbawa, Chitedze, Bvumbwe and Makoka Research Stations and replicated four times. Soil was sampled at harvest for the 0–15, 15–30 and 30–45 cm depths. In addition, a national survey, soil samples were collected based on soil FAO classification and covered major soil types in Malawi. Selenium was analysed in all soil samples collected using ICP-MS (Thermo-Fisher Scientific X-Series^{II}) after digesting soil samples in 40% AR hydrofluoric acid, 70% TAG nitric acid and 70% AR perchloric acid. KH_2PO_4 was used to determine extractable Seric acid.

3 RESULTS AND DISCUSSION

3.1 *Soil Se concentrations across soil types*

Total and extractable soil Se concentrations in the soil were higher in Humic Alisols, Ferralsols and Haplic Luvisols than in the other soil types. Extractable Se concentration was also high in vertisols suggesting that KH_2PO_4 extractant is a good indicator of Se availability only in vertisols, in which the values were correlated with grain Se (Chilimba *et al.*, 2011). Humic Alisols and Ferralsols are strongly acidic and selenate is reduced to selenite at low pH levels, which is likely the form of Se to be strongly absorbed by hydrous secondary iron oxides. By contrast, Se fixation would have been lower in the other, less acidic soils, thereby explaining the differences in available soil Se concentration. Haplic Lixisols are highly weathered, are dominated by low activity clays, and have a moderate to high base saturation with the result that Se fixation is likely to be lower than in Humic Alisols and Ferralsols. Eutric Cambisols and Chromic Luvisols have a higher base saturation and pH than Humic Alisols and Ferralsols and would therefore exhibit lower Se fixation.

3.2 *Relationship of soil Se with soil properties*

The total variation of 71%, 66% and 87% in soil Se can be explained by the linear relationship between soil organic matter, clay content and extractable Se. Soil organic matter and texture are among the most important soil properties associated with high soil fertility and so may form a component of soil management to increase soil Se through enhancement of soil organic matter in low input cropping systems in Malawi.

Table 1. Total soil Se, KH_2PO_4-extractable soil Se, and soil pH for the major soil types in Malawi. Data shown are means and standard errors.

Soil type (FAO)	Total Se (mg/kg)	KH_2PO_4 Se (μg/kg)	pH (H_2O)
Chromic Cambisols	0.20 ± 0.035	3.62 ± 0.52	4.70 ± 0.12
Chromic Luvisols	0.20 ± 0.021	4.68 ± 0.34	5.50 ± 0.12
Eutric Cambisols	0.14 ± 0.013	3.26 ± 0.19	5.97 ± 0.14
Eutric Planosols	0.12 ± 0.010	3.60 ± 0.31	5.61 ± 0.25
Ferralic Cambisols	0.09 ± 0.019	2.84 ± 0.47	5.56 ± 0.14
Ferralsols	0.45 ± 0.091	5.92 ± 0.67	4.62 ± 0.23
Haplic Lixisols	0.23 ± 0.020	4.21 ± 0.33	5.07 ± 0.08
Haplic Luvisols	0.41 ± 0.120	4.97 ± 0.70	6.19 ± 0.28
Humic Alisols	0.58 ± 0.077	6.46 ± 0.40	4.41 ± 0.04
Rhodic Ferralsols	0.15 ± 0.028	4.24 ± 0.47	4.98 ± 0.26
Vertisols	0.21 ± 0.006	6.28 ± 0.24	7.68 ± 0.10

3.3 *A variation in native and applied Se with soil depth*

At Mbawa, Chitedze, Bvumbwe and Makoka either native or applied Se increased with depth. At Ngabu, Se application increased soil Se at all depths ($P < 0.001$), and the values decreased with depth in the applied Se treatment ($P = 0.018$) but not in the control treatment. The behaviour of soil Se at Chitedze and Bvumbwe was similar as concentrations increased with depth after Se application ($P < 0.003$), while native Se was unaffected by depth at both sites. At Mbawa, Se increased in the topsoil, suggesting that applied Se was more available than the native Se. Chilimba *et al.* (2012b) also reported that maize grain at Mbawa accumulated a greater proportion of its Se from applied Se than at Chitedze, suggesting that soil type influences the behaviour of native and applied Se at these sites. The soils at Chitedze and Mbawa are respectively a Chromic Luvisol and a Haplic Lixisol, which confirms the earlier observation that applied Se is more available in Haplic Lixisols.

4 CONCLUSIONS

The behaviour of applied and native Se was affected by soil types and soil properties, such as clay and organic matter content, but further studies are needed to establish the precise nature of the correlation.

REFERENCES

Chilimba, A.D.C., Young, S.D., Black, C.R., Rogerson, K.B., Louise Ander, E., Watts, M.J. *et al.* 2011. Maize grain and soil surveys reveal suboptimal dietary selenium intake is widespread in Malawi. *Scientific Reports:* 1: 72.

Chilimba, A.D.C., Young, S.D., Black, C.R., Meacham, M.C., Lammel, J. & Broadley, M.R. 2012a. Agronomic biofortification of maize with selenium (Se) in Malawi. *Field Crops Research* 125: 118–128.

Chilimba, A.D.C. Young, S.D., Black, C.R. Meacham, M.C., Lammel, J. & Broadley, M.R. 2012b. Assessing residual availability of selenium applied to maize crops in Malawi. *Field Crops Research* 134: 11–18.

Fairweather-Tait, S.J., Bao, Y., Broadley, M.R., Collings, R., Ford, D. & Hesketh, J. *et al.* 2011. Selenium in human health and disease. *Antioxidants and Redox Signaling* 14: 1337–1383.

Selenium in the Environment and Human Health – Bañuelos, Lin & Yin (Eds)
© 2014 Taylor & Francis Group, London, ISBN 978-1-138-00017-9

Use of stable enriched selenite tracer in field experiments to investigate short- and long-term fate in soil of selenium

P. Di Tullo, M. Bueno, I. Le Hécho & F. Pannier
Laboratory of Bioinorganic Analytical and Environmental Chemistry (LCABIE), University of Pau et des Pays de l'Adour/CNRS, UMR 5254, IPREM, Hélioparc, Pau Cedex, France

Y. Thiry
Andra, Research and Development Division, Châtenay, Malabry Cedex, France

1 INTRODUCTION

Selenium (Se) as an essential trace element for living organisms is of great worldwide environmental concern because its concentration range between deficiency and toxicity is very narrow. Furthermore, among different radionuclides, ^{79}Se is of environmental concern for safety assessment of nuclear waste repositories because of its presence in high level long-lived nuclear waste. Thus, to prevent future health hazards—both related to Se excess or deficiency in food chain and possible radiological contamination—Se transformations and distribution among the different components of soil have to be clearly understood.

The purpose of this study was therefore to investigate and compare the dynamics of added and ambient Se in soil and thereby acquire insights into the processes involved in Se ageing and possible accumulation, migration to waters or transfer to vegetation. It is particularly important to link fundamental knowledge of the biogeochemistry of Se, obtained from instance from laboratory experiments or thermodynamic modeling, to field experiments in order to enhance understanding of Se in natural environment and consequently improve our ability to predict its behavior. To achieve this goal, we developed a new methodology involving the use of isotopic enriched tracers in field experiment that enable accurate and precise Se measurements at trace level by HPLC-ICP-MS. Simultaneous monitor of the reactivity of added and ambient Se in soils was based on selective extraction.

2 MATERIALS AND METHODS

Three soils types (agricultural, forest and grassland) were collected (5–15 cm depth) at the monitoring sites managed by Andra and located in the region of Meuse/Haute-Marne (France), a site selected for the future underground repository for nuclear wastes. The soils were spiked with isotopically enriched selenite (^{77}Se IV) at concentrations close to natural levels (µg/kg). Bags (10 × 10 cm size) prepared from a nylon tissue (20 µm stitch NYTAL 25TII) were filled with the spiked soils. The bags were buried on May 2012 in each soil type (*i.e.*, agricultural, forest and grassland) at 15 cm depth in order to monitor changes in distribution/speciation of ambient and added Se after different times of incubation over a 2-year period.

For both ambient and added Se, we evaluated the time course of solid/liquid partition coefficient (Kd), of distribution and speciation in different methodologically defined fractions (water soluble, exchangeable and organic-matter associated). Total and Se species concentration, both for ambient and added Se were determined by reverse isotope dilution (RID), with natural abundance Se standard used as a spike and ^{78}Se/^{77}Se ratio selected as measurement isotope pair (Hintelmann *et al.*, 1997). Operating condition and instrumental parameters of ICP-MS have been firstly optimized to overcome the spectral interferences encountered for accurate and precise Se isotopic ratio determination (Tolu *et al.*, submitted).

3 RESULTS AND DISCUSSION

3.1 *Ambient and added Se soil distribution*

For grassland and forest soils, added $Se_{water\text{-}soluble}$ fraction reached values close to the ambient one after 3 months. In contrast, added $Se_{water\text{-}soluble}$ fraction showed a different pattern for agricultural soil, decreasing along the six months incubation period without reaching Se ambient state (Fig. 1). The observed decrease of added $Se_{water\text{-}soluble}$ in agricultural soil may be partly due to Se losses in the system since added Se_{total} has decreased (23 ± 3% less than T_0 value) during the in situ experiment. In contrast, for forest soil no Se losses were observed. The decreasing solubility of Se is assumed to mainly result from Se adsorption by oxides and/or

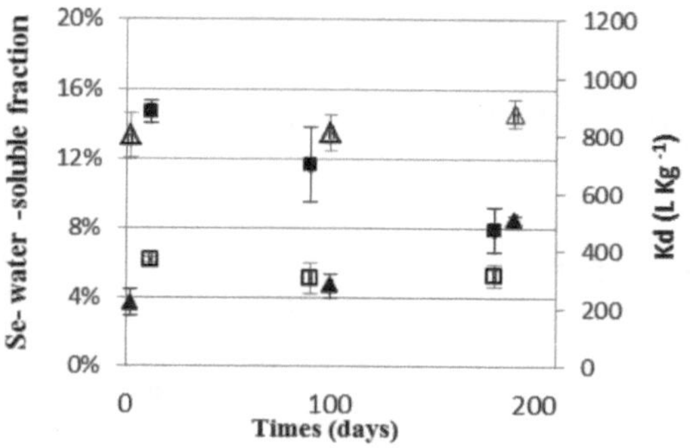

Figure 1. $Se_{water\text{-}soluble}$ fraction (square symbols) and Kd values (triangle symbols) of ambient (open symbols) and added Se (filled symbols) in the agricultural soil.

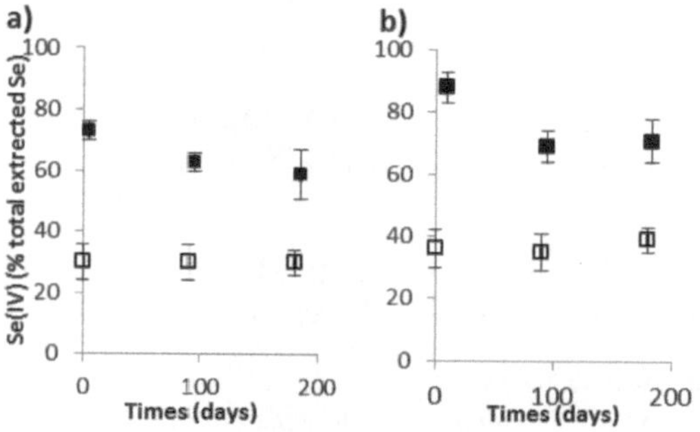

Figure 2. Ambient (open symbols) and added (filled symbols) Se(IV) proportion in soil extracts of forest soil. (a) exchangeable fraction; (b) OM fraction.

clay mineral phases. Added $Se_{exchangeable}$ fraction showed similar values for forest and grassland soils (32 ± 3%; 36 ± 1% respectively) without significant change during the considered time of incubation. For the agricultural soil this fraction decreased regularly but it remains still higher than the one of ambient Se after six months (57 ± 10% and 24 ± 4%, respectively). Moreover, for the three soils no modification of added Se fraction associated with OM was observed, showing extractabilities two or three times superior to that of ambient Se. Thus, no diffusion of added Se towards more thermodynamically stable binding sites (refractory to NaOH) was noticed till now.

3.2 *Ambient and added Se soil speciation*

For the three soils, ambient Se speciation was found to be stable over the incubation period, with only Se(IV) detected in the three soil extracts.

After the spiking-homogenization step, added Se(IV) was rapidly transformed into unidentified Se species and partially into Se(VI), as detected in the water-soluble fraction of the forest and grassland soil, and to a lesser extent in the agricultural soil. This modification was probably promoted during air-drying of soils after spiking. Furthermore, additional ^{77}Se IV chemical transformation into unidentified compounds occurred during the incubation period and they were observed in the three soils extracts with different kinetics as a function of soil type (Fig. 2).

It is interesting to note that even if ^{77}Se IV proportion in extracts decreased with time, speciation of ambient and added Se remained very different. Actually, a large reservoir of ambient Se is associated with a pool of unidentified Se species, *i.e.*, Se bound/incorporated into organic and/or organo-mineral colloids that could affect Se mobility or bio-availability in the long term.

4 CONCLUSIONS

The use of isotopically enriched tracer demonstrated to be very promising for a better monitoring and understanding of processes involved in long term Se cycling in soils. After six months of soils incubation under real environmental conditions, added Se and ambient Se showed different pattern in both speciation in and distribution between soils methodologically defined fractions, indicating of lower strength retention of added Se. First results thus evidence slow processes controlling Se fate in soils. These finding are relevant for managing both contamination and Se soil fertilization issues.

REFERENCES

Hintelmann, H. & Evans, R.D. 1997. Application of stable isotopes in environmental tracer studies-Measurement of monomethylmercury (CH3Hg+) by isotope dilution ICP-MS and detection of species transformation. *Fresenius Journal of Analytical Chemistry* 358: 378–385.

Tolu, J., Di Tullo, P., Le Hécho, I., Thiry, Y., Pannier, F., Potin-Gautier, M. & Bueno, M. 2013. A new methodology involving stable isotope tracer to simultaneously compare short and long-term selenium mobility in soils. *Analytical & Bioanalytical Chemistry* (submitted).

Selenium in the Environment and Human Health – Bañuelos, Lin & Yin (Eds)
 ISBN 978-1-138-00017-9

Use of stable isotope signatures in plants as a tool to explore the selenium cycle in the "critical zone"

H.M. Banning, M. Stelling, E. Eiche, A.K. Nothstein & T. Neumann
Karlsruhe Institute of Technology, Institute of Mineralogy and Geochemistry, Karlsruhe, Germany

R. Schoenberg
University of Tübingen, Geochemistry Group, Tübingen, Germany

M. Riemann & P. Nick
Karlsruhe Institute of Technology, Molecular Cell Biology, Botanical Institute, and Center for Functional Nanostructures (CFN), Karlsruhe, Germany

1 INTRODUCTION

The trace element selenium (Se) is of high environmental significance, because it is an essential nutrient and a toxin for humans and mammals. Selenium is predominantly transported to the organisms via the food chain. Hence, a deep understanding of the Se cycle in the so called "critical zone" (soil-plant-groundwater system) is crucial and is highly relevant for environmental issues such as remediation. Furthermore, the understanding of processes controlling, promoting or reducing the uptake, transport, accumulation and volatilization of Se in agricultural crops is key to avoid Se deficiency and toxic concentrations (Ellis & Salt, 2003). Based on that, the natural biogeochemical conditions as well as the applied agricultural methods in problematic areas can be monitored and adapted with the aim of an appropriate Se supply.

Stable Se isotopes are a promising tool to explore the Se cycle within the critical zone. Several studies (Schilling *et al.*, 2011; Mitchell *et al.*, 2011; Herbel *et al.*, 2002; Johnson, 2004; Johnson & Bullen, 2003; Ellis *et al.*, 2003; Clark & Johnson, 2008; Johnson *et al.*, 1999) examined the relation between isotopic signatures and biotic/abiotic processes taking place in the critical zone, *e.g.*, redox reactions (Fig. 1), sorption, and volatilization. Most of the processes within the plants and at the interfaces between plant and soil, groundwater and atmosphere are not explored regarding their isotope fractionation (Fig. 2). Under controlled conditions we will determine the isotope patterns in different plant parts induced by assessable or known processes (*e.g.*, uptake from water and atmosphere, metabolic and transformation processes within the plant, storage in grains, emission and volatilization). The aim is to acquire knowledge of characteristic isotope patterns and thereby reconstruct the underlying processes from in situ plant material by its isotope signatures.

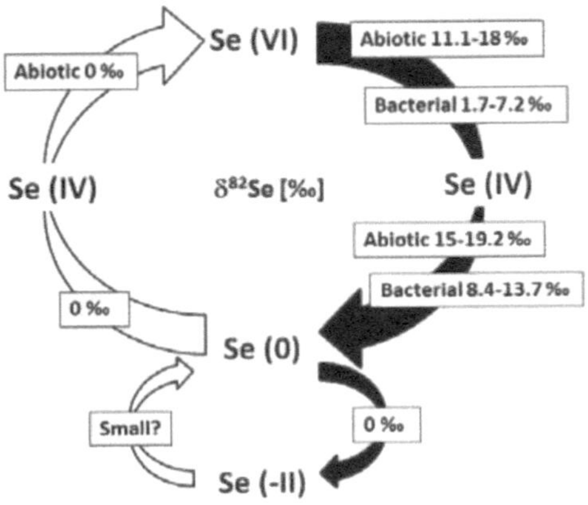

Figure 1. Selenium isotope fractionation during redox processes within the critical zone (white—oxidation; black—reduction) (modified after Johnson, 2004) (Schilling *et al.*, 2011; Herbel *et al.*, 2002; Ellis *et al.*, 2003).

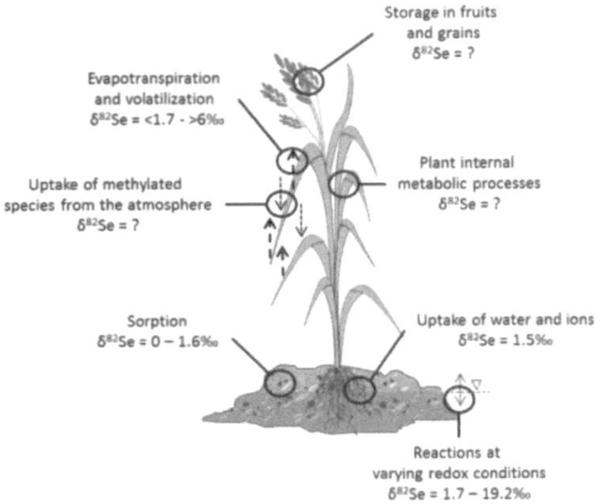

Figure 2. Selenium isotope fractionation during plant related processes (Mitchell *et al.*, 2003; see Fig. 1).

2 SELENIUM ISOTOPE ANALYTICS

Multi-Collector Inductively Coupled Plasma Mass Spectrometry (MC-ICP-MS) enables highly accurate measurements of the stable Se isotopic composition of environmental samples (*e.g.*, Zhu *et al.*, 2008). For isotope measurements we use a ThermoScientific Neptune Plus MC-ICP-MS. Isotope fractionation during instrumental mass bias and sample preparation is corrected by a $^{74}Se/^{77}Se$ double spike (Zhu *et al.*, 2008). An on-line Hydride Generator (HG) (Elemental Scientific hydride ICP system) is connected to exclusively carry the hydride generating elements (Se, arsenic, germanium and others) into the spectrometer. This step amplifies the Se signal and separates the main matrix elements from the target element Se.

3 SAMPLE PURIFICATION

The basic prerequisite for a precise isotope measurement is that the Se analyte is highly purified from the sample matrix in order to minimize isobaric interferences. Two different Se separation methods are prominently described in literature, named as the Method I (Clark & Johnson, 2008) and the Method II (Elwaer & Hintelmann, 2008). Both methods are based on anion exchange techniques in flow-through columns packed with different materials, such as BioRad AG1-X8 anion exchange resin for the Method I and thiol activated cellulose powder for the Method II. We tested and optimized these methods for different matrices with regard to purity and Se yield, as well as with time and cost. For this purpose, artificial test samples and digested plant tissue material has been used. Both methods achieved good results for Se purification (removal of matrix elements >99%). Yields ranged from 45 to 100% and were strongly dependent on the kind and quantity of matrix elements. Method I is superior regarding the elimination of matrix elements, especially concerning arsenic and copper. On the average, higher yields have been reached with Method II, which is more suitable for high matrix element concentration (*e.g.*, soils). Each method has its strengths and weaknesses, mainly depending on the sample matrix and the element concentrations.

4 CULTIVATION EXPERIMENTS

To investigate the Se isotope composition of plants we will conduct pot experiments using nutrient solution (hydroponics) and artificial soils (Nothstein *et al.*, 2013). In a first step (hydroponic) we will map the isotope signature of whole plants, plant parts and in Se volatiles as a function of Se species in nutrient solution and plant material. In a second step (artificial soils) we want to determine out if these isotope signatures in plants change as a function of land use and identify the conclusions that can be drawn on underlying processes.

REFERENCES

Clark, S.K. & Johnson, T.M. 2008. Effective isotopic fractionation factors for solute removal by reactive sediments: A laboratory microcosm and slurry study. *Environ. Sci. Technol.* 42: 7850–7855.

Ellis, D.R. & Salt, D.E. 2003. Plants, selenium and human health. *Curr. Opin. Plant Biol.* 6: 273–279.

Ellis, A.S., Johnson, T.M., Herbel, M.J. & Bullen, T.D. 2003. Stable isotope fractionation of selenium by natural microbial consortia. *Chem. Geol.* 195: 119–129.

Elwaer, N. & Hintelmann, H. 2008. Selective separation of selenium (IV) by thiol cellulose powder and subsequent selenium isotope ratio determination using multicollector inductively coupled plasma mass spectrometry. *J. Anal. At. Spectrom.*, 23: 733–743.

Herbel, M.J., Johnson, T.M., Tanji, K.K., Gao, S.D. & Bullen, T.D. 2002. Selenium stable isotope ratios in California agricultural drainage water management systems. *J. Environ. Qual.* 31: 1146–1156.

Johnson, T.M., Herbel, M.J., Bullen, T.D. & Zawislanski, P.T. 1999. Selenium isotope ratios as indicators of selenium sources and oxyanion reduction. *Geochim. Cosmochim. Acta* 63: 2775–2783.

Johnson, T.M. & Bullen, T.D. 2003. Selenium isotope fractionation during reduction by Fe(II)-Fe(III) hydroxide-sulfate (green rust). *Geochim. Cosmochim. Acta* 67(3): 413–419.

Johnson, T.M. 2004. A review of mass-dependent fractionation of selenium isotopes and implications for other heavy stable isotopes. *Chem. Geol.* 204: 201–214.

Mitchell, K., Couture, R.M., Johnson, T.M., Mason, P.R.D. & van Cappellen, P. 2011. Selenium sorption and isotope fractionation: iron(III) oxides versus iron(II) sulfides. *Chem. Geol.* 342: 21–28.

Nothstein, A.K., Eiche, E., Stelling, M., Kosmala, R., Eiche, E., Stelling, M. *et al.* 2013. Speciation-dependent uptake of Se by rice (Oryza sativa) from varying substrates. In G.S. Banuelos, Z.-Q. Lin & X.B. Yin (eds), *Selenium in the Environment and Human Health.* Boca Raton, FL: CRC Press.

Rouxel, O., Ludden, J., Carignan, J., Marin, L. & Fouquet, Y. 2002. Natural variations of Se isotopic composition determinded by hydride generation multiple collector inductively coupled plasma mass spectrometry. *Geochim. Cosmochim. Acta* 66(18): 3191–3199.

Schilling, K., Johnson, T.M. & Wilcke, W. 2011. Isotope fractionation of selenium during fungal biomethylation by *Alternaria alternate. Environ. Sci. Technol.* 45: 2670–2676.

Zhu, J.-M., Johnson, T.M., Clark, S.K. & Zhu, X.-K. 2008. High precision measurement of selenium isotopic composition by hydride generation multiple collector inductively coupled plasma mass spectrometry with a ^{74}Se-^{77}Se double spike. *Chin. J. Anal. Chem.* 36(10): 1385–1390.

Selenium in the Environment and Human Health – Bañuelos, Lin & Yin (Eds)
© 2014 Taylor & Francis Group, London, ISBN 978-1-138-00017-9

The source-partitioning of selenium volatilization in soil-*Stanleya pinnata* and *Brassica juncea* systems

L. Jones, V. Sever & Z.-Q. Lin
Southern Illinois University, Edwardsville, Illinois, USA

G.S. Bañuelos
USDA-ARS, Parlier, California, USA

1 INTRODUCTION

Prince's Plume (*Stanleya pinnata*) is a selenium (Se) hyperaccumulator known to accumulate over 1% of Se in its dry matter (Parker *et al.*, 2003). For comparison, non-hyperaccumulators (such as *Brassica juncea*) accumulate only about 0.001% of Se in its dry matter (Brown & Shrift, 1981). Because of a large amount of Se accumulated in shoot tissues, *S. pinnata* has been used to phytoextract Se from contaminated soil and water. It was proposed that the superior ability of *S. pinnata* in accumulating Se may be in part linked to the unique metabolism pathways of Se in these plants (Freeman & Bañuelos, 2011). LeDuc *et al.* (2004) indicated that in hyperaccumulator plants toxic inorganic selenate (SeO_4^{2-}) can be biotransformed to methylselenocysteine (MetSeCys), which can prevent the mis-incorporation of selenocysteine into proteins and therefore reduces Se phytotoxicity. MetSeCys can further be transformed to volatile dimethyl diselenide (DMDSe). The metabolism pathway of selenate to MetSeCys and DMDSe could be unique in Se hyperaccumulator plants, because the dominant volatile Se compounds produced by other non-hyperaccumulator species and soil microbes is dimethylselenide (DMSe), followed by small proportions of DMDSe (<5%) along with trace amounts of other volatile Se compounds (Mason, 2012; Gao *et al.*, 2000). With non-hyperaccumulator plants and soil microbes, previous studies suggest that the soil-root system be the major source of Se volatilization (Lin, 2008).

Biogenic Se volatilization in soil-plant systems represents an environmentally sound and sustainable remediation technology of Se-contaminated soil and water. Although high rates of Se volatilization in the *S. pinnata* field has been observed under field conditions by Freeman & Bañuelos (2011), the source-partitioning of Se volatilization in the soil-*S. pinnata* system has not been investigated. We speculated that the unique metabolism pathway of Se volatilization in *S. pinnata* might result in a different source-partitioning of Se volatilization, compared with non-hyperaccumulator plants. Thus, the specific objective of this study was to determine the amount of volatile Se dissipated from the soil and roots of *S. pinnata* versus the amount of Se volatilized from the shoots in the soil-plant system. For direct comparison, the source-partitioning of Se volatilization in the soil-Indian mustard (*Brassica juncea*) system was also determined under the same laboratory conditions.

2 METHODS AND MATERIALS

Stanleya pinnata seed was obtained from Western Native Seed (Colorado, USA). The seeds first underwent a 21-day cold-wet stratification prior to seed germination in 6-inch diameter plastic pots containing 1.7 kg of a mixture of agricultural soil, sand and peat moss in a ratio of 1:2:0.5. The soil was then further amended with 1 mg/kg (DW) Se from Na_2SeO_4 because Se-hyperaccumulator species grow better in soil with low levels of Se. After germination, *S. pinnata* seedlings were thinned to one seedling per pot, and continued growing for another 10 weeks in a plant growth chamber at 23°C, 70% relative humidity, and 75 $\mu M/m^2/s$ PAR with a 16-hour photo-period. *Brassica juncea* seeds were germinated directly in 6-inch diameter pots containing the same growing medium, but without the 1 mg/kg Se soil amendment. The brassica plants were grown under the same growth chamber conditions as *S. pinnata*.

When the plant seedlings reached the pre-mature growth stage (about 15 weeks for *S. pinnata* and 7 weeks for *B. juncea*), three pots of similar shoot biomass were selected. Each pot was treated with Se from Na_2SeO_4 solution for a soil Se treatment of 4 mg/kg (DW). The collection of volatile Se from the soil-plant system started immediately after the final Se treatment. Volatile Se was collected separately from the shoots and the soil-root compartment using double shoot versus soil/root volatilization chambers (Zayed & Terry, 1994). Volatile Se compounds from shoots and the soil-root compartment were collected using a series of gas-washing bottles containing H_2O_2 and NaOH trap solution (Lin & Terry, 2003). The trap solution was replaced every 24 hours, the total volume of trap solution in each washing bottle was recorded, and a trap solution sample was collected for total Se concentration analyses each time. The Se volatilization measurement continued for a time period of six days for each plant species under the same laboratory conditions of 24–26°C and 175–207 μM/m²/s PAR from a LED light source. The chemical analysis of Se was carried out using inductively-coupled plasma/mass spectrometry (ICP-MS). The cumulative Se mass volatilized from shoots versus from the soil-root compartment was calculated for each pot of a plant species.

3 RESULTS AND DISCUSSION

The cumulative Se mass volatilized from two different sources of the soil-*S. pinnata* system during a 6-day experimental period is shown in Figure 1. At the end of Day 6, more volatile Se was dissipated from shoots (15.2 μg/pot), compared with the soil-root compartment (3.5 μg/pot). On average *S. pinnata* shoots volatilized about 77% of the total Se (Fig. 1). A large variation in Se volatilization flux from shoots was observed, compared with the soil-root compartment that showed a range of 2.3–5.3 μg/pot among the three replicates. The ratio of Se volatilized from the shoots to the soil-root compartment also increased with increasing time.

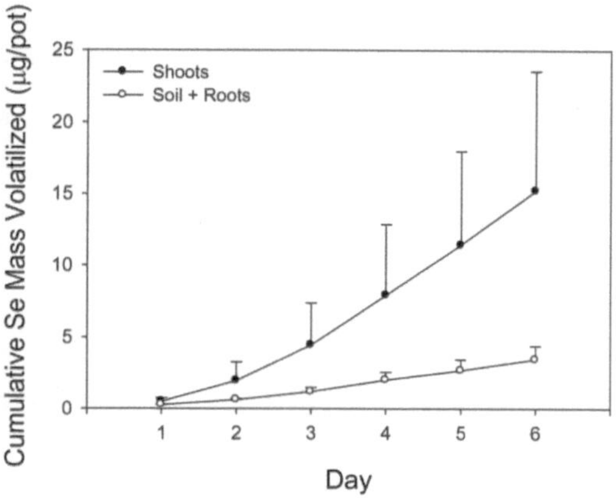

Figure 1. Cumulative Se mass volatilized from *S. pinnata* shoots and the root-soil compartment. Data shown are the means ± one standard error (n = 3).

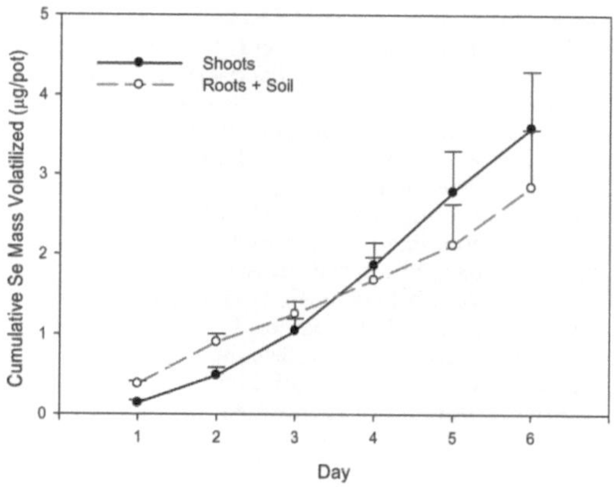

Figure 2. Cumulative Se mass volatilized from *B. juncea* shoots and from the soil-root compartment. Data shown are means ± on standard error (n = 3).

The total Se mass volatilized from the soil-*B. juncea* system was much less than the soil-*S. pinnata* system (Fig. 2). The shoots have slightly higher rates of Se volatilization (about 56%), compared to the soil-roots system. Zayed and Terry (1994) previously indicated that the plant root of broccoli in a hydroponic experimental system is the major source of volatile Se. Lin *et al.* (unpubl.) also observed that the shoots of *Salicornia bigelovii* only accounts for about 10% of the total Se mass volatilized from the soil-plant system. Clearly, the Se hyperaccumulator plant, *S. pinnata*, shows a different source partitioning of Se volatilization, compared with other non-hyper-accumulator species.

4 CONCLUSIONS

Different soil-plant systems may show significant source-partitioning of Se volatilization. While compared with a non-Se hyperaccumulator species, *B. juncea*, the shoot of *S. pinnata* volatilized the majority of volatile Se in the soil-plant system. Future research needs to focus on the effects of different chemical forms of Se on biogenic Se volatilization in the soil-*S. pinnata* system and the role of soil microbes in the source-partitioning of Se volatilization by Se-hyperaccumulator plants.

REFERENCES

Freeman, J.L. & Bañuelos, G.S. 2011. Selection of salt and boron tolerant selenium hyperaccumulator *Stanleya pinnata* genotypes and characterization of se phytoremediation from agricultural drainage sediments. *Environmental Science and Technology* 45: 9703–9710.

Gao, S., Tanji, K.K., Peters, D.W. & Herbel, M.J. 2000. Water selenium and sediment fractionation in a California flow-through wetland system. *J. Environ. Qual.* 29: 1275–1283.
LeDuc, D.L., Tarun, A.S., Montes-Bayon, M., Meija, J., Malit, M.F., Wu, C.P. *et al.* 2004. Overexpression of selenocysteine methyltransferase in Arabidopsis and Indian mustard increases selenium tolerance and accumulation. *Plant Physiology* 135(1): 377–383.
Lin, Z.-Q. 2008. Ecological process: volatilization. In S.E. Jorgensen & B.D. Fath (eds.), *Encyclopedia of Ecology*, 1st edition, 3700–3705. Oxford, U. Elsevier B.V.K.
Lin, Z.-Q. & Terry, N. 2003. Selenium removal by constructed wetlands: Quantitative importance of biological volatilization in the treatment of selenium-laden agricultural drainage water. *Environmental Science and Technology* 3: 606–615.
Mason, R.P. 2012. The methylation of metals and metalloids in aquatic systems. In A. Dricu (ed.), *Methylation—from DNA, RNA and Histones to Diseases and Treatments.* 271–302. ???: InTech.
Parker, D.R., Feist, L.J., Varvel, T.W., Thomason, D.N. & Zhang, Y. 2003. Selenium phytoremediation potential of *Stanleya pinnata*. *Plant and Soil* 249: 157–165.
Zayed, A.M. & Terry, N. 1994. Selenium volatilization in roots and shoots: Effects of shoot removal and sulfate level. *Journal of Plant Physiology* 143(1): 8–14.

2 *Effects of selenium on animal and human health*

Selenium in the Environment and Human Health – Bañuelos, Lin & Yin (Eds)
© 2014 Taylor & Francis Group, London, ISBN 978-1-138-00017-9

Selenium status and widespread endocrine diseases—a critical view

Lutz Schomburg
Institute for Experimental Endocrinology, Charité—Medical University Berlin, Germany

1 INTRODUCTION

Endocrinology is the science of endocrine glands, the secretion of hormones and their molecular and metabolic actions. Endocrine signaling is a universal and ubiquitous process. Already the desire to produce offspring involves a number of neuroendocrine and endocrine signals. Similarly, conception, growth and development are hormone-controlled processes. Defects in the endocrine system can lead to infertility, growth or developmental defects highlighting the central importance of hormones, endocrine glands and hormone receptors for human life and reproduction. However, not only development but also diseases and ageing are affected by endocrine glands and hormonal signals. Among the most prevalent endocrine diseases are diabetes mellitus, autoimmune thyroid diseases and osteoporosis. They all have in common a strong increase in incidence and prevalence during the last decades. Oddly enough, this tendency is partially due to an ever-improving health system and the resulting rise in life expectancy. Diabetes mellitus and osteoporosis are especially well-established diseases of the elderly. As ageing is often viewed as a result of an increased tone of reactive oxygen species causing molecular damage, measures aimed at improving a subject's defense system are considered as meaningful strategy for prolonging youth and avoiding age-related decline. Consequently, an increasing number of experimental and clinical studies are testing the importance of selenium (Se) status and Se supplementation for these endocrine diseases. However, the published study results are often over-interpreted and extrapolated in a non-scientific way, causing contradictions, misconceptions or unjustified hopes.

2 MATERIALS AND METHODS

The publications, insights and data used for this presentation were obtained either from own work or upon searching the PubMed database. Search key words included "selenium", "selenoprotein", "endocrinology", "autoimmune disease", "diabetes" and "osteoporosis" in different combinations. References of the identified articles were further analyzed.

3 RESULTS AND DISCUSSION

At the time of writing this abstract, a total of >400 articles on the topic "selenium AND diabetes" were published and available via PubMed, including 86 reviews. When studying this literature, a major bottom line emerges claiming that high Se intake causes type 2 diabetes mellitus (Rayman & Stranges, 2013; Rocourt & Cheng, 2013). This conclusion is mainly based on US American Association Studies, which by definition do not provide causal links. The most recent respective cross-sectional analysis reports that serum Se concentrations are on average higher in diabetic participants compared to non-diabetic participants by 7.3 μg/L (Laclaustra *et al.*, 2009). Interestingly, the average serum Se concentrations of these subjects were around 140 μg/l, and thus outside the range in which the expression of selenoproteins are affected by Se intake (Combs *et al.*, 2011). It is thus more than doubtful that the reported diabetes rates are related to selenoproteins. Moreover, these findings are of marginal importance for the majority of humans have Se concentrations below 100 μg/L. Well-controlled Se supplementation trials were conducted in the US to study causal relations. Analysis of the NPC data 7.7 years after initiation of the study (where by an extra 200 μg Se per day was provided to the participants for an average of 4.5 years) indicated that among the women participating, there were 9 new cases of self-reported diabetes in the Se group and 8 respective cases in the control group, *i.e.*, no difference (Stranges *et al.*, 2007). Similarly, the SELECT study did not observe a statistically increased risk for diabetes upon Se supplementation (Lippman *et al.*, 2009). Despite these inconclusive data, a common fear of diabetes induction has spread even in populations in which a selenium supplement would reduce a number of health risks thereby hindering a meaningful supplementation of selenium-poor subjects.

A similar belief instead of clear scientific rationale is currently dominating the area of Se supplementation for improving autoimmune thyroid diseases. Avoiding a Se deficit condition is clearly of importance for maintaining thyroid health and for reducing goiter and nodule risk in selenium-poor subjects (Derumeaux *et al.*, 2003;

Rasmussen *et al.*, 2011). However, the clinical results of supplementation trials are currently often interpreted as if they indicated that Se can treat the diseases, while in fact selenium's effects in Hashimoto's thyroiditis is relatively small and not uniformly observed in all trials alike. Moreover, the promising effects of Se in Graves eye disease have not been replicated yet (Schomburg, 2012). At present, it is thus definitely false to praise Se supplementation as an efficient thyroid disease treatment. Data on the interaction of Se and bone quality and osteoporosis risk are similarly sparse, fragmentary but providing a promising tendency.

4 CONCLUSIONS

Despite a number of published clinical studies on the importance of Se for wide-spread endocrinopathies, substantial knowledge gaps remain and premature findings of single trial results should not be extrapolated into medical advice. It is mandatory to discriminate between Se supplementation effects observed with well-supplied individuals and substitutions of selenium-poor subjects in need of the trace element, otherwise even more contradictions may arise.

ACKNOWLEDGEMENTS

Research in the author's laboratories is supported by the DFG, BMBF, and BMWi.

REFERENCES

Combs, G.F., Jr., Watts, J.C., Jackson, M.I., Johnson, L.K., Zeng, H., Scheett, A.J. et al. 2011. Determinants of selenium status in healthy adults. *Nutr. J.* 10: 75.

Derumeaux, H., Valeix, P., Castetbon, K., Bensimon, M., Boutron-Ruault, M.C., Arnau, J. *et al.* 2003. Association of selenium with thyroid volume and echostructure in 35- to 60-year-old French adults. *Eur. J. Endocrinol.* 148(3): 309–315.

Laclaustra, M., Navas-Acien, A., Stranges, S., Ordovas, J.M. & Guallar, E. 2009. Serum selenium concentrations and diabetes in U.S. adults: National Health and Nutrition Examination Survey (NHANES) 2003–2004. *Environ. Health Perspect.* 117(9): 1409–1413.

Lippman, S.M., Klein, E.A., Goodman, P.J., Lucia, M.S., Thompson, I.M., Ford, L.G. *et al.* 2009. Effect of selenium and vitamin E on risk of prostate cancer and other cancers: the Selenium and Vitamin E Cancer Prevention Trial (SELECT). *JAMA* 301(1): 39–51.

Rasmussen, L.B., Schomburg, L., Kohrle, J., Pedersen, I.B., Hollenbach, B., Hog, A. *et al.* 2011. Selenium status, thyroid volume, and multiple nodule formation in an area with mild iodine deficiency. *Eur. J. Endocrinol.* 164(4): 585–590.

Rayman, M.P. & Stranges, S. 2013. Epidemiology of selenium and type 2 diabetes: Can we make sense of it? *Free Radic. Biol. Med.* http://dx.doi.org/10.1016/j.freeradbiomed.2013.04.003

Rocourt, C.R. & Cheng, W.H. 2013. Selenium supranutrition: are the potential benefits of chemoprevention outweighed by the promotion of diabetes and insulin resistance? *Nutrients* 5(4): 1349–1365.

Schomburg, L. 2012. Selenium, selenoproteins and the thyroid gland: interactions in health and disease. *Nat. Rev. Endocrinol.* 8(3): 160–171.

Stranges, S., Marshall, J.R., Natarajan, R., Donahue, R.P., Trevisan, M., Combs, G.F. *et al.* 2007. Effects of long-term selenium supplementation on the incidence of type 2 diabetes: a randomized trial. *Ann. Intern. Med.* 147(4): 217–223.

Selenium in the Environment and Human Health – Bañuelos, Lin & Yin (Eds)
© 2014 Taylor & Francis Group, London, ISBN 978-1-138-00017-9

The effect of mycotoxins and selenium on the modification of cartilage proteoglycan sulphation

Junling Cao, Mingxiu Luo, Jinghong Chen, Qiang Fu & Zengtie Zhang
Institute of Endemic Diseases, College of Medicine, Xi'an Jiaotong University, Xi'an, Shaanxi, China
Key Laboratory of Environment and Genes Related to Diseases (Xi'an Jiaotong University), Ministry of Education, China
Key Laboratory of Microelement and Endemic Disease (Xi'an Jiaotong University), Ministry of Health, China

Bruce Caterson
Laboratory of Connective Tissue Biology, School of Biosciences, Cardiff University, Cardiff, UK

1 INTRODUCTION

Kashin-Beck Disease (KBD) is an endemic and chronic disease and causes severe and progressive osteoarthropathy in China. Clinical symptoms of the disease involve limb joint enlargement and pain and movement restriction. The KBD impacts human health primarily in the Se-deficiency areas from northeast to southwest of China. The characteristic pathological changes in KBD patients show an over-apoptosis in epiphyseal plate chondrocyte, developmental disorder of secondary ossification endochondrale, and chondrocyte's necrosis in the deep layer of articular cartilage. The etiology of KBD is still not clear. Recently, the etiologic studies of KBD were mainly focused on low environmental selenium (Se) and its interaction with mycotoxins, such as nivalenol (NIV), deoxynivalenol (DON) and T-2 toxin in food.

Sulfation of extracellular matrix proteoglycan (PG) in articular cartilage is an important step in maintaining normal cartilage metabolism. However, the sulfation degree of Chondroitin Sulfate (CS) in the cartilage matrix of KBD patients was decreased and the sulfate metabolism of cartilage was also abnormal. These observations were found earlier in 1984 from the scientific investigation of KBD in Yongshou county in China. Guo *et al.* (1997) found that the serum from KBD children could significantly inhibited ^{35}S incorporation rate in cultured chondrocytes. A series study showed that the sulfation degree of cartilage chondroitin sulfate-proteoglycan (CS-PG) was decreased in experimental animals fed with water and grain from KBD areas. The abnormal metabolism of cartilage matrix in experimental animals and KBD patient's body was related to low Se. Consequently, Se supplementation can improve cartilage matrix metabolism. The serum of KBD patients had contained many factors that interfered with the metabolism of PG, which could reduce molecular weight of aggrecan monomer and the capacity of aggrecan aggregates forming. Relatedly, the biological activity of cytokines promoting PG decomposition and degradative enzyme in synovial fluid and serum of KBD patients were increased.

In this study, we investigated the potential role of the sulfate modification of cartilage aggrecan on the pathogenesis of KBD patients. In addition, the effects of mycotoxins or presence of Se or absence to sulphate modification of aggrecan in extracelluar matrix on human cartilage cell line C28/I2 was evaluated. These in vitro chondrocytes and cartilage cultures mimic the KBD environments and elucidate the potential factors that cause the cartilage development degeneration and ageing.

2 MATERIALS AND METHODS

Samples of articular cartilage were divided into 2 groups: control children (5 samples), and children with KBD (5 samples). The morphology and pathology of hand joint cartilage of KBD children were examined by light microscopy. The expression of proteoglycans was determined by histochemical staining. The aggrecan, PAPSS2, PAPST1, CHST15, ARSB and GALNS localization were performed by inmmunohistochemistry. Articular cartilage was classified in 3 zones, and positive findings were counted by light microscopy for cytoplasmic staining by antibodies of PAPSS2, PAPST1, CHST15, ARSB and GALNS.

The C28/I2 chondrocyte was monolayer cultures in vitro. The MTT assay was used to assess the chondrocyte proliferation with different concentration of mycotoxins; and Real-time PCR and Western Blot methods were used to detect the expression of mRNA and protein of aggrecan, PAPSS2, PAPST1, CHST15 and ARSB respectively.

3 RESULTS AND DISCUSSION

The chondrocytes staining percentages of PAPSS2, PAPST1 and CHST15 in KBD patients were significantly lower than in controls ($p < 0.05$), and in the upper and middle zones of KBD samples. The chondrocytes staining percentage of the ARSB and GALNS were significantly higher than in controls. Staining of PAPSS2, PAPST1 and CHST15 in KBD samples significantly decreased in the upper zone (44.61 ± 5.65%: 18.38 ± 3.22%, 52.21 ± 9.12%: 16.79 ± 3.42%, and 61.87 ± 13.38%: 13.14 ± 1.17%, respectively) and the middle zone (21.53 ± 6.24%: 12.47 ± 1.47%, 23.49 ± 6.07%: 7.35 ± 1.42%, and 33.47 ± 4.74%: 6.75 ± 0.61%, respectively) of articular cartilage. Staining ARSB and GALNS in KBD samples significantly increased in the upper zone (0.00 ± 0.00%: 43.79 ± 11.05%, 0.00 ± 0.00%: 35.53 ± 4.24%, respectively) and the middle zone (0.00 ± 0.00%: 27.12 ± 3.45%, 0.00 ± 0.00%: 14.45 ± 1.53%, respectively) of articular cartilage.

This study showed that the different level mycotoxins could inhibit the proliferation of chondrocyte, and affect the expression of enzymes related to sulfate modification of Aggrecan. The synthetic interrelated enzymes PAPSS2, PAPST1 and CHST15 were decreased and the catabolism related enzymes ARSB were increased. Supplementing with physiological Se could partly inhibit those effects of mycotoxins, but the antagonism was incomplete.

4 CONCLUSIONS

Our results suggest that the altered sulfate modification of cartilage proteoglycan has an important effect on the pathogenesis of KBD. Mycotoxins could obviously inhibit the growth of chondrocyte, and avert an effect similar to the sulphate modification of cartilage proteoglycan by altering the expression of some related enzymes. Supplementing the physiological selenium could partly rivalry those effects of mycotoxins. Our study provides strong evidence for further studying the etiology and pathogenesis of KBD.

ACKNOWLEDGEMENTS

This research was supported by NSFC (No. 31070725, 30872187, 30471499 and 30170831).

REFERENCES

Cao, Junling. 2012. Relationship of proteoglycan with the structure and function of cartilage and osteoarthropathy. *Journal of XI'an Jiaotong University (Medical Sciences)*. 33(2): 131–136.

Cao, J.-L., Li, S.-Y., Shi, Z.-L., Yue, Y., Sun, J., Chen, J. *et al.* 2008. Articular cartilage metabolism in patients with Kashin-Beck Disease: an endemic osteoarthropathy in China. *Osteoarthritis and Cartilage* 16(6): 680–688.

Chen, J.-H, Chu, Y.-L., Cao, J.-L., Wang, W., Liu, J.-Y. & Wang, J.-L. 2011. Effects of T-2 toxin and selenium on chondrocyte expression of matrix metalloproteinases (MMP-1, MMP-13), α 2-macroglobulin (α2M) and TIMPs. *Toxicology in Vitro* 25: 492–499.

Guo, C.Y., Cao, J.L. & Zhang, S.Y. 1997. Effect of serum from children with Kashin-Beck disease on proteoglycan metabolism of bovine articular cartilage in steady-state culture. *China. J. Epidem.* 16(3): 143–144.

Li, S.-Y, Cao, J.-L., Shi, Z.-L., Zhang, Z.T., Hughes, C.E. et al. 2008. Promotion of the articular cartilage proteoglycan degradation by T-2 toxin and selenium protective effect. *Journal Zhejiang University Science B* 9(1): 22–33.

Minling Lu, Jinling Cao, Fuqiang Liu, Siyuan Li, Jinghong Chen, Qiang Fu *et al.* 2012. The Effects of mycotoxins and selenium deficiency on tissue engineered cartilage. *Cells Tissues Organs* 196(3): 241–250.

Siyuan Li, Jinling Cao, Caterson, B. & Hughes, C.E. 2012. Proteoglycan metabolism, cell death and Kashin-Beck Disease. *Glycoconjugate J.* 29: 241–248.

Stone, R. 2009. A medical mystery in middle China—China has launched a massive effort to stamp out Kashin-Beck disease, including moving populations from affected areas, but the cause of this crippling ailment remains elusive. *Science* 324: 1378–1381.

Zhang, A., Cao, J.-L., Yang, B., Chen, J.H., Zhang, Z.T., Li, S.Y. *et al.* 2010. The effects of moniliformin and selenium on human articular cartilage metabolism and their potential relationships to the pathogenesis of Kashin-Beck Disease. *Journal of Zhejiang University Science B* 11(3): 200–208.

Zhongli Shi, Junling Cao, Jinghong Chen, Zengtie Zhang, Bo Yang & Shuangqing Peng. 2009. Butenolide induced cytotoxicity by disturbing the prooxidant-antioxidant balance, and antioxidants partly quench in human chondrocytes. *Toxicology in Vitro* 23: 99–104.

Selenium in the Environment and Human Health – Bañuelos, Lin & Yin (Eds)
© 2014 Taylor & Francis Group, London, ISBN 978-1-138-00017-9

Selenium nutrition in Africa

E.J.M. Joy, C.R. Black, S.D. Young & M.R. Broadley
School of Biosciences, University of Nottingham, Sutton Bonington, Leicestershire, UK

E.L. Ander & M.J. Watts
British Geological Survey, Keyworth, Nottingham, UK

A.D.C. Chilimba
Ministry of Agriculture and Food Security, Department of Agricultural Research Services, Lunyangwa Research Station, Mzuzu, Malawi

1 INTRODUCTION

Selenium (Se) is an essential element with many roles in human health (Fairweather-Tait *et al.*, 2011). The risk of Se deficiency can be assessed reliably by measuring the concentration of Se in whole-blood or plasma fractions, or through direct dietary intake analysis. However, the collection of blood and diet samples is an invasive and time-consuming process. Furthermore, the preparation and analysis of material for Se analyses requires access to reliable and expensive analytical equipment. Thus, many countries have no nationally-representative dietary or clinical surveys of Se intakes or status, and therefore the prevalence of Se deficiency risks remains difficult to quantify in many regions.

Published Food Balance Sheets (FBSs), representing net food availability at a retail level, and food composition data can be used to estimate the daily *per capita* supply of mineral elements. These supply data can then be combined with an Estimated Average Requirement (EAR) cut-point to estimate population deficiency risks. This method is based on the assumption that net food supply is consumed in full with no storage losses or household waste. Inter- and intra-household economic, social or gender inequalities are captured by a generic 25% inter-individual coefficient of variation, which is assumed to be normally-distributed. Whilst these assumptions are major simplifications of reality, this EAR cut-point method can be used to estimate deficiency risks in broad and comparative terms. Such an approach has been used previously to estimate global deficiencies of Zn (Wessells and Brow, 2012), and Mg/Ca deficiencies in Africa (Broadley *et al.*, 2012; Joy *et al.*, 2013).

The aim of this study was to determine the risk of dietary Se inadequacy in Africa using net food availability and food Se composition data, and builds on recent published work on Se food composition and food supply data (Hurst *et al.*, 2013).

2 MATERIALS AND METHODS

We estimated the *per capita* supply of dietary Se available for human consumption in 46 African countries using an approach described previously for Mg (Joy *et al.*, 2013), and further developed for Se by Hurst *et al.* (2013). For this, we used published Food and Agriculture Organization (FAO) FBSs for 2009 (FAO, 2012; N.B. 2007 FAO data were used for Democratic Republic of Congo) and food Se composition data which captured some of the broad regional differences in food composition in Africa (Hurst *et al.*, 2013). The risk of inadequate Se intake was estimated using an EAR cut-point method described previously (Wuehler *et al.*, 2005; Joy *et al.*, 2013). An EAR for Se was estimated for each country, using United Nations 2010 population data distributed by age groups, which was within the range 36 to 41 µg Se per day.

3 RESULTS AND DISCUSSION

3.1 *Per capita energy and selenium supplies*

Median *per capita* energy supply among 46 African countries from 2009 FBSs was 2637 kcal d^{-1}, range 1553 (Burundi) to 3842 (Morocco). Median *per capita* Se supply at a national scale, which was based on combining FBSs with published food Se composition data, was 62 µg d^{-1}, ranging from 24 (Djibouti) to 216 (Ghana) and 219 (United Republic of Tanzania) (Fig. 1).

3.2 *Selenium deficiency risks in Africa*

The prevalence of Se deficiency in Africa (*i.e.*, the total number of people with access to Se supply below their national mean EAR, divided by the total population of Africa) was 19%. The median national-level deficiency risk among the

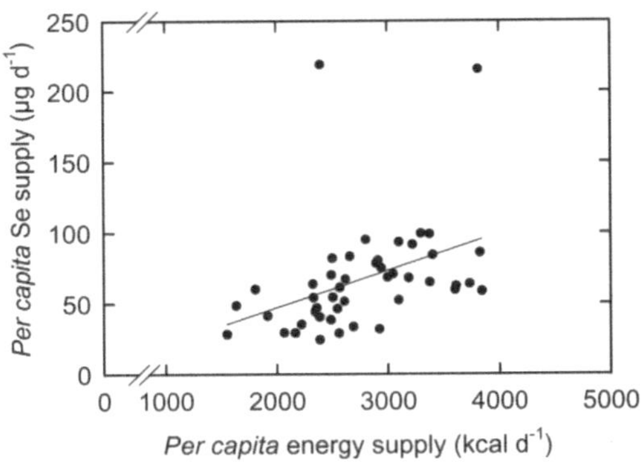

Figure 1. *Per capita* Se supply as a function of energy supply among 46 African countries. Data are estimates based on published food balance sheets, combined with food Se composition data.

46 countries was 6.5%, with UR Tanzania (<1%) and Djibouti (99%) the counties with the lowest and highest Se deficiency risk, respectively. The prevalence of Se deficiency risks was estimated to be lower in W. and N. Africa, but >60% in Burundi (92%), Democratic Republic of the Congo (83%), Djibouti, Lesotho (79%), Malawi (64%), Zambia (84%) and Zimbabwe (89%).

4 CONCLUSIONS

These data show that Se deficiency is likely to be widespread in Africa, albeit variable from country-to-country. Our analysis is highly sensitive to (often very limited) food Se composition data and other methodological assumptions regarding the distribution of food supply and its intake at an individual level. However, it is expected that identifying Se deficiency risks based on food supply is more likely to underestimate rather than overestimate deficiency risks, given that not all food will be consumed and given that inter-individual variation in consumption is likely to be high based on socio-economic and other factors. Despite these caveats, direct observations of Se intake and status among small groups within Malawi and Ghana are consistent with our more general country-level observations (Hurst *et al.*, 2013). Nevertheless, further data are required to assess the true scope of Se deficiency risks in Africa.

In countries/regions where Se deficiency risks are high, it may be feasible to adopt agriculturally-based public health programmes to enhance the mineral composition of foodstuffs, for example, by applying fertilisers enriched with Se to staple crops (Chilimba *et al.*, 2012).

REFERENCES

Broadley, M.R., Chilimba, A.D.C., Joy, E.J.M., Young, S.D., Black, C.R., Ander, L.E. *et al.* 2012. Dietary requirements for magnesium, but not calcium, are likely to be met in Malawi based on national food supply data. *International Journal for Vitamin and Nutrition Research* 82: 192–199.

Chilimba, A.D.C., Young, S.D., Black, C.R., Meacham, M.C., Lammel, J. & Broadley, M.R. 2012. Agronomic biofortification of maize with selenium (Se) in Malawi. *Field Crops Research* 125: 118–128.

Fairweather-Tait, S.J., Bao, Y.P., Broadley, M.R., Collings, R., Ford, D., Hesketh, J. & Hurst, R. 2011. Selenium in human health and disease. *Antioxidants and Redox Signaling* 14: 1337–1383.

Food and Agriculture Organization (FAO). 2012. Food Balance Sheet data. http://faostat.fao.org. Accessed 1st March 2012.

Hurst, R., Siyame, E.W.P., Young, S.D., Chilimba, A.D., Joy, E.J., Black, C.R. *et al.* 2013. Soil-type influences human selenium status and underlies widespread selenium deficiency risks in Malawi. *Scientific Reports* 3: 1425.

Joy, E.J.M., Young, S.D., Black, C.R., Ander, E.L., Watts, M.J. & Broadley, M.R. 2013. Risk of dietary magnesium deficiency is low in most African countries based on food supply data. *Plant and Soil* 368(1–2): 129–137.

Wessells, K.R. & Brown, K.H. 2012. Estimating the global prevalence of zinc deficiency: results based on zinc availability in national food supplies and the prevalence of stunting. *PLoS ONE* 7: 11.

Wuehler, S.E., Peerson, J.M. & Brown, K.H. 2005. Use of national food balance data to estimate the adequacy of zinc in national food supplies: methodology and regional estimates. *Public Health Nutrition* 8: 812–819.

Selenium in the Environment and Human Health – Bañuelos, Lin & Yin (Eds)
© 2014 Taylor & Francis Group, London, ISBN 978-1-138-00017-9

Dietary selenium intake in Zangtang county, a Kashin-Beck disease area in Sichuan Province, China

T. Yu, Z.F. Yang, Y.Y. Lv & Q.Y. Hou
China University of Geosciences, Beijing, China

L.X. Jin
Sichuan Institute of Geological Survey, Chengdu, Sichuan, China

1 INTRODUCTION

The trace mineral element selenium (Se) is an essential nutrient of fundamental importance to human and other animals' health, but greater amounts can be toxic (Combs & Combs, 1986; Wang, 1993; Fordyce, 2007; Huang *et al.*, 2013). Of all the elements, Se has one of the narrowest ranges between dietary deficiency (<40 μg/day) and toxic levels (>400 μg/day). For this reason it is necessary to carefully control its intake by humans and other animals and understand its relationship between environmental exposure and health (Yang & Xia, 1995; WHO, 1996). The link between the environmental biogeochemistry of Se and health outcomes have been documented for many years, as well as the relationship between Se deficiency and the Kaschin-Beck Disease (KBD) (Tan & Huang, 1991; Tan *et al.*, 2002; Fordyce, 2005). In this paper, dietary selenium intake in a typical KBD area will be discussed.

2 MATERIALS AND METHODS

2.1 *Study areas*

Zangtang County, where KBD is still active and severe, is located in the northwest of Sichuan Province and in the southeast of the Tibetan Plateau. The annual average temperature is 5°C, and the annual precipitation is more than 750 mm. Due to the diverse climate, Highland barley is the predominant crop for the residents of this county. The farmland in Zangtang County mostly consists of small barley patches dotted around the villages. Brown and dark brown soil are the main soil types in the study area.

2.2 *Methods*

A total of 7 towns were involved in this study, and we collected samples including Highland barley grains, tsamba, beef, wheat, sugar, milk, potatoes, tea and table salt. All samples were marked either as KBD group or control group. Drinking water was well bottled and kept under ambient temperature. Fresh food samples were washed with deionized water to prevent contamination. The edible parts of food samples were dried and concentrations of Se were determined by Hydride Generation Atomic Fluorescence Spectrometry.

Daily intake of Se was estimated from the food balance sheet for adults reported by Zhu *et al.* (1997) and from a 24-h food questionnaire filled out by 380 subjects in Zangtang. Based on the relative proportions of consumed food in their diet and concentrations of Se in the respective foods, mean daily intake of Se was calculated from the formula below:

$$I = \sum C_j \cdot d_j$$

where I = daily intake of Se; C_j = Se concentration in food; and d_j = daily consumption of food.

3 RESULTS AND DISCUSSION

3.1 *Selenium concentration in foods*

Mean concentrations of Se (μg/g, dryt weight) in different foods and food ingredients typically collected in Zangtang diets are reported with the range and standard error of means.

The highest Se concentration was found in table salts (1.08 ± 0.52 μg/g). Teas contained moderate concentrations of Se, with a mean of 0.20 ± 0.18 μg/g, whereas leafy vegetables, non-leafy vegetables and meats contained similar concentrations of Se with mean concentrations of approximately 0.14 μg/g. Highland barley grains, tsamba, wheat, sugar, milk and potatoes contained mean concentrations of Se <0.10 Se μg/g. There was significant difference between KBD and control group. The table salts were Se-iodized products, which were not available around KBD group or control group.

Concentrations of Se in foods vary among geographical regions, mainly due to variations in total concentrations of Se in soils (even within the same town). Bioavailability of Se from soils into plants is also a factor that determines accumulation of Se into the food diets (EC, 2000).

3.2 *Estimation of the daily dietary intake of Se*

The daily average Se intake was calculated using the estimated quantities of each food consumed, per person per day. The major sources of Se intake by KBD group of Zangtang were as follows in%: meat and meat products (35%), cereal and table salt cereal products (20%), milk (14%), teas (10%), potatoes (11%), table salts (5%), vegetables (3%) and sugars (1%). The sources of Se intake by control group of Zangtang were as follows: meat and meat products (18%), cereal and cereal products (39%), milk (7%), teas (5%), potatoes (6%), table salts (3%), vegetables (21%) and sugars (1%). Taking into account the mean Se concentrations measured in the tested samples, the total average intake of Se from food is 20 μg per person per day for the KBD group and 39 μg per person per day for the control group in Zangtang. This average Se daily intake is lower than that previously reported (40 μg per person per day) that satisfies the dietary selenium requirement (Yang & Xia, 1995). The aforementioned differences can be due to the lower Se content in some foods and the consumption amount found in this study. Dietary supplementation of Se could likely reduce the incidence of KBD.

4 CONCLUSIONS

The results indicate: (a) the levels of Se of foods were very low and the study area belongs to a Se-deficient ecological landscape; (b) the KBD becomes much more severe with decreasing intake of Se under the Se-deficient condition. Namely, the lower intake of Se is, the more severe the disease is; (c) Se deficiency plays a critical role for the prevalence of local KBD.

ACKNOWLEDGEMENTS

This work was supported by the Major Programs of the Geological Survey of Land Resources, China Geological Survey (No. 1212010560101 and No. 1212010511218) and by the *Fundamental Research Funds for the Central Universities* (No. 2010ZY54).

The authors would like to thank Dr. Wenjun Ma of Peking University Health Science Center for suggestions and analytical support.

REFERENCES

Combs, G.F. & Combs, S.B. 1986. The biological availability of selenium in foods and feeds. In G.F. Combs & S.B. Combs (eds), *The role of selenium in nutrition*. New York: Academic Press.

EC (European Commission). 2000. Scientific Committee on Food. *Opinion of the Scientific Committee on food on the tolerable upper intake level of Selenium; CF/CS/NUT/UPPLEV/25 Final, 28 November*. Bruxelles.

Fordyce, F.M. 2005. Selenium deficiency and toxicity in the environment. In O. Selinus (ed.), *Essentials of medical geology: 373–415*. Waltham: Academic Press.

Fordyce, F.M. 2007. Selenium geochemistry and health. *Ambio* 36: 94–97.

Huang, Y., Wang, Q., Gao, J., Lin, Z.-Q., Bañuelos, G.S., Yuan, L.X. *et al.* 2013. Daily dietary selenium intake in a high selenium area of Enshi, China. *Nutrients* 5: 700–710.

Tan, J.A. & Huang, Y.J. 1991. Selenium in geo-ecosystem and its relation to endemic diseases in China. *Water, Air, and Soil Pollution* 57–58: 59–68.

Tan, J.A., Zhu, W.Y., Wang, W.Y., Li, R., Hou, S., Wang, D. *et al.* 2002. Selenium in soil and endemic diseases in China. *The Science of the Total Environment* 284: 227–235.

Wang, Z.J. 1993. Some biogeochemical aspects of selenium in Chinese low-selenium belt. *Environmental Chemistry* 12: 237. (in Chinese)

WHO (World Health Organization). 1996. *Trace elements in human nutrition and health. Geneva*. World Health Organization.

Yang, G.Q. & Xia, Y.M. 1995. Studies on human dietary requirements and safe range of dietary intakes of selenium in China and their application to the prevention of related endemic diseases. *Biomedical and Environmental Sciences* 8(3): 187–201.

Yang, X.E., Chen, W.R. & Feng, Y. 2007. Improving human micronutrient nutrition through biofortification in the soil-plant system: China as a case study. *Environmental Geochemistry Health* 29: 413–428.

Selenium in the Environment and Human Health – Bañuelos, Lin & Yin (Eds)
© 2014 Taylor & Francis Group, London, ISBN 978-1-138-00017-9

Selenium compounds as antioxidants and radioprotectors

K. Indira Priyadarsini, B.G. Singh & A. Kunwar
Radiation and Photochemistry Division, Bhabha Atomic Research Centre, Trombay, Mumbai, India

P. Prabhu & V.K. Jain
Chemistry Division, Bhabha Atomic Research Centre, Trombay, Mumbai, India

1 INTRODUCTION

Selenium is an essential trace element for both animals and humans. Selenium deficiency has been implicated in several diseases and some studies have also correlated it with incidence of cancer. Inside plants, inorganic selenium is converted to low molecular weight amino acids and finally into selenoproteins. These proteins are responsible for most of the physiological functions mediated by selenium such as antioxidative action, redox regulation, immune function, *etc.* The most important selenoprotein is Glutathione Peroxidase (GPx). It is an antioxidant enzyme that detoxifies peroxides. In addition to enzyme activity another important requirement for selenium antioxidants is their ability to scavenge reactive oxygen free radicals in oxidative stress. Another closely associated activity of selenium compounds is in development of radioprotectors, which are agents employed to minimize radiation induced damage to living systems (Priyadarsini *et al.*, 2013).

With our recent objective of developing low molecular weight organoselenium compounds as antioxidants and radioprotectors, a number of selenium compounds have been synthesized and examined, out of which diselenodipropionic acid (DSePA) was found to exhibit promising radioprotection at non-toxic doses. The results leading to this promising achievement will be presented in the conference.

2 MATERIALS AND METHODS

DSePA and selenonicotinamide (SNA) were synthesized by literature methods with suitable modifications. Antioxidant activity was examined by free radical scavenging reactions and in vitro biochemical assays. *In vivo* radioprotection was examined in mice models after exposing to lethal and sublethal doses of radiation from a Cobalt-60 γ-source and the parameters were evaluated as per the protocols standardized in our laboratory.

3 RESULTS & DISCUSSION

3.1 *Free radical reactions and GPx activity*

Initially we performed studies on free radical reactions, GPx mimicking ability and antioxidant activity of a number of aliphatic symmetric mono and diselenides having functional groups like -OH, $-NH_2$, -COOH. The results indicated that organic diselenides substituted with carboxylic acid functional group are more effective as antioxidants than the others due to less toxicity, ability to scavenge oxidizing free radicals and preventing membrane peroxidation (Kunwar *et al.*, 2007; Prabhu *et al.*, 2011). From this, DSePA, a diselenide (RSeSeR), was selected for *in vitro* and *in vivo* studies. DSePA showed a reaction with reactive oxygen free radicals like hydroxyl and peroxyl radicals with rate constants comparable to standard antioxidants. It exhibits GPx activity by hydroperoxide reactions and forms a seleninic acid intermediate that is reduced by glutathione. DSePA also inhibited peroxyl radical and radiation induced lipid peroxidation in several membrane models, which demonstrates its antioxidant activity.

3.2 *Toxicity and in vivo radioprotection studies*

The maximum tolerable dose of DSePA in mice was 8.82 mg/kg body weight. Therefore, for *in vivo* radioprotection studies of DSePA (2 mg/kg body weight) was injected into mice for five days and they were exposed to whole body γ-radiation of ~9 Gy. Their survival was monitored daily for 30 days and compared with that of mice exposed to radiation without DSePA (as control animals). There was a significant 35% improvement in the survival of animals treated with DSePA and exposed to radiation as compared to radiation control animals. No death was observed in animals treated with DSePA alone. DSePA showed significant protection to radiosensitive organs like gastrointestinal (GI) tract, and hematopietic system. In the irradiated mice, DSePA maintained

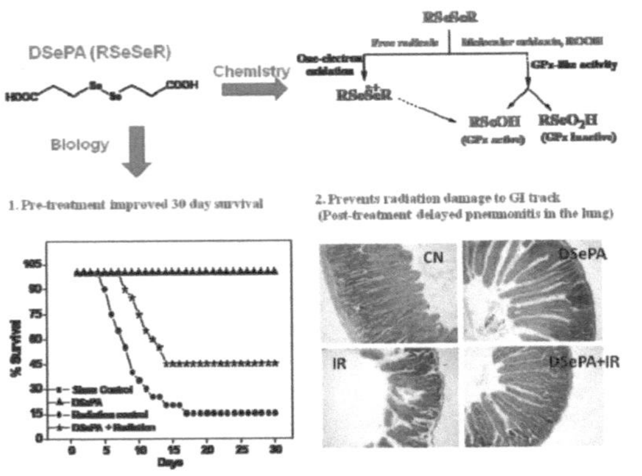

Figure 1. Summary of *in vitro* and *in vivo* studies on DSePA.

the villi height and villi/circumference of the small intestine and prevented epithelial cells (intestinal lining) from undergoing apoptosis. DSePA also exhibited prophylactic action in the hepatic system of irradiated mice through maintenance of the antioxidant enzymes (*e.g.*, GPx) and hepatic architecture. Similarly, DSePA prevented DNA damage in peripheral leukocytes in mice exposed to radiation. Additionally post-irradiative administration of DSePA significantly reduced the radiation induced peneumonitis in irradiated mice (Kunwar *et al.*, 2010, 2011). All these results on DSePA are summarized in Figure. 1

3.3 *New selenium compounds under investigation*

Recently we have synthesized a diselenide derivative of nicotinamide, 2,2'-diselenobis[3-amidopyridine] (SNA) that showed promising activity. GPx activity of SNA was found to be more than that of well known GPx mimic ebselen. The higher activity of SNA is attributed to the formation of a stable selone intermediate in the enzyme cycle. SNA also showed *in vitro* antioxidant activity by inhibiting lipid peroxidation and DNA damage induced by radiation and other free radical initiators. The compound will be studied for *its in vivo* activity in future.

4 CONCLUSIONS

Our *in vitro* and *in vivo* antioxidant and radioprotection studies on organoselenium compounds revealed that selenium compounds exhibit promising activity. DSePA, a low-molecular weight water-soluble compound with a very low cytotoxicity, is a promising candidate as a radioprotector. Many other selenium compounds exhibit antioxidant activity, they can also act as pro-oxidants depending on dosage, and other cellular conditions. It is therefore necessary to evaluate the complete cytotoxicity parameters of selenium compounds in the context of study and interest of its application.

REFERENCES

Kunwar, A., Mishra, B., Barik, A., Kumbhare, L.B., Pandey, R., Jain, V.K. *et al.* 2007. 3,3'-Diselenodipropionic acid, an efficient peroxyl radical scavenger and a GPx mimic, protects erythrocytes (RBCs) from AAPH-induced hemolysis. *Chem. Res. Toxicol.* 20: 1482–1487.

Kunwar, A., Bansal, P., Kumar, S.J., Bag, P.P., Paul, P., Reddy, N.D. *et al.* 2010. In vivo radioprotection studies of 3,3' -diselenodipropionic acid, a selenocystine derivative. *Free Radic. Biol. Med.* 48: 399–410.

Kunwar, A., Bag, P.P., Chattopadhyay, S., Jain, V.K. & Priyadarsini, K.I. 2011. Anti-apoptotic, anti-inflammatory and immunomodulatory activities of 3,3'- diselenodipropionic acid in mice exposed to whole body radiation. *Archives of Toxicology* 85: 1395–1405.

Prabhu, P., Bag, P.P., Singh, B.G., Hodage, A., Jain, V.K., Iwaoka, M. *et al.* 2011. Effect of functional groups on antioxidant properties of substituted selenoethers. *Free Rad. Res.* 45: 461–468.

Priyadarsini, K.I., Singh, B.G. & Kunwar, A. 2013. Current developments on synthesis, redox reactions and biochemical studies of selenium antioxidants. *Curr. Chem. Biolo.* 7: 37–46.

Selenium in the Environment and Human Health – Bañuelos, Lin & Yin (Eds)
© 2014 Taylor & Francis Group, London, ISBN 978-1-138-00017-9

Difference of geographic region in plasma selenium level with risk of type 2 diabetes patients: Meta-analysis of cross-section studies

Xia Jiang & Li-Qiang Qin
Department of Nutrition and Food Hygiene, School of Public Health, Soochow University, China

Bo Wang
Suzhou Center for Disease Prevention and Control, Suzhou, China

Xue-Bin Yin
Suzhou Key Laboratory for Selenium and Human Health, USTC, China

1 INTRODUCTION

Selenium (Se) is an essential trace element in human nutrition. In a previous study, we reported that Se supplements improve GPx activities (Jiang *et al.*, 2011). Since the decrease of GPx activities is a risk factor for Type 2 Diabetes Mellitus (T2DM), we considered that there is association of Se level with T2DM risk. In fact, this relationship has been considered since the end of last century, but studies have yielded inconsistent results. Thus, we conducted a meta-analysis to examine the association between the plasma Se level and T2DM.

2 MATERIALS AND METHODS

We searched published studies in the PubMed and CNKI databases through February 2013, using the search keywords Se in combination with diabetes, and examined the reference lists of the obtained articles. Studies were included in this meta-analysis if they met the following criteria: (1) subjects were diagnosed with T2DM; and (2) plasma selenium and Standard Deviation (SD) were reported. The following information was extracted: the first author's last name, publication year, study population, location, and the number of participants.

The data for individual study were converted to Standardized Mean Difference (SMD). SMD refers to the difference of selenium level between T2DM and non-T2DM divided by the combined SD. The heterogeneity of the effect size among studies was tested using the Q test based on χ^2 distribution. Either a fixed-effects or, in the presence of heterogeneity, a random-effect model was used to calculate the combined effect size. We also calculated the I^2 statistic, a quantitative measure of inconsistency across studies. An I^2 value > 50% was considered to indicate substantial heterogeneity between trials. Stratified analysis and sensitivity analysis was applied according to study characteristics. Analyses were performed using Stata version 11 (Stata Corp, College Station, TX).

3 RESULTS AND DISCUSSION

We identified 17 studies that met the criteria for this meta-analysis. The included trials were published between 1998 and 2011. There were four studies conducted in the Americas, five in Europe, and eight in Asia. Among a total of 12016 participants, 2429 were T2DM patients. Due to the wide heterogeneity across the studies ($P < 0.1$, $I^2 >$ 50%), a random effect model was used to calculate combined SMD. Compared with the non-T2DM, total SMD (95%CI) in T2DM patients were –0.16 (–0.33, 0.001). Subgroup analyses were performed according to geographic region. The combined SMD of the studies conducted in Americas and Europe were 0.04 (–0.37, 0.44) and –0.18 (–0.5, 0.14), respectively. On the other hand, the combined SMD in Asia was –0.25 (–0.48, –0.03) (Fig. 1). The sensitivity analyses omitting one study at a time and calculating the overall pooled estimate for the remaining studies showed a range from –0.10 (–0.18, –0.02) and –0.11 (–0.06, 0.03). Begg's tests and Egger's tests did not indicate the evidence of publication bias (Begg P = 0.20, Egger P = 0.44).

In general, T2DM patients have lower plasma selenium level, compared with non-T2DM subjects. Subgroup analysis shows that lower level of plasma selenium in T2DM patients was found in subjects from Asia, not from America. Selenium status greatly varies in different countries worldwide (Rayman, 2000). Geology exerts a fundamental control on the concentrations of selenium in soils, crops and animals. When there are relatively high Se concentrations in soils, the plasma Se level is correspondingly higher in subjects from Americas than from Europe and Asia. The main limitation

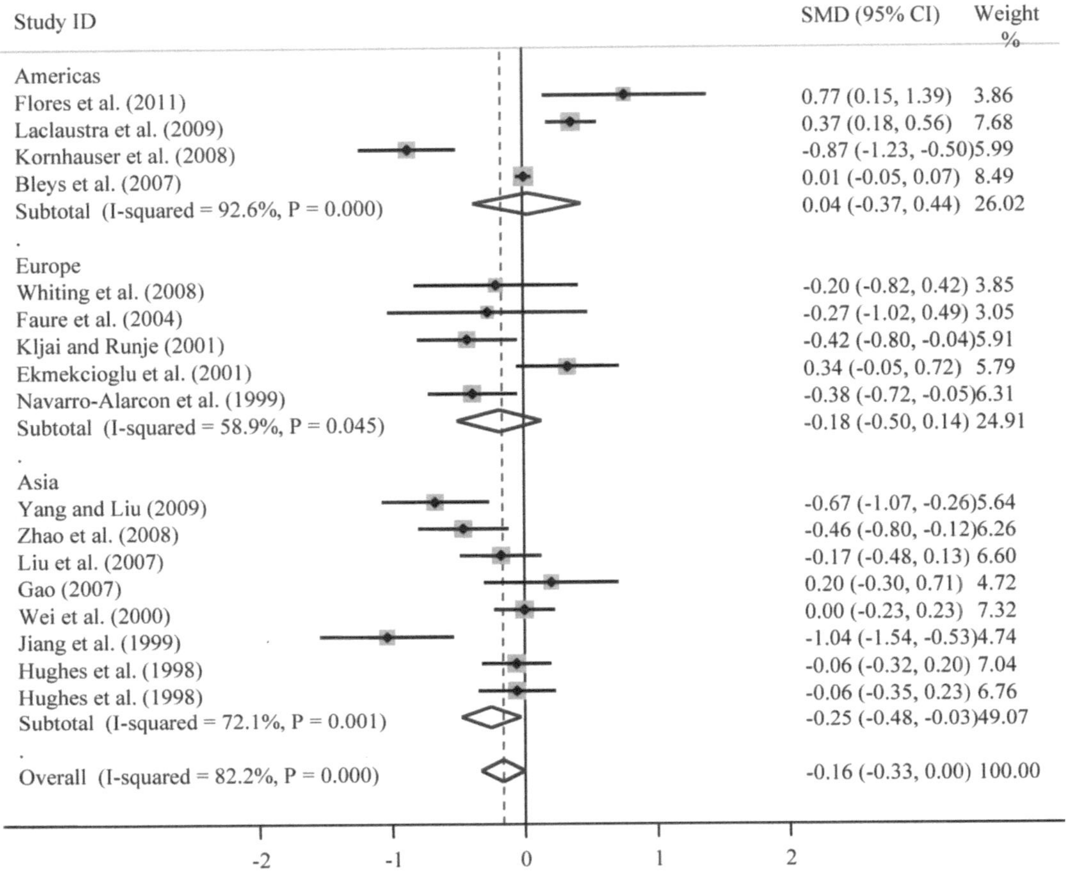

Figure 1. Combined and subgroup SMD of plasma selenium level in T2DM patients compared with the non-T2DM subjects. Note: Weights are from random effects analysis.

of our meta-analysis is the design of cross-section studies, which cannot ascertain whether the altered plasma selenium level is a consequence or a causative factor for diabetes.

4 CONCLUSIONS

In conclusion, our finding indicated that low plasma selenium level in T2DM patients is dependent on geographic region. More evidences are needed to reveal the relationship between selenium status and T2DM risk.

ACKNOWLEDGEMENTS

This research was funded by Dongwu Scholarship by Soochow University and Jiangsu Province Department of Health (H201040).

REFERENCES

Bleys, J., Navas-Acien, A. & Guallar, E. 2007. Serum selenium and diabetes in U.S. adults. *Diabetes Care* 30(4): 829–834.

Ekmekcioglu, C., Prohaska, C., Pomazal, K., Steffan, I., Schernthaner, G. & Marktl, W. 2001. Concentrations of seven trace elements in different hematological matrices in patients with type 2 diabetes as compared to healthy controls. *Biol. Trace Elem. Res*, 79(3): 205–219.

Faure, P., Ramon, O., Favier, A. & Halimi, S. 2004. Selenium supplementation decreases nuclear factor-Kappa B activity in peripheral blood mononuclear cells from type 2 diabetic patients. *European J. Clin. Invest.* 34(7): 475–481.

Flores, C.R., Puga, M.P., Wrobel, K., Garay Sevilla, M.E. & Wrobel, K. 2011. Trace elements status in diabetes mellitus type 2: Possible role of the interaction between molybdenum and copper in the progress of typical complications. *Diabetes Res. Clin. Pract.* 91(3): 333–341.

Gao, R. 2007. Analysis of the mineral elements in type 2 diabetes in combined with coronary heart disease. *J. Chengde Med. College* 24(1): 27–29.

Hughes, K., Choo, M., Kuperan, P., Ong, C.N. & Aw, T.C. 1998. Cardiovascular risk factors in non-insulin-dependent diabetics compared to non-diabetic controls: A population-based survey among Asians in Singapore. *Atherosclerosis*, 136(1): 25–31.

Jiang, W.L., Zhang, Z. & Yu, D.M. 1999. Study of serum trace elements in 2 type 2 diabetes. *J Tianjin Med College* 4: 60–62.

Jiang, X. & Qin, L.Q. 2011. Selenium supplement and glutathione peroxidase activities in infants: a meta-analysis of randomized controlled trials. In G.S. Banuelos, Z.Q. Lin, X.B. Yin, N. Duan (eds), *Selenium: Global Perspectives of Impacts on Humans, Animals and the Environment*, 59–60. Hefei, China: USTC Press.

Kljai, K. & Runje, R. 2001. Selenium and glycogen levels in diabetic patients. *Biol. Trace Elem. Res.* 83(3): 223–229.

Kornhauser, C., Garcia-Ramirez, J.R., Wrobel, K., Perez-Luque, E.L. & Garay-Sevilla, M.E. 2008. Serum selenium and glutathione peroxidase concentrations in type 2 diabetes mellitus patients. *Prim. Care Diabetes* 2(2): 81–85.

Laclaustra, M., Navas-Acien, A., Stranges, S., Ordovas, J.M. & Guallar, E. 2009. Serum selenium concentrations and diabetes in U.S. adults: National Health and Nutrition Examination Survey (Nhanes) 2003–2004. *Environ. Health Perspect.* 117(9): 1409–1413.

Liu, Xiaoyun, Yang, Shuhong, Niu, Yan & Li, Hongmei. 2007. Analysis of the mineral elements in type 2 diabetes. *Ningxia Med. J.* 5: 405–407.

Navarro-Alarcon, M., de la Serrana, G., Lopez, H., Perez-Valero, V. & Lopez-Martinez, C. 1999. Serum and urine selenium concentrations as indicators of body status in patients with diabetes mellitus. *Sci. Total Environ.* 228(1): 79–85.

Rayman, M.P. 2000. The importance of selenium to human health. *The Lancet* 356(9225): 233–241.

Whiting, P.H., Kalansooriya, A., Holbrook, I., Haddad, F. & Jennings, P.E. 2008. The Relationship between chronic glycaemic control and oxidative stress in type 2 diabetes mellitus. *Br. J. Biomed. Sci.* 65(2): 71–74.

Wei, Jianhua, Luo, Jin, Ji, Limin, Qi, Rongxiang & Chen, Peiying. 2000. Determination of the contents of eight kinds of essential trace element in the sera of patients with diabetes. *J. Chinese Phys.* 8: 464–465.

Yang, Guang, & Liu, Xuejun. 2009. Diabetes and trace elements. *China Prac. Med.* 12: 113–114.

Zhao, Changfeng, Wang, Huimin, Zhang, Junli, Feng, Li, Liu Yanxun, Zou, Linyan, & Hu, Qinjing. 2008. Correlations of trace elements, glucose and body compositions in type 2 diabetes. *J. Hygiene Res.* 5: 600–601.

Selenium in the Environment and Human Health – Bañuelos, Lin & Yin (Eds)
 ISBN 978-1-138-00017-9

A preliminary study on inhibitory effect of selenium against *Penicillium expansum* and *Botrytis cinerea*

Zhilin Wu, Qian Wang & Miao Li
Key Laboratory of Agri-Food Safety of Anhui Province, School of Plant Protection, Anhui Agriculture University, Hefei, Anhui, China

Linxi Yuan
Jiangsu Bio-Engineering Research Centre on Selenium, Suzhou, Jiangsu, China

Xuebin Yin & Ying Liu
Advanced Lab for Selenium and Human Health, Suzhou Institute for Advanced Study, University of Science and Technology of China, Suzhou, Jiangsu, China

1 INTRODUCTION

Recent reports have indicated that certain organic and inorganic salts inhibit the growth of many pathogens (Millis *et al.*, 2004). Selenium (Se), as an essential micronutrient, was effective in improving the plant resistance and antioxidant capacity at a low concentration (Zhu *et al.*, 2009). *P. expansum*, a widespread filamentous fungus, is a major causative agent of fruit and vegetables decay. While *B. cinerea*, a ubiquitous fungal pathogen, causes gray mold rot on a large number of economically important agricultural crops (Qin *et al.*, 2010; Cao *et al.*, 2012). In the present study, the inhibitory effect of Se against *P. expansum* and *B. cinerea* were evaluated. The results showed that selenium could inhibite spore germination, germ tube length, and mycelial growth of *P. expansum* and *B.cinerea*. These results suggest that Se could serve as a potential alternative to synthetic fungicides for the control of the postharvest disease of fruit and vegetables caused by *P. expansum* and *B.cinerea*.

2 MATERIALS AND METHODS

P. expansum and *B. cinerea* were obtained from Key Laboratory of Plant Resources in Northern China, Institute of Botany, Chinese Academy of Sciences. Cultures of *P. expansum* and *B. cinerea* were placed on Potato Dextrose Agar (PDA), and Colonies grown on PDA plates at 23 ± 2°C produced cylindrical conidia 12–16 × 6–8 μm. Selenium in the form of sodium selenite was purchased from Sigma-Aldrich (St. Louis, MO).

The effects of Se on conidial germination and germ tube elongation of *P. expansum* and *B. cinerea* were evaluated in Potato Dextrose Broth (PDB) according to the method by Qin *et al.* (2010). The experiment was divided into two parts: firstly, fungal spores were obtained from 2-week-old PDA cultures incubated at 23°C. Spore suspensions were collected by adding sterile distilled water to the surface of the culture and obtained by filtering the solution through four layers of sterile cheesecloth. They were added to several triangular flasks of 20 ml PDB culture with a spore suspension of 5.0×10^5 spores/ml in different concentrations (0, 4, 8, 12 and 16 mg/L) of Se solution and incubated at 25°C on a rotary shaker at 200 rpm. Spore germination and germ tube elongation were assayed microscopically after a 7–11h incubation period.

Secondly, we applied 100 μl spore suspensions of 5×10^5 spores/ml to the surface of PDA culture and incubated at 23°C. A 5 mm diameter plug of mycelial agar obtained from the growing edge of 7-day-old cultures was placed in the center of a 9 cm diameter Petri dish containing PDA medium with sodium selenite at different concentrations (0, 12 mg/L). Colony diameter was measured after 2–5d incubation period. The method of *B. cinerea* was similar to the one described above.

3 RESULTS AND DISCUSSION

The effect of Se on spore germination and germ tube elongation in PDB medium was shown in Figures 1 and 2. The inhibition was positively correlated with the concentration of Se especially when the Se concentration was greater than 4 mg/L. The spore germination ratio of control was about 90%, while the treatment group was lower than 30% at 12 mg/L, and less than 10% at 16 mg/L. The germ tube elongation of *P. expansum* and *B. cinerea* were degraded to 15.8% and 16.7%, respectively, at 12 mg/L. *B. cinerea* was slightly less sensitive to Se than *P. expansum*. At 12 mg/L, *P. expansum* spores began to germinate after incubation for 9 h at 25°C under Se treatment, whereas extensive

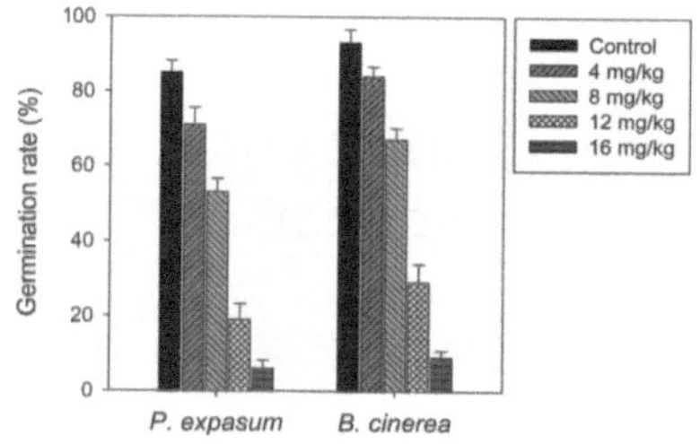

Figure 1. Effects of Se concentrations on spore germination.

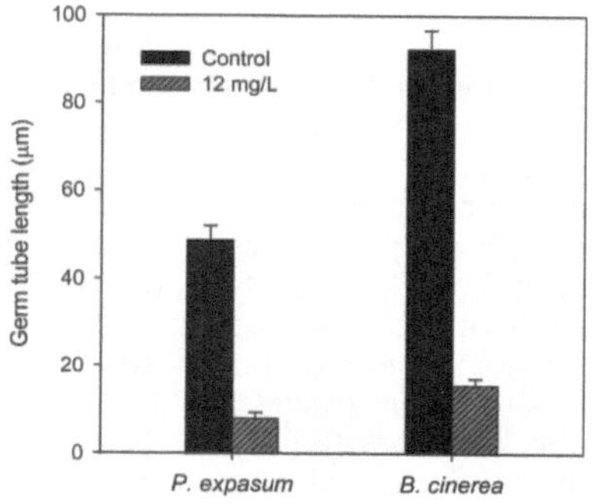

Figure 2. Effects of 12 mg/L Se on germ tube elongation at 9h. Values with different letters are statistically different ($P < 0.05$).

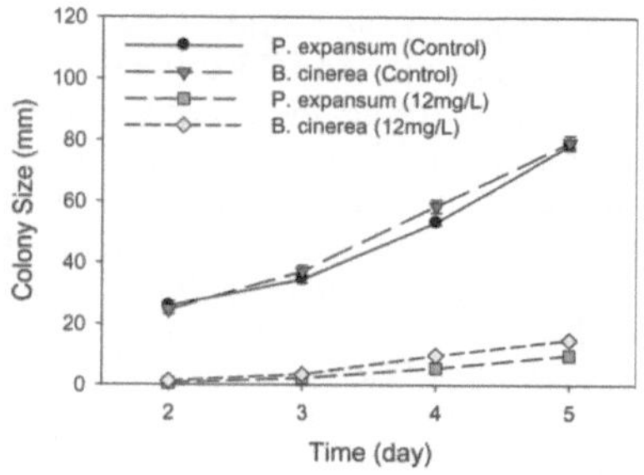

Figure 3. Effects of Se on spore morphology of *P. expansum*. Top: control; Bottom: 12 mg/L Se treatment.

mycelial growth was observed in culture medium without Se at the same time (Fig. 3).

The effect of Se on development of the mycelium was shown in Figure 4. The colony diameters of treatment groups were 34.6 mm (*P. expansum*) and 37.1 mm (*B. cinerea*) while the control were 2.2 mm (*P. expansum*) and 3.4 mm (*B. cinerea*) after 3-d incubation time period. Fungus colony treated by Se not only was smaller in diameter but also appeared to be decreased in spores and distorted in mycelium. From the growth curve, we could conclude that Se significantly inhibited fungal growth rate.

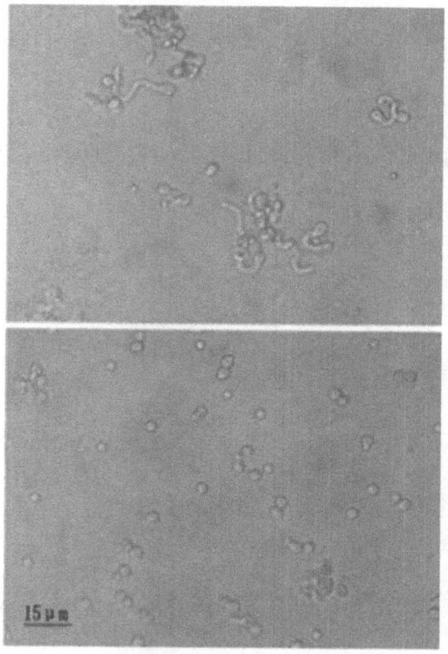

Figure 4. Effects of Se on mycelial growth during a 5-d incubation time period.

4 CONCLUSIONS

This study demonstrated that Se had significant inhibition effects on the growth of *P. expansum* and *B. cinerea*, and the response was dose and time dependent. The possible mechanisms in the inhibition of *P. expansum* and *B. cinerea* by Se need to be explored by conducting further experiments.

ACKNOWLEDGEMENTS

The research funding was provided by Natural Science Youth Foundation of Jiangsu Province for Youth, China (BK 2012195, BK 2012202), Jiangsu Research-Industry Joint Innovation Project (BY 2011186).

REFERENCES

Cao, B.H., Li, H., Tian, S.P. & Qin, G.Z. 2012. Boron improves the biocontrol activity of *Cryptococcus laurentii* against *Penicillium expansum* in jujube fruit. *Postharv. Biol. Technol.* 68: 16–21.

Mills, A.A.S., Platt, H.W. & Hurta, R.A.R. 2004. Effect of salt compounds on mycelial growth, sporulation and spore germination of various potato pathogens. *Postharv. Biol. Technol.* 34: 341–350.

Qin, G.Z., Zong, Y.Y., Chen, Q., Hua, D. & Tian, S. 2010. Inhibitory effect of boron against *Botrytis cinerea* on table grapes and its possible mechanisms of action. *International Journal of Food Microbiology* 138: 145–150.

Zhu, Y.G., Pilon-Smits, E.A.H., Zhao, F.J., Williams, P.N. & Meharg, A.A. 2009. Selenium in higher plants: understanding mechanisms for biofortification and phytoremediation. *Trends in Plant Sci.* 14(8): 436–442.

Selenium in the Environment and Human Health – Bañuelos, Lin & Yin (Eds)
© 2014 Taylor & Francis Group, London, ISBN 978-1-138-00017-9

Bioaccessibility and transformations of selenium in the human intestine: Selenium-enriched crops versus food supplements

R.V.S. Lavu, T. Van de Wiele, V.L. Pratti, F.M.G. Tack & G. Du Laing
Ghent University, Ghent, Belgium

1 INTRODUCTION

It is well-known that the effective bioavailability of Se in Se-enriched products may depend on the chemical forms in which Se occurs, i.e. its speciation, as well as the matrix in which Se is embedded. In recent years, the (potential) bioavailability of Se in different food sources has been investigated by various *in vitro* intestinal digestion procedures to assess bioaccessibility (Reyes *et al.*, 2006). The term bioaccessibility is defined as the fraction of a compound that is released from its matrix in the gastrointestinal tract and thus becomes available for intestinal absorption (Fernandez-Garcia *et al.*, 2009).

We aimed to assess and compare the bioaccessibility of Se-enriched food crops and commercial Se supplements. Three Se supplements that include SelenoPrecise, Se + vit ACE and yoghurt-based Se-containing tablets, as well as Se-enriched leek (*Allium ampeloprasum*) and kenaf (*Hibiscus cannabinus*) were tested. Moreover, we also assessed the role of microorganisms in Se bioaccessibility and bio-transformations (using reference Se compounds) during gastrointestinal digestion, with focus on the colon. Fate of Se in the colon has hardly been studied before although some studies suggested relationships between Se deficiency and incidence of colon cancer (Dworkin *et al.*, 1988; Psathakis *et al.*, 1998).

This information should give insights into the effect of food matrix and speciation on Se bioaccessibility in both the small and large intestine. Moreover, the role of microbial Se biotransformations during intestinal digestion, will be described, which may contribute to developing more effective Se supplementation strategies in populations with Se deficient as well as sub-optimal Se levels.

2 MATERIALS AND METHODS

Selenium-enriched samples were as follows: two lyophilized and powdered samples of Se-enriched food crops (leek and kenaf) grown on soil fertilized with selenate (Na_2SeO_4), leek grown on soil fertilized with selenite (Na_2SeO_3) and three food supplements (tablets), *i.e.* SelenoPrecise tablets (SP, Se-enriched yeast), a Se+ACE-vitamins mixture (ACE), and a Se-enriched Yoghurt–Based tablet (YB), obtained from commercially available sources were used for assessment of Se bioaccessibility. The speciation of Se in the leek and kenaf were reported by Lavu *et al.* (2012, 2013).

Three gram of lyophilized and powdered samples of Se-enriched kenaf and leek were boiled with 10 mL of deionized water for 3 min to mimic the food preparation process applied prior to human consumption. After cooling, 3 g of the obtained suspension were transferred into 100 mL amber colored bottles. Thirty mL of simulated gastric juice (10 g/L pepsin adjusted to pH 2.0 with 2M HCl) were added to the penicillin bottles. The bottles were capped and shaken (100 rpm) in an incubator (37°C) for 1 h. Five mL sample was taken and considered to represent the gastric phase. Afterwards, 12.5 mL of small intestine fluid (7.5 g/L bile powder, 5.0 g/L pancreatin and 15 g/L sodium bicarbonate) were added and the mixture was shaken in the incubator for 2 h again. After 2 h, 5 mL of sample, representing the small intestine, was taken. Subsequently, 25 mL of colon suspension, sampled from the colon compartments of the SHIME® system (Simulator of the Human Intestinal Microbial Ecosystem) (Van de Wiele *et al.*, 2004), was added. The bottles were capped, flushed with nitrogen gas to create anaerobic conditions, shaken in the incubator (100 rpm, 37°C) and sampled after 0 h (T0), 2 h (T2), 24 h (T24) and 48 h (T48). The samples collected in each step were placed in 10 mL polypropylene tubes and centrifuged at 10,000 g for 10 minutes. The supernatant was filtered (0.45 μm) and stored at –80°C. Collected filtrates and pellets (residues) were analyzed for total Se using ICP-MS after microwave digestion with concentrated HNO_3 and H_2O_2. Speciation of Se was measured in the supernatants using HPLC-ICP-MS. A similar procedure was conducted for Se supplements (tablets) by weighing 0.3 g powdered samples, which were prepared by crushing 20 tablets using mortar and pestle to obtain sample homogeneity. The entire procedure was repeated using pure Na_2SeO_4

and SeMet spiked into the gastric solution to a concentration of 2.4 mg/L. All experiments were conducted in triplicate.

The relative bioaccessibility of Se was calculated for each digestion phase as the ratio (%) between the amount of Se in the supernatant and the total amount of Se in the suspension.

The procedure described above was repeated for SelenoPrecise and pure SeMet. However, the active SHIME suspension collected for use in the colon phase was now autoclaved twice for 30 min (121°C, 1 bar overpressure) to inactivate colon microbiota. This step allowed us to assess the role of microbiota in biotransformation of Se in the colon.

3 RESULTS AND DISCUSSIONS

3.1 *Bioaccessibility of Se from Se-enriched food crops, Se-enriched food supplements, and pure Se reference compounds*

The relative bioaccessibility of the pure Se reference compounds SeMet and selenate is high (almost 100%) and does not vary significantly in upper intestinal phases (data not shown). However, significant differences were observed between these two Se forms in the colon, with the bioaccessibility being much lower when SeMet is used. The bioaccessibility decreases to below 40% and 70% after 48 hours of colon incubation for SeMet and selenate, respectively. Following the early physical processes when moving from small intestine to colon, Se is probably actively taken up by the bacterial cell fractions of gut contents and feces. The uptake of Se by colon microbiota should be attributed to the fact that microbes also require Se as micronutrient for their metabolic processes. Our results indicate that the uptake of SeMet by colon microbiota is much more efficient compared to the uptake of selenate.

Compared to food supplements and food crops (Fig. 1), the bioaccessibility in upper intestinal tract is higher for pure compounds. This is due to the absence of a matrix, allowing instantaneous solubilization of these reference compounds when subjected to gastrointestinal conditions. For both crops, the bioaccessibility is higher in the small intestine compared to the stomach and it generally decreases with time in the colon (Fig. 1). Bioaccessibility does not differ between the crops in the stomach and small intestine, but it clearly differs in the colon, where a higher bioaccessibility was observed for leek as compared to Kenaf.

Bioaccessibility in the Yoghurt-Base Se supplement (YB) is significantly lower compared to the bioaccessibility in the other Se supplements (SP and ACE) (Fig. 1). This may be due to the presence

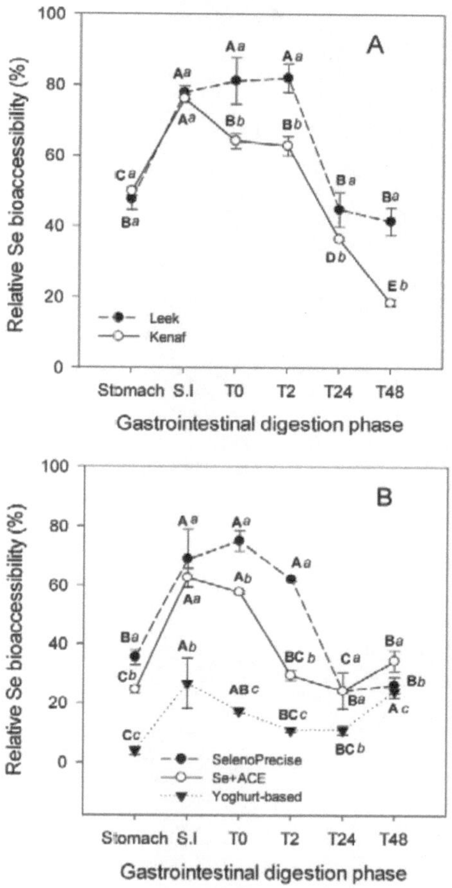

Figure 1. Relative bioaccessibility of Se for (A) two Se-enriched crops and (B) three Se-enriched food supplements in different steps of an in vitro gastrointestinal digestion (T0, T2, T24 and T48 refer to 0, 2, 24 and 48 hours after starting colon incubation, respectively; S.I. refers to small intestine).

of elemental nano or microparticles of elemental Se formed by the microbiota in the yoghurt. Solubilisation of these particles in the intestinal environment is difficult.

Colon incubation of the Se-ACE supplement and selenized yeast (SelenoPrecise) resulted in a decrease of less than 35% and about 50%, respectively, whereas incubation of the yoghurt-based supplement did not result in significant Se uptake by the microbial biomass. The Se-ACE supplement results in a significant decrease of bioaccessibility already after 2 hours of incubation, whereas a significant drop was observed for SelenoPrecise only after 24 hours. Presence of vitamins in the Se-ACE supplement probably promoted microbial

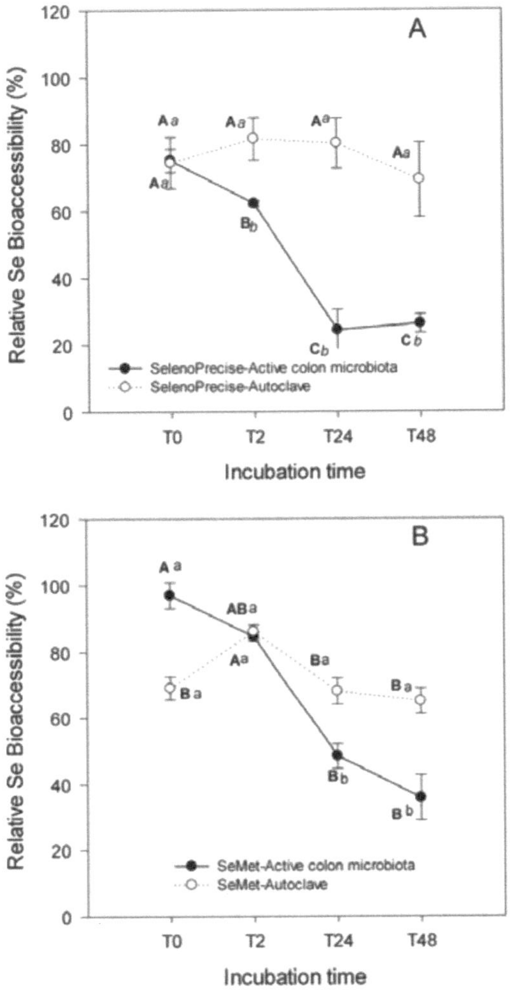

Figure 2. Comparison of relative Se bioaccessibility in the colon phase at various incubation times (T0, T2, T24 and T48) for (A) SelenoPrecise supplement (SP) and (B) SeMet, in presence of active and inactivated (autoclaved) colon microbiota.

Se uptake. However, all food supplements resulted in a similar residual bioaccessibility after 48 hours of colon incubation.

3.2 *Role of colon microbiota*

In a second experiment, SP and SeMet were incubated with inactivated colon microbiota, using autoclaved colon suspension. Colon microbiota were found to play a prominent role in reducing Se bioaccessibility in the colon environment (Fig. 2). The presence of inactivated microbiota resulted in a higher bioaccessibility compared to the presence of an active microbial community. For the inactivated microbiota, no significant difference is observed between SeMet and SP.

3.3 *Stability of Se species during gastrointestinal digestion*

Biotransformation of SeMet was observed in the colon. SeMet concentrations decreased, whereas SeMeCys was formed to some extent (data not shown). Moreover, Se-methionine oxide (SeMetO) was detected in the small intestine when incubating SeMet. In contrast, selenate was highly stable. Only selenate was identified in the bioaccessible fraction of the colon at various time points when incubating this compound. No unknown peaks were observed.

4 CONCLUSIONS

We observed a difference in bioaccessibility between different Se supplements, with a yoghurt-based supplement having the lowest bioaccessibility and a yeast-based supplement (SelenoPrecise) having the highest bioaccessibility. Compared to other digestion phases, the bioaccessibility of Se was significantly lower in the colon phase due to presence and activity of colon microbiota. In the colon phase, the bioaccessibility of Se was higher for Se-enriched food crops compared to Se-enriched food supplements. Moreover, Se biotransformations were also identified during intestinal digestion. These observations highlight the need for assessing Se speciation and bioaccessibility when evaluating the efficacy of new food supplements and fortified food products launched in the market.

Although the majority of Se is bioaccessible in the small intestine, a significant fraction still reaches the colon, where it seems to be taken up by the microbial community and assumed to also induce positive health effects. Various clinical studies have shown that supplementing Se may reduce the incidence of colon cancer and tumor growth. However, the mechanism behind this is yet unclear. Therefore, further research is needed to assess the fate and role of Se in the colon, including its effect on microbial activity and community composition, and the health benefits it may induce in this environment.

REFERENCES

Dworkin, B.M., Rosenthal, W.S., Mittelman, A., Weiss, L., Applebee-Brady, L. & Arlin, Z. 1988. Selenium status and the polyp-cancer sequence—a colonoscopically controlled-study. *American Journal of Gastroenterology* 83: 748–751.

Fernandez-Garcia, E., Carvajal-Lerida, I. & Perez-Galvez, A. 2009. In vitro bioaccessibility assessment as a prediction tool of nutritional efficiency. *Nutrition Research* 29: 751–760.

Lavu, R.V., Du Laing, G., Van de Wiele, T., Pratti, V.L., Willekens, K., Vandecasteele, B. *et al.* 2012. Fertilizing soil with selenium fertilizers: impact on concentration, speciation, and bioaccessibility of selenium in leek (*Allium ampeloprasum*). *Journal of Agriculture and Food Chemistry* 60: 10930–10935.

Lavu, R.V., De Schepper, V., Steppe, K., Majeti, P.N.V., Tack, F., Du Laing, G. 2013. Use of selenium fertilizers for production of Se-enriched Kenaf (*Hibiscus cannabinus*): Effect on Se concentration and plant productivity. *Journal of Plant Nutrition and Soil Science*, DOI: 10.1002/jpln.201200339.

Psathakis, D., Wedemeyer, N., Oevermann, E., Krug, F., Krug, F., Siegers, C.P. & Bruch, H.P. 1998. Blood selenium and glutathione peroxidase status in patients with colorectal cancer. *Diseases of the Colon & Rectum* 41: 328–335.

Reyes, L.H., Encinar, J.R., Marchante-Gayon, J.M., Alonso, J.I. & Sanz-Medel, A. 2006. Selenium bioaccessibility assessment in selenized yeast after "in vitro" gastrointestinal digestion using two-dimensional chromatography and mass spectrometry. *Journal of Chromatography A* 1110: 108–116.

Van de Wiele, T., Boon, N., Possemiers, S., Jacobs, H. & Verstraete, W. 2004. Prebiotic effects of chicory inulin in the simulator of the human intestinal microbial ecosystem. *FEMS Microbiology Ecology* 51: 143–153.

Selenium in the Environment and Human Health – Bañuelos, Lin & Yin (Eds)
© 2014 Taylor & Francis Group, London, ISBN 978-1-138-00017-9

Food selenium and the prevention of colorectal cancer

G.H. McIntosh, Y. Hu & G.P. Young
Flinders Centre for Innovation in Cancer, Flinders University of South Australia, Bedford Park, South Australia, Australia

1 INTRODUCTION

We have studied a number of dietary components for their ability to influence colon cancer expression in a rodent model. Selenium (Se) has been the most effective chemopreventive agent examined so far in reducing azoxymethane (AOM) induced tumours (McIntosh *et al.*, 2006); Shamberger (1970) reported an inhibitory effect of selenium on carcinogenesis. Clark *et al.* (1996) showed this trace element in organic form prevented expression of colorectal cancers (CRC) and some other cancers (prostate, lung) in senior Americans attending skin cancer clinics. In this intervention study Se was provided at 200 μg Se/day as yeast Se over 4.5 years in a double blind placebo control study. Relative risk of CRC was reduced to 0.42 and total cancer incidence to 0.63. Epidemiological evidence showed that highest versus lowest blood Se quartiles had an odds ratio of 0.66 for colon adenomas in 3 disparate American studies (Jacobs *et al.*, 2004). The source and form of Se can have a significant influence on cancer prevention (Medina *et al.*, 2001). Using a breast cancer model Medina et al reported a dose effect for Se-methylselenocysteine one fifth that of selenomethionine which required 4–5 mg Se/kg diet to achieve 50% inhibition of expression.

We undertook a series of studies examining some differing food Se sources and their influence on tumorigenesis in a colon cancer model and/or on early biomarkers, with an interest in the likely mechanism(s) involved.

2 MATERIALS AND METHODS

Food sources rich in Se were examined, including milk proteins from yeast Se-fed cows, brewers yeast and torula yeast Se, and Brazil nuts. Human clinical and rodent cancer model studies were used. We used laboratory mice and rats in which colon Aberrant Crypt Foci (ACF) and tumors were induced with azoxymethane (AOM) (Hu *et al.*, 2008; Hu *et al.*, 2013). ACF were assessed at 10 weeks post initiation, and tumors at 20–24 weeks. The latter were classified histologically into adenomas and adenocarcinomas. Human clinical studies were undertaken with Se supplements at 50–150 μg Se/day over 6 weeks, with a 6 week follow up (Hu *et al.*, 2011). Blood samples were taken fortnightly and rectal pinch biopsies were undertaken at 0, 6 and 12 weeks.

3 RESULTS AND DISCUSSION

In both rat and mouse studies we have observed that dairy protein Se in 1 mg Se/kg diet significantly reduced ACF expression at 10 weeks (*e.g.*, total ACF by 54% and large ACF by 65%) and for tumors, incidence was reduced by 34%, multiplicity by 78%, and size by 73%. In contrast, yeast Se at 1 mg/kg had no effects. Only yeast Se at 4 mg/kg exhibited a significant reduction of similar magnitude to that of milk protein Se in 1 mg/kg Se diet. We interpret this difference in effect to the form of Se, which influenced the distribution of Se to critical sites via relevant selenoproteins (*e.g.*, selenoprotein P, and SePP).

We tested brewers yeast and milk protein Se sources as supplements to humans at a rate of 150 μg Se/day for 6 weeks, and examined potential biomarkers of Se status and function in plasma and rectal mucosa biopsies. Biomarkers glutathione peroxidase (GPx) 1 and 2 and SePP were monitored at 0, 6, and 12 weeks. Plasma Se increased with both supplements over 30% ($P < 0.05$). All three rectal markers were significantly increased (about double) with Se supplement as dairy protein but not with yeast Se. SePP remained elevated at 12 weeks with the dairy Se supplement (Hu *et al.*, 2011). Selenium was supplemented to subjects as Brazil nuts (2.7 mg Se/kg) which are a rich but variable Se source (1 to 20 mg Se/kg). Six nut kernels per day provided about 50 μg Se/day over 6 weeks. We observed a significant increase in plasma Se levels. Rectal mucosa selenoproteins expression is not yet available.

Finally, we have compared the single and combined influences of green tea extract (0.5%) and milk protein Se (1 mg Se/kg) in rat diets on early colon cancer markers (ACF), and tumor expression in rats induced with AOM (7). Rats fed green tea extract alone significantly reduced the number of large ACF (but not small nor total), by 38% ($P < 0.05$), compared with those fed a control diet. With the combination diet the number of all sizes of ACF, including total, small and large, was significantly reduced (81%) ($P < 0.05$) and by Se alone, (45% for large) compared with those fed a control diet. Large ACF correlate closely with developing invasive tumors. Overall, there was 42.5% inhibition of tumor incidence with Se, and 80% with combined tea and Se. Tumor multiplicity and size were halved with the combined treatment relative to Se alone. The combination of green tea and Se is effective additively in tumor reduction in the rodent model. Plasma Se is a useful indicator of Se status, but there are other biomarkers that appear to offer more predictability with regard to risk of CRC, and may be useful in ongoing human clinical trials.

The preventive effect was associated with regulating genetic and epigenetic biomarkers implicated in colonic carcinogenesis, as evidenced by restoring *SFRP5* gene expression, increasing histone H3 acetylation and reducing DNMT1 expression, inhibiting ß-catenin nuclear accumulation, reducing *Cyclin D1* expression and cell proliferation in normal-appearing crypts.

4 CONCLUSIONS

Selenium supplementing of rodent diet with 1 mg Se/kg as dairy protein Se significantly reduced preneoplastic leions (ACF) and colon tumors. Se supplements to an average Australian human diet of 50–150 µg Se/day over 6 weeks leads to significantly increased plasma Se concentrations. The dairy protein Se supplement but not yeast Se also increased colorectal mucosa mRNA expression of selenoproteins (i.e., SePP, GPx1 and GPx2), potentially useful biomarkers of protection against sporadic colorectal tumors. However, the food form of Se determines the amount necessary to effectively influence rectal mucosa selenoproteins. Used expeditiously Se can achieve an anticancer effect, and this may be augmented by dietary components such as green tea, which together influence relevant biochemical mechanisms offering protection.

REFERENCES

Clark, L.C., Combs, G.F., Jr., Turnbull, B.W., Slate, E.H., Chalker, D.K., Chow, J. *et al.* 1996. Effects of selenium supplementation for cancer prevention in patients with carcinoma of the skin. A randomized controlled trial. Nutritional Prevention of Cancer Study Group. *JAMA* 276: 1957–1963.

Hu, Y., McIntosh, G.H., Le Leu, R.K., Woodman, R. & Young, G.P. 2008. Suppression of colorectal oncogenesis by selenium-enriched milk proteins: apoptosis and K-ras mutations. *Cancer Res.* 68: 4936–4944.

Hu, Y., McIntosh, G.H., Le Leu, R.K., Nyskohus, L.S., Woodman, R.J. & Young, G.P. 2013. Combination of selenium and green tea improves the efficacy of chemoprevention in a rat colorectal cancer model by modulating genetic and epigenetic biomarkers. *PLoS One* 8, e64362.

Hu, Y., McIntosh, G.H., Le Leu, R.K., Upton, J.M., Upton, J.M., Woodman, R.J. & Young, G.P. 2011. The influence of selenium-enriched milk proteins and selenium yeast on plasma selenium levels and rectal selenoprotein gene expression in human subjects. *Brit. J. Nutr.* 106(4): 572–582.

Jacobs, E.T., Jiang, R., Alberts, D.S., Greenberg, E.R., Gunter, E.W., Karagas, M.R. *et al.* 2004. Selenium and colorectal adenoma: results of a pooled analysis. *J. Natl. Cancer Inst.* 96: 1669–1675.

McIntosh, G.H., Royle, P.J., Lesno, S. & Scherer, B.L. 2006. Selenised casein protects against AOM-induced colon tumors in Sprague Dawley rats. *Nutr. Cancer* 54: 209–2152.

Medina, D., Thompson, H., Ganther, H. & Ip, C. 2001. Se-methylselenocysteine: a new compound for chemoprevention of breast cancer. *Nutr. Cancer* 40: 12–17.

Shamberger, R.J. 1970. Relationship of selenium to cancer.1. Inhibitory effect of selenium on carcinogenesis. *J. Nat. Cancer Inst.* 44: 931–936.

Selenium in the Environment and Human Health – Bañuelos, Lin & Yin (Eds)
© 2014 Taylor & Francis Group, London, ISBN 978-1-138-00017-9

Optimizing selenylation of seleno-astragali polysaccharide by Box-Behnken design, its characterization and antioxidant activity

Yu-Bin Ji, Fang Dong & Miao Yu
Research Center on Life Sciences and Environmental Sciences, Harbin University of Commerce, Harbin, China
Engineering Research Center of Natural Anticancer Drugs, Ministry of Education, Harbin, China

Dan Liu & Long Qin
Research Center on Life Sciences and Environmental Sciences, Harbin University of Commerce, Harbin, China

1 INTRODUCTION

Selenium (Se) is a necessary micronutrient for human beings and animals. It plays a key role in the field of immunity intervention and tumor prevention (Micke *et al.*, 2009). In China, Chinese Nutrition Society recommends that adult intake should be 50~250 μg/d, but about 700 million people live in low Se areas. Therefore, Se supplementation is necessary and a feasible strategy to consider.

Selenium is usually toxic at elevated concentration in inorganics forms. As organic Se compound, seleno-polysaccharide maintains the basic configuration of the polysaccharide and physiological functions. Moreover, seleno-polysaccharide can improve the bio-availability of Se and reduce the toxicity and side effects of inorganic Se. Many studies show that the biological activity of seleno-polysaccharide is higher than that of polysaccharide and Se (Shang *et al.*, 2011).

Meanwhile, the natural seleno-polysaccharide is insufficient in plants, animals and microorganisms, and it is necessary to increase the yield of seleno-polysaccharide by artificial methods. Compared with biotransformation, chemical synthesis is the most effective approach to harvest seleno-polysaccharide.

Radix Astragali is cultivated as a traditional functional plant for thousands of years to prevent and treat various diseases (Wu & Chen, 2004). Based on its extensive activity and safety, Ministry of Health of China has approved that the Radix Astragali could be utilized as functional food. Additionally, the studies confirm that Astragalus polysaccharide (APS) is one of active ingredients (Xu *et al.*, 2008).

Therefore, this paper employed a Box-Behnken Design (BBD) (3 factors and 3 levels) to optimize the effects of reaction time, reaction temperature and ratio of Na_2SeO_3 to APS on the Se content in Se-APS.

Furthermore, the characterization and antioxidant activities of Se-APS were evaluated for seeking the higher biological functional seleno-polysaccharide as a medicine or as a Se supplement.

2 EXPERIMENTS AND RESULTS

2.1 *Optimization of selenylation*

The ratios of Na_2SeO_3 and APS (0.8–1.2 g/g) were dissolved by HNO_3 (100 mL, 0.05%) in Erlenmeyer flasks. Then, the mixed solution was reacted for 7–9 h, at a reaction temperature of 60–80°C and shook in a water bath at a rate of 40 r/min. After the reaction, Na_2CO_3 was added into the reacted solution to adjust pH to 5–6. The solution was centrifuged to remove insoluble residues (3000 × g for 5 min, at 20°C), and then dialyzed with dialysis membrane in running water. Ascorbic acid was used to detect whether free SeO_3^{2-} was dialyzed out from the solution. When ascorbic acid did not redden the solution, the Se-APS solution was collected, and freeze-dried.

BBD designed seventeen experimental points for optimizing the three individual parameters. Five replicates at the center of the design were used to allow for estimation of a pure error sum of squares. The response surface analysis predicted that the optimal reaction time, reaction temperature and the ratio of Na_2SeO_3 and APS were 8.01 h, 80°C and 0.99 g/g, respectively. The model predicted the optimized Se content in Se-APS was 8.409 mg/g. For the actual operating convenience, the confirmatory experimental verification was tested under the conditions: reaction time 8 h, reaction temperature 80°C and the ratio of Na_2SeO_3 to APS 1 g/g. A value of 8.462 mg/g was obtained from the experiments. The results indicate that the actual value is close to the predicted result, and the optimization parameters are reliable.

2.2 *Characterization of Se-APS*

The solution of Se-APS and APS in water (0.02 mg/mL) was detected by Zetasizer Nano ZS90 nano-particle size and zeta potential analyzer. The results showed that the particle size of Se-APS (85.65 nm) was significantly smaller than that of APS (159 nm). The reason was that APS was hydrolyzed into many smaller pieces by acidic water in selenylation. Moreover, it also implied that absorption and utilization of Se-APS were better than that of APS at the same dosing. Additionally, zeta potential of APS (–0.237 mV) was higher than that of Se-APS (–0.190 mV). This result shows that Se-APS was prone to clump and agglomerate and that the stability of Se-APS was lower than that of APS in solution.

Pyris 6-thermal gravimetric analysis instrument was used to analyze the thermal weight loss of Se-APS and APS (air flow rate: 5 mL/min, heating rate: 10°C/min, and scope of temperature: 20–760°C). The differences between Se-APS and APS could be observed from the TG-DTG curves. On the one hand, the temperature of dehydration peak (5% loss, around 70°C) of Se-APS was significantly lower than that of APS (5% loss, around 78°C). On the other hand, in the decomposition peak (200–400°C), the fastest weight loss temperatures of Se-APS and APS were 267°C and 296°C, respectively. This demonstrated that stability of Se-APS was lower than that of APS.

Meanwhile, in order to reveal the configuration after selenylation, Se-APS and APS were detected by PW1700 X-ray diffraction instrument (Radiation: CuKα, $2\theta = 10°–80°$). The results showed that Se-APS and APS were amorphous, and indicates that Se introduction had little effect on amorphous configuration of polysaccharide. Consequently, this also explained the regularity of polysaccharide molecules was poor and the configuration of polysaccharide was complex.

Besides, Se-APS and APS (5 μL, 0.01 mg/mL) were dropped on the freshly cleaved mica and Atomic Force Microscopy (AFM) was used to observe the morphology of them. The results showed that the length of Se-APS was shorter than that of APS and the shape of Se-APS was spherical. Meanwhile, APS had some long chains. The reason was that the chains of Se-APS were cut in acidic water, and Se-APS was intertwined by the hydrogen bond association and the van der Waals force interactions.

2.3 *Antioxidant activity of Se-APS*

Kunming mice were transplanted with tumor S_{180} 0.2 mL (living cells number ≥ 97%) in the armpit. The weighed mice after inoculation were randomly divided into six groups (n = 8), half male and half female: (1) negative group: mice were treated with normal saline. (2) positive group: mice were treated with lentinan at a dosage of 100 mg/kg/d. (3) APS group: mice were treated with APS at a dosage of 80 mg/kg/d. (4) Se-APS groups: mice of low, medium, high-dose groups were treated with Se-APS at the dosage of 40, 80, 120 mg/kg/d, respectively.

After 24 h of inoculation, the mice were treated in peritoneal injection for 10 days. Then, catalase (CAT), malondialdehyde (MDA), superoxide dismutase (SOD) and glutathione peroxidase (GSH-Px) of mice bearing tumor S_{180} in blood, heart and liver were detected by antioxidant assay kits. The results showed that compared with negative group, Se-APS could significantly improve the CAT, SOD and GSH-Px activity and reduce MDA level in blood, heart and liver. Moreover, the result also showed that antioxidant activity of Se-APS was better than that of APS at the same doses.

3 CONCLUSION

In this paper, the selenylation conditions for Se-APS were optimized by BBD. The experiments confirmed that optimum conditions consisted at a reaction time of 8 h, reaction temperature of 80°C and the ratio of Na_2SeO_3 to APS of 1 g/g. The actual experimental Se content of 8.462 mg/g closely matched the predicted value. Furthermore, particle size, zeta potential, the thermal stability, and shape of Se-APS were significantly different from those of APS. The stability of Se-APS was lower than that of APS. In the antioxidant activity assays, besides decreasing MDA, Se-APS also could enhance CAT, SOD and GSH-Px activity. These results demonstrated that Se-APS could significantly enhance antioxidant system and selenylation was a feasible approach to seek high biological functional seleno-polysaccharide.

REFERENCES

Micke, O., Schomburg, L., Buentzel, J., Kisters, K. & Muecke, R. 2009. Selenium in oncology: from chemistry to clinics. *Molecules* 14: 3975–3988.

Shang, D., Li, Y. & Wang, C.A. 2011. Novel polysaccharide from Se-enriched *Ganoderma lucidum* induces apoptosis of human breast cancer cells. *Oncology Reports* 25: 267–272.

Wu, F. & Chen, X. 2004. A review of pharmacological study on *Astragalus membranaceus* (Fisch) Bge. *Zhong Yao Cai* 27: 232–234. (in Chiese)

Xu, D.J., Xia, Q., Wang, J.J. & Wang, P.P. 2008. Molecular weight and monosaccharide composition of *Astragalus polysaccharides*. *Molecules* 13: 2408–2415.

Selenium in the Environment and Human Health – Bañuelos, Lin & Yin (Eds)
© 2014 Taylor & Francis Group, London, ISBN 978-1-138-00017-9

How can we understand the epidemiology of selenium and type-2 diabetes?

M.P. Rayman
Faculty of Health and Medical Sciences, University of Surrey, Guildford, UK

S. Stranges
Division of Health Sciences, University of Warwick Medical School, Medical School Building, Gibbet Hill Campus, Coventry, UK

1 INTRODUCTION

The trace mineral, selenium (Se), is essential for human health (Rayman, 2012). As selenocysteine, it is the key component of a number of selenoproteins with essential enzymatic functions that include redox homeostasis, thyroid hormone metabolism and protection from oxidative stress and inflammation (Rayman, 2012). We know that selenoproteins play a role in certain health conditions because Single Nucleotide Polymorphisms (SNPs) in a number of selenoprotein genes affect the risk of those conditions and indeed this is the case for inflammation and diabetes-related conditions (Rayman, 2012). However, not all Se in the body is present as selenoproteins; a proportion is present as selenomethionine in body proteins, while some is present as low-molecular-weight species which can give rise to the unstable, reactive methyl selenol (Rayman, 2012).

The potential of the selenoproteins to protect against oxidative stress led to the expectation that Se would be protective against type-2 diabetes, and indeed in the 1990s, Se (as selenate) was shown to have anti-diabetic and insulin-mimetic effects (Steinbrenner *et al.*, 2011). However, more recently, findings from observational epidemiological studies and randomized clinical trials have raised concern that high Se exposure may lead to type-2 diabetes or insulin resistance, at least in well-nourished populations (Rayman & Stranges, 2013).

2 EPIDEMIOLOGY OF SELENIUM AND TYPE-2 DIABETES RISK

2.1 *Cross-sectional studies of Se and type-2 diabetes*

Five out of eight studies identified significant positive associations between serum/plasma selenium and type-2 diabetes or fasting plasma glucose, two have shown no significant association and one has shown an inverse association (Rayman & Stranges, 2013). However, cross-sectional studies cannot show causality and do not allow us to determine whether high Se is a cause or a consequence of the disease process.

2.2 *Longitudinal studies of Se and type-2 diabetes*

Longitudinal studies have generally not supported a causal role for Se in type-2 diabetes (Rayman & Stranges, 2013). In the Olivetti Heart Study, baseline Se status was not prospectively associated with type-2 diabetes, while prospective findings from the EVA study in France showed that high plasma Se was associated with a marginally significantly reduced risk of hyperglycaemia in men. Recently, a pooled longitudinal analysis from two large US cohorts actually showed inverse associations between toenail Se and incident type-2 diabetes; diabetes risk reduced as quintile of toenail Se increased (Park *et al.*, 2012).

2.3 *Randomized Controlled Trials (RCTs) of Se and type-2 diabetes risk*

A number of RCTs have examined the effect of Se on the risk of type-2 diabetes (Rayman & Stranges, 2013). These were generally cancer trials where type-2 diabetes was only a secondary outcome.

In a *post-hoc* analysis of the Nutritional Prevention of Cancer (NPC) Trial in 1312 US participants, the risk of developing type-2 diabetes was significantly higher in the Se-supplemented group than in the placebo group (Hazard Ratio, HR 1.55; 95% CI, 1.03 to 2.33). When subjects were divided into tertiles by baseline plasma Se, a significantly increased risk of type-2 diabetes was found only in the highest Se tertile, *i.e.* plasma Se > 121.6 μg/L (HR 2.70; 95% CI, 1.30 to 5.61) (Stranges *et al.*, 2007).

By contrast, SELECT (Selenium and Vitamin E Cancer Prevention Trial) found no adverse effect of supplementation with 200 μg/d Se, as selenomethionine, on type-2 diabetes risk in 35,533 N. American men (Lippmann *et al.*, 2009). Given the difference in effect seen on type-2 diabetes risk in this trial and the NPC trial, it should be noted that SELECT men had quite high Se status at baseline (serum Se 136 μg/L), considerably higher than NPC-trial participants (114 μg/L) (Rayman & Stranges, 2013).

A number of smaller trials including PRECISE (500 participants), and the Watchful Waiting Trial (140 men) have also found no evidence for an adverse effect of Se supplementation on type-2 diabetes risk (Rayman & Stranges, 2013).

3 POSSIBLE EXPLANATIONS FOR THE EPIDEMIOLOGICAL FINDINGS

3.1 *Cross-sectional associations between Se and type-2 diabetes*

The mainly positive cross-sectional associations between serum/plasma Se and type-2 diabetes have a number of possible explanations (Rayman & Stranges, 2013). We know that the expression of SeP, a major component of Se in plasma, and gluconeogenic enzymes is linked in that increased expression of SeP will raise plasma Se, while increased expression of gluconeogenic enzymes will raise plasma glucose by a number of plausible mechanisms (Steinbrenner *et al.* 2011). This linked expression could explain the significant correlations found between SeP, and fasting plasma glucose, and between SeP and glycosylated haemoglobin (HbA1c), a marker of type-2 diabetes (Misu *et al.*, 2012) and the significantly higher serum SeP concentrations in people with type-2 diabetes or pre-diabetes than in those with normal glucose tolerance (Yang *et al.*, 2011).

Of course the cross-sectional associations could result from Se, as SeP, causing type-2 diabetes. However, longitudinal and prospective studies have not supported such a causal role nor have trial results been as consistent as would be expected if this were the case.

3.2 *The inconsistent results of Randomized Controlled Trials*

There appears to be no adverse effect of Se supplementation when Se status is low. However, the effect of supplementation with Se in those of higher Se status is not consistent. Possible explanations for this inconsistency include the following.

1. Could the NPC trial results be a chance finding? Results derive from a *post-hoc* analysis of a small trial. SELECT was a much larger study and its findings are therefore more robust.
2. Was the Se status of SELECT men at baseline already above a threshold of risk because of their higher Se status? Baseline Se status was already high enough at baseline for SeP (which may impair insulin signalling) to be maximised (Hurst *et al.*, 2010).
3. SELECT men were supplemented with selenmethionine rather than Se-yeast as used in the NPC trial.
4. Was arsenic exposure, which increases the risk of type-2 diabetes, an additional risk factor in NPC subjects (Rayman & Stranges, 2013)?

4 CONCLUSIONS

The relationship between Se and type-2 diabetes is undoubtedly complex. It is clear that higher concentrations of SeP have been found in type-2 diabetic than in non-diabetic subjects (Misu *et al.*, 2010; Yang *et al.*, 2011). By contrast, significantly lower plasma GPx3 protein concentrations have been found in such subjects (Chung *et al.*, 2009). As major contributors to plasma Se, these two factors will have opposing effects on apparent Se status. A further complication is that plasma Se is lowered by the systemic inflammatory response (Renko *et al.*, 2009) which is a feature of type-2 diabetes. Stage of disease and associated level of oxidative stress may influence levels of these selenoproteins in the plasma.

Finally, it may be that the association between Se and cardio-metabolic outcomes is U-shaped with potential harm occurring both below and above the physiological range for optimal activity of some or all selenoproteins (Rayman, 2012).

REFERENCES

Chung, S.S., Kim, M., Youn, B.S., Lee, N.S., Park, J.W., Lee, I.K. *et al.* 2009. Glutathione peroxidase 3 mediates the antioxidant effect of peroxisome proliferator-activated receptor gamma in human skeletal muscle cells. *Mol. Cell Biol.* 29: 20–30.

Hurst, R., Armah, C.N., Dainty, J.R., Hart, D.J., Teucher, B., Goldson, A.J. *et al.* 2010. Establishing optimal selenium status: results of a randomized, double-blind, placebo-controlled trial. *Am. J. Clin. Nutr.* 91: 923–931.

Lippman, S.M., Klein, E.A., Goodman, P.J., Lucia, M.S., Thompson, I.M., Ford, L.G. *et al.* 2009. Effect of selenium and vitamin E on risk of prostate cancer and other cancers: the Selenium and Vitamin E Cancer Prevention Trial (SELECT). *JAMA.* 301: 39–51.

Misu, H., Takamura, T., Takayama, H., Hayashi, H., Matsuzawa-Nagata, N., Kurita, S. *et al.* 2010. A liver-derived secretory protein, selenoprotein p, causes insulin resistance. *Cell Metabolism* 12: 483–495.

Park, K., Rimm, E.B., Siscovick, D.S., Spiegelman, D., Manson, J.E., Morris, J.S. *et al.* 2012. Toenail selenium and incidence of type 2 diabetes in U.S. men and women. *Diabetes Care* 35: 1544–1551.

Rayman, M.P. & Stranges, S. 2013. Epidemiology of selenium and type 2 diabetes: can we make sense of it? *Free Rad. Biol. Med.* DOI 10.1016/j.freeradbiomed.2013.04.003.

Rayman, M.P. 2012. Selenium and human health. *Lancet* 379: 1256–1268.

Renko, K., Hofmann, P.J., Stoedter, M., Hollenbach, B., Behrends, T., Köhrle, J. *et al.* 2009. Down-regulation of the hepatic selenoprotein biosynthesis machinery impairs selenium metabolism during the acute phase response in mice. *FASEB J.* 23: 1758–1765.

Steinbrenner, H., Speckmann, B., Pinto, A. & Sies, H. 2011. High selenium intake and increased diabetes risk: experimental evidence for interplay between selenium and carbohydrate metabolism. *J. Clin. Biochem. Nutr.* 48: 40–45.

Stranges, S., Marshall, J.R., Natarajan, R., Donahue, R.P., Trevisan, M., Combs, G.F. *et al.* 2007. Effects of long-term selenium supplementation on the incidence of type 2 diabetes: a randomized trial. *Ann. Intern. Med.* 147: 217–23.

Yang, S.J., Hwang, S.Y., Choi, H.Y., Yoo, H.J., Seo, J.A., Kim, S.G. *et al.* 2011. Serum selenoprotein P levels in patients with type 2 diabetes and prediabetes: implications for insulin resistance, inflammation, and atherosclerosis. *J. Clin. Endocrinol. Metab.* 96: E1325–1329.

Selenium in the Environment and Human Health – Bañuelos, Lin & Yin (Eds)
© 2014 Taylor & Francis Group, London, ISBN 978-1-138-00017-9

Selenium in cancer prevention: Did we move too quickly?

James R. Marshall
Cancer Prevention and Control, Roswell Park Cancer Institute, Buffalo, NY, USA

1 INTRODUCTION

Selenium (Se) has strong biologic actions in humans and animals. Based on experimental animal research, it is clearly known that inadequate Se intake can have significant consequences. Clark's findings (Clark *et al.*, 1996) generated tremendous excitement among cancer prevention researchers on his relatively small chemoprevention trial. The findings were seen as quite reasonable, in light of strong basic scientific understanding of the biological importance of Se nutriture.

2 HOPE FROM BASIC SCIENCE

The expectation that Se might provide a chemopreventive benefit was based at least partly on its role as an antioxidant. Selenium does play a critical role, as a constituent of glutathione peroxidase, in protection against oxidative stress. However, it is not clear that selenium's antioxidant function is its most critical for protection against disease. Absorbed and metabolized much as sulfur and substituting in many compounds for sulfur (S), Se may provoke a perturbation of the metabolism of sulfurous compounds. As Ip and others have shown, Se may sensitize the endoplasmic reticulum and the androgen receptor (Zu & Ip, 2003; Lee *et al.*, 2006; Zu *et al.*, 2006) to disturbances in cellular function, and may activate cell repair or the initiation of apoptosis.

3 PROVOCATIVE EVIDENCE FROM A SINGLE TRIAL

The trial led by Dr. Larry Clark (Clark *et al.*, 1996), published in 1996 (JAMA), probably did more than any other agent to increase interest in Se as a chemopreventive agent. That trial, with subjects randomized to placebo or to 200 µg Se in selenized yeast, showed that assignment to Se treatment was associated with a 50% decrease in prostate cancer diagnosis and other less dramatic effects on other cancers. The precise form of the Se was, not well documented, although it is believed that it was largely selenomethionine. Moreover, the effect of Se appeared to be strongest among those with the lowest baseline blood Se levels. This striking finding led the noted Se biochemist, Clement Ip, to note the distinct contributions of clinical trials, epidemiology and basic research in selenium-based chemoprevention (Ip, 1998).

4 PIVOTOL AND DEFINITIVE TRIALS

One outgrowth of the Clark study was a trial initiated among men diagnosed with High Grade Prostatic Intraepithelial Neoplasia (HGPIN). This condition is recognized to be a predictor of increased risk of prostate cancer. Subjects were randomized to a selenium supplementation 200 µg/day or to placebo, with the selenium delivered in the form of selenomethionine. The result of the trial, with some 460 men randomized, was that Se supplementation had no effect on the development of prostate cancer (Marshall *et al.*, 2011).

The much larger SELECT, a 2 × 2 factorial design trial which studied supplementation by both selenium 200 µg/day and vitamin E 400 IU/day, enrolled 35,000 average risk US men. Selenium was again in the form of selenomethionine, and prostate cancer was the primary endpoint. The trial was ended prematurely, as a futility analysis showed it was extremely unlikely that the result of the trial would be anything other than null (Lippman *et al.*, 2005).

5 SELENOMETHIONINE AS THE AGENT?

An issue that remains concerns the form of Se used in both the HGPIN and SELECT studies. Selenomethionine, found in a number of foodstuffs, is known to be relatively nontoxic in even supra-nutritional doses. However, cell-based research has suggests that selenomethionine is a relatively inefficient means of generating the methylated Se species that may be most pertinent other forms. Methyl selenocysteine or methyl selenic acid may be much more efficient sources of methylated Se (Marshall *et al.*, 2011). Thus, we are engaged in an 84-day trial in which we compare the pharmacokinetics of selenomethionine and methyl selenocysteine at doses of 400, 800 and 1,200 µg/day. We will present preliminary data from that trial in our key note presentation.

REFERENCES

Clark, L.C., Combs, G.F., Turnbull, B.W., Slate, E.H., Chalker, D.K., Chow, J. *et al.* 1996. Effects of selenium supplementation for cancer prevention in patients with carcinoma of the skin. A randomized controlled trial. *J. Am. Med. Assoc.* 276: 1957–1963.

Ip, C. 1998. Lessons from basic research in selenium and cancer prevention. *J. Nutr.* 128: 1845–1854.

Lee, S.O., Yeon, C.J., Nadiminty, N., Trump, D.L., Ip, C., Dong, Y. *et al.* 2006. Monomethylated selenium inhibits growth of LNCaP human prostate cancer xenograft accompanied by a decrease in the expression of androgen receptor and prostate-specific antigen (PSA). *Prostate* 66: 1070–1075.

Lippman, S.M., Goodman, P.J., Klein, E.A., Parnes, H.L., Thompson, I.M. Jr., Kristal, A.R. *et al.* 2005. Designing the selenium and vitamin E cancer prevention trial (SELECT). *J. Natl. Cancer Inst.* 97: 94–102.

Marshall, J.R., Ip, C., Romano, K., Fetterly, G., Fakih, M., Jovanovic, B. *et al.* 2011. Methyl selenocysteine: Single-dose pharmacokinetics in men. *Cancer Prevention Research* 4: 1938–1944.

Marshall, J.R., Tangen, C.M., Sakr, WA., Wood, D.P. Jr, Berry, D.L., Klein, E.A. *et al.* 2011. Phase III trial of selenium to prevent prostate cancer in men with high-grade prostatic intraepithelial neoplasia: SWOG S9917. *Cancer Prevention Research* 4: 1761–1769.

Zu, K. & Ip, C. 2003. Synergy between selenium and vitamin E in apoptosis induction is associated with activation of distinctive initiator caspases in human prostate cancer cells. *Cancer Res.* 63: 6988–6995.

Zu, K., Wu, Y., Park, Y. & Ip, C. 2006. Peering down the kaleidoscope of thiol proteomics and unfolded protein response in studying the anticancer action of selenium. In M.J. Nerry, V.N. Gladyshev & D. Hatfield (eds.), *Selenium, its Molecular Biology and Role in Human Health*, 265–276. Norwell, MA: Springer.

Selenium in the Environment and Human Health – Bañuelos, Lin & Yin (Eds)
© 2014 Taylor & Francis Group, London, ISBN 978-1-138-00017-9

Interaction between selenoprotein P and tubulin and the implications in Alzheimer's disease

Zhi Wang & Xiubo Du
College of Life Sciences, Shenzhen Key Laboratory of Microbial Genetic Engineering, Shenzhen University, Shenzhen, China

Shi Qiu, Chao Wang & Qiong Liu
Department of Marine Biology, Shenzhen Key Laboratory of Marine Biotechnology and Ecology, Shenzhen, China

1 INSTRUCTION

Selenoproteins are a group of special proteins that contain the 21st amino acid, selenocysteine (Sec, U) in their sequence. Sec is encoded by TGA, which is generally used as a terminal signal in protein translation. Among the 25 selenoproteins identified in human, selenoprotein P (SelP) is the only one containing multiple (ten) Sec residues. SelP was suggested to function as the carrier and transporter of selenium (Se) that delivers Se from liver to peripheral tissues. Expression of SelP is upregulated in an age-dependent manner (Lu *et al.*, 2004). Moreover, SelP was found to be co-localized with Aβ plaques in the postmortem tissue from individuals with the hallmark lesions of AD (Bellinger *et al.*, 2008). The direct association of SelP expression with the pathology of AD suggests that this protein is involved in the response or progression of the disorder. However, the exact function and mechanism of SelP in AD prevention remain unknown. In this study, the interaction between SelP and tubulin was investigated and the implications in AD were discussed.

2 MATERIALS AND METHODS

2.1 *Yeast two-hybrid screening*

To facilitate the expression of SelP in yeast and mammalian cells, the ten Sec-encoding TGAs were mutated to Cys-encoding TGCs and named as SelP′. SelP′ was used to screen the human fetal brain cDNA library via the yeast two-hybrid system. The screened positive prey plasmid was co-transformed with the bait plasmid into yeast for re-transformation verification.

2.2 *FRET Analyses*

HEK293T cells were co-transfected with pEYFP-C1-*Tub* and pECFP-C1-*SelP′* or pEYFP-C1-*Tub-C* and pECFP-C1-*SelP-H* for FRET analysis using laser confocal microscope (Olympus FV1000, Tokyo, Japan). Two types of FRET methods, sensitized emission and receptor bleaching, were used to measure FRET efficiency and the distance between the interactive proteins.

2.3 *Isothermal Titration Calorimetry (ITC)*

ITC measurements were performed on a MicroCal iTC-200 microcalorimeter (Northampton, MA) at 25°C. Briefly, 2 μl of 3 mM Tub-C were titrated into 200 μl of 500 μM proteins over 4 s with a 3-min interval between each injection. Twenty injections were made in total. The reaction solution was stirred at 1000 rpm. The results were analyzed by Origin 7.0 (Microcal) using one-site binding model.

3 RESULTS AND DISCUSSIONS

3.1 *Screening the interacting protein of SelP from the human fetal-brain cDNA library*

NpGBKT7-SelP′-containing yeast was used to screen the fetal brain cDNA library. On the corresponding selection plates, sixteen colonies were grown, the plasmids of which were extracted and transformed into the E. coli Top10 cells separately, to screen for colonies carrying the gene of the interactive protein (AD plasmid). The BD plasmid and the screened AD plasmid were then re-transformed into AH109 yeast cells. The AD plasmids in two positive colonies following re-transformation were subjected to sequencing and bioinformatics analysis with NCBI's non-redundant (nr) protein database. One of them was identified to be Homo sapiens tubulin, alpha 1a (TUBA1 A).

3.2 *Verification the iteraction between SelP′ and tubulin by FRET*

To verify the interaction between SelP′ and tubulin, FRET methods of sensitized emission and receptor photobleaching were performed. For the sensitized emission assay, the energy transfer efficiency between CFP-SelP′ (donor) and YFP-Tub (receptor) was calculated to be 21.8 ± 7.2% (n = 11), and the distance between the donor and receptor was calculated to be 7.5 ± 0.4 nm (n = 11).

Results from the acceptor photobleaching experiments showed that fluorescence of the CFP-SelP′ donor was significantly increased after the receptor was bleached, which was not observed in control cells. The distance between CFP-SelP′donor and YFP-Tub receptor was estimated to be 6.8 ± 0.5 nm (n = 5). The energy transfer efficiency between the two proteins was calculated to be 21.3 ± 5.6% (n = 5). Results from FRET assays confirmed the interaction between SelP′ and tubulin.

3.3 *Studying the interaction domain between SelP′ and tubulin by FRET*

The C-terminus of tubulin is involved in multiple aspects of the regulation of microtubule assembly. A recent study revealed that this region (the last 42 residues) interacted with several cationic molecules, including Tau, polyamines and calcium (Lefevre *et al.*, 2011). SelP encodes two His-rich regions, located at residue 204–217 and residue 244–250, respectively. The His-rich domain of SelP (188–263) was predicted to be on the surface of the protein via PROFace program. We proposed tubulin might interact with the His-rich domain of SelP (named as SelP-H) via its C-terminal region (named as Tub-C).

The interaction between SelP-H and Tub-C was first tested by FRET with the method of acceptor photobleaching. The energy transfer efficiency and distance between CFP-SelP′ donor and YFP-Tub receptor were estimated to be 22.8 ± 5.4% and 7.4 ± 0.5 nm (n = 12), respectively. The FRET experiments confirmed the interaction between SelP-H and Tub-C preliminarily.

3.4 *Verification of the interaction between SelP′ and tubulin by ITC*

SelP-H was expressed and purified as described previously (Du *et al.*, 2013). To express Tub-C, DNA fragment encoding the last 42 residues of tubulin was amplified and inserted into pET15b vector to generate pET15b-Tub-C plasmid. Tub-C protein was expressed from BL21 transformed with pET15b-Tub-C and purified with the aid of Ni^{2+}-IMAC chromatography.

Titration of Tub-C into SelP-H resulted in exothermic peaks, which eventually diminished to just the heat of dilution after 12 injections. Analysis of the binding isotherm using one set of sites binding model approximated the association constant to be 2.8E6 ± 5.2E6 /M, and the binding stoichiometry to be 0.4 ± 0.1. The ΔH and ΔS were estimated to be −398.6 ± 41.78 cal/mol and 28.1 cal/mol K, respectively. Therefore, Tub-C binding to SelP-H is an enthalpically and entropically driven process.

4 CONCLUSIONS

By yeast two-hybrid screening, SelP was found to interact with tubulin, which was confirmed by FRET with the methods of sensitized emission and receptor photobleaching. Next, we found the C-terminus of tubulin bound directly to the His-rich domain of SelP through ITC study. In AD, microtubule associated protein tau was hyperphosphorylated, resulting in the collapse of cytoskeleton and the formation of neurofibrillary tangles (NFT). Previously, van Eersel reported sodium selenate reduced tau phosphorylation both in vitro and in vivo (van Eersel *et al.*, 2010). The interaction between SelP and tubulin suggested that SelP may provide Se for tau to inhibit its abnormal phosphorylation or function as a stabilizer of microtubule. It would be of interest for future work to investigate the exact mechanism of the function of SelP in AD.

ACKNOWLEDGEMENTS

The two authors Zhi Wang and Shi Qiu contributed equally to this work.

REFERENCES

Bellinger, F.P., He, Q.P., Bellinger, M.T., Lin, Y., Raman, A.V., White, L.R. *et al.* 2008. Association of selenoprotein p with Alzheimer's pathology in human cortex. *J. Alzheimers Dis.* 15(3): 465–472.

Du, X., Li, H., Wang, Z., Qiu, S., Liu, Q., Ni, J. 2013. Selenoprotein P and selenoprotein M block Zn(2+)-mediated Abeta42 aggregation and toxicity. *Metallomics* 5(7): 861–870.

Lefèvre, J., Chernov, K.G., Joshi, V., Delga, S., Toma, F., Pastré DLefevre, J. *et al.* 2011. The C terminus of tubulin, a versatile partner for cationic molecules: binding of Tau, polyamines, and calcium. *J. Biol. Chem.* 286(4): 3065–3078.

Lu, T., Pan, Y., Kao, S.Y., Li, C., Kohane, I., Chan, J. et al. 2004. Gene regulation and DNA damage in the ageing human brain. *Nature* 429(6994): 883–891.

van Eersel, J., Ke, Y.D., Liu, X., Delerue, F., Kril, J.J., Götz, J. et al. 2010. Sodium selenate mitigates tau pathology, neurodegeneration, and functional deficits in Alzheimer's disease models. *Proc. Natl. Acad. Sci.* 107(31): 13888–13893.

Selenium in the Environment and Human Health – Bañuelos, Lin & Yin (Eds)
© 2014 Taylor & Francis Group, London, ISBN 978-1-138-00017-9

Effect of sodium selenate on the level of extracellular homocysteine in Alzheimer's disease model cells

L.X. Wang & Q. Liu
Department of Marine Biology, Shenzhen Key Laboratory of Marine Biotechnology and Ecology, Shenzhen, China

P. Chen
College of Life Sciences, Shenzhen Key Laboratory of Microbial Genetic Engineering, Shenzhen University, Shenzhen, China

1 INTRODUCTION

Alzheimer's Disease (AD) is a progressive neurodegenerative disorder affecting over 20 million people worldwide and is the most common form of dementia and cognitive impairment. AD is characterized by progressive loss of memory, intellectuality and emotion, as well as further executive, attentional or behavioral function (Loef *et al.*, 2011).

Extracellular Senile Plaques (SP) and intracellular neurofibrillary tangles (NFT) are two core histopathological hallmarkers of this disease (Selkoe, 2001). SP is composed of the β-amyloid peptide (Aβ) formed by the sequence cleavage of Amyloid Precursor Protein (APP) via β- and γ-secretase enzymes (Selkoe, 1996). NFTs in the neuron and dystrophic neuritis are composed of hyperphosphorylated Microtubule Associated Protein (MAP) tau (Corcoran *et al.*, 2010).

High concentration of homocysteine (Hcy) seems to be an independent risk factor for many age related diseases, including dementia and AD. Elevated level of plasma total Hcy among the old people is strongly related to global cognitive decline in AD patients (Oulhaj *et al.*, 2010). The association between Hcy and the development of AD is mediated by several potential biological pathways. Firstly, Hcy serves as the methyl donor for methylation reaction within neurons. Secondly, Hcy plays an important role in post-translational modification of proteins and syntheses of nucleic acids, phospholipids, and neurotransmitters. Hyperhomocysteinemia changes the structure and function of cerebral blood vessels and neurons, and oxidative stress appears to play a major role in mediating these changes (Seshadri, 2006; Morris, 2003).

Recently, it was found that low doses of sodium selenate specifically activate Protein Phosphatase 2 A (PP2 A), a family member of heterotrimeric enzymes and a major brain Ser/Thr phosphatase, which is regulated by methylation involved in AD pathogenesis. Chronic oral treatment of transgenic mice that over-express tau protein with sodium selenate; (1) reduced tau phosphorylation; (2) completely prevented NFT formation in the hippocampus and amygdale; and (3) improved spatial learning and motor per-formance. Subsequent experiments indicated that this effect was mediated uniquely by PP2 A (Corcoran *et al.*, 2010; Van Eersel *et al.*, 2010).

Serum selenium (Se) concentration and plasma Hcy level have been found to be inversely associated in elderly human population (Gonzalez *et al.*, 2004). Both Hcy and Se are involved in the development of AD (Liu *et al.*, 2012). Here, we found that sodium selenate can increase the proliferation of AD model cells N2a_sw, while lower the Hcy levels in the culture medium.

2 MATERIALS AND METHODS

2.1 *Cell culture*

N2a_sw cell line was derived from N2a neuroblastoma cell stably expressing APPsw, a Swedish mutant of APP protein. Those cells were cultured in DMEM/Opti-MEMI (Thermo/Invitrogen, USA) medium supplemented with 5% FBS (Invitrogen, USA), G418 (Sigma-Aldrich, USA) and penicillin/streptomycin at 37°C / 5% CO_2.

2.2 *Cell viability assay*

Sodium selenate (Sigma-Aldrich, USA) was dissolved in sterile ddH_2O to prepare 100 mM stock solution. N2a_sw cells (5×10^3 per well) were plated in 96-well microplates. After treatment with a series of dosages of sodium selenate (0, 0.00001, 0.0001, 0.001, 0.01, 0.1, 1, 10 mM Se final concentrations) for 24 h. 10% Cell Counting Kit-8 (CCK-8)

(Beyotime, China) was added to each well for an additional 2 h cell culture. The absorbance of sample was measured at 450 nm wavelength with a reference at 650 nm on iMark Microplate Reader (Bio-Rad, USA). Cells without sodium selenate treatment served as the control.

2.3 *Cell proliferation assay*

N2a_sw cells (1 × 10^6 per well) were plated in 6-well microplates, incubated with sodium selenate (0, 0.01, 0.1 mM at final concentrations) for 24 h, harvested and washed three times with PBS, labeled with carboxyfluorescein diacetate succinimidyl ester (CFDA SE, often called CFSE) according to the instructions of CFDA SE Cell Proliferation Assay and Tracking Kit (Beyotime, China). Cell proliferation was measured by flow cytometry and data were analyzed by the WinMDI 2.9 software.

2.4 *Sample preparation and homocysteine analysis*

Cell culture media were collected and spinned to remove pellets. Protein concentrations of the culture medium were determined using the bicinchoninic acid (BCA) assay (Pierce, USA). Hcy level was determined with Hcy assay Kit (AUSA, China) following the manufacturer's instructions.

3 RESULTS AND DISCUSSION

3.1 *Effect of sodium selenate on cell viability*

Cell viability was evaluated by CCK-8 assay. As shown in Fig. 1, the growth of N2a_sw cells was increased by sodium selenate in dose-dependent manner. Only at a concentration greater than 0.1 mM were they inhibited significantly by selenate. The optimum concentration for selenate to increase cell viability was 0.01 mM.

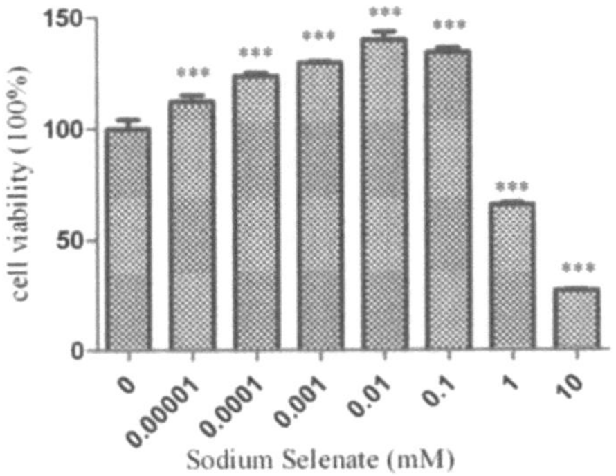

Figure 1. Dose-dependent effect of sodium selenate on the viability of N2a_sw cells after 24 h treatment. ***$P < 0.001$ represents very significant difference between the selenate-treated and control cells.

3.2 *Effect of sodium selenate on cell proliferation*

To further investigate the effect of 0.01 mM sodium selenate on cell proliferation, the treated cells were labeled with CFDA and analyzed by flow cytometry. The result indicated that sodium selenate can promote cell proliferation compared with the control cells (untreated with selenate). These results on cell viability and proliferation are in accordance with previous reports where sodium selenate was found to have no cytotoxicity with hippocampal neurons and reducing tau protein phosphorylation, preventing NFT formation, improving spatial learning and motor performance (Corcoran *et al.*, 2010; van Eersel *et al.*, 2010). Thus 0.01 mM sodium selenate was selected for the mechanism studies described below.

3.3 *Extracellular Hcy level altered by selenate*

Recently, Hcy level is emerging as a risk factor for Alzheimer's disease. However, study on the correlation of Se status with Hcy level seems to be conflict with one another. One hand, serum Se level and plasma Hcy concentration have been found to be independent inversely associated in elderly human population (Gonzalez *et al.*, 2004). Plasma total Hcy concentration decreased in rats fed a low-Se diet and increased with Se supplementation (Uthus *et al.*, 2002). On the other hand, though low Se status was significantly associated with high plasma

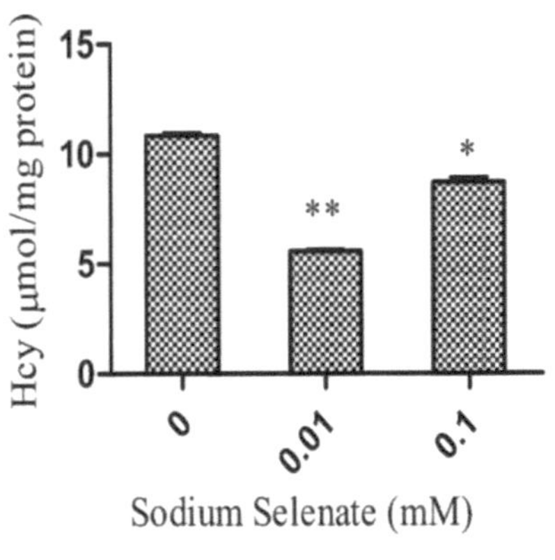

Figure 2. Supplementation of sodium selenate on extracellular Hcy level. N2a sw cell culture supernatants were collected for detecting the Hcy levels. *$P < 0.05$; **$P < 0.01$ represents significant and very significant differences between the selenate-treated and control cells, respectively.

Hcy level at baseline, Se supplement couldn't decrease Hcy in plasma (Bekaert *et al.*, 2008). In order to clarify whether supplement Se can affect the alteration of Hcy, sodium selenate was selected to affect the AD model cell, followed by the detection of Hcy levels in cell culture. As showed in Figure 2, N2a_sw cells treated with relatively low level of selenate caused a decline of Hcy level in the culture medium. Those are only primary results, and further study remains to be carried out.

4 CONCLUSIONS

The present study demonstrated that sodium selenate has low toxicity on AD model cell N2a_sw. 0.01 mM selente can promote the cell proliferation significantly, down-regulate Hcy levels of the culture medium. These results indicate that Hcy levels can be mediated by sodium selenate and thus alter the progression of AD. However, a clear relationship between Se and Hcy need to be further investigated.

ACKNOWLEDGEMENTS

This work was financially supported by the National Natural Science Foundation of China (No. 31070731 and 21271131), the Natural Science Foundation of Guangdong Province (No. 10151806001000023). We are grateful to Prof. Huaxi Xu and Prof. Yunwu Zhang (Institute for Biomedical Science, Xiamen University, Xiamen, China) for providing the N2a_sw cell line, and to the AUSA Pharmed Ltd in China for their help with Hcy measurement.

REFERENCES

Bekaert, B., Cooper, M.L., Green, F.R., McNulty, H., Pentieva, K., Scott, J.M. *et al.* 2008. Effect of selenium status and supplementation with high-selenium yeast on plasma homocysteine and B vitamin concentrations in the UK elderly. *Mol. Nutr. Food. Res.* 52: 1324–1333.

Corcoran, N.M., Martin, D., Hutter-Paier, B., Windisch, M., Nguyen, T., Nheu, L. *et al.* 2010. Sodium selenate specifically activates PP2 A phosphatase, dephosphorylates tau and reverses memory deficits in an Alzheimer's disease model. *Journal of Clinical Neuroscience* 17: 1025–1033.

Gonzalez, S., Huerta, J.M., Alvarez-Uria, J., Fernandez, S., Patterson, A.M. & Lasheras, C. 2004. Serum selenium is associated with plasma homocysteine concentrations in elderly humans. *J. Nutr.* 134: 1736–1740.

Liu, Q., Jing, T., Chen, P., Lin, Y.S. & Yun, S. 2012. Selenium deficiency and Alzheimer's disease. *Chinese Bulletin of Life Sciences* 24(8): 892–900.

Loef, M., Schrauzer, G.N. & Walach, H. 2011. Selenium and Alzheimer's disease: A systematic review. *J. Alzheimers Dis.* 25: 1–24.

Morris, M.S. 2003. Homocysteine and Alzheimer's disease. *Lancet Neurology* 2: 425–428.

Oulhaj, A., Refsum, H., Beaumont, H., Williams, J., King, E., Jacoby, R. *et al.* 2010. Homocysteine as a predictor of cognitive decline in Alzheimer's disease. *Int. J. Geriatr. Psychiatry.* 25: 82–90.

Selkoe, D.J. 2001. Alzheimer's disease: genes, proteins, and therapy. *Physiol. Rev.* 81: 741–766.

Selkoe, D.J. 1996. Amyloid beta-protein and the genetics of Alzheimer's disease. *J. Biol. Chem.* 271: 18295–18298.

Seshadri, S. 2006. Elevated plasma homocysteine levels: risk factor or risk marker for the development of dementia and Alzheimer's disease? *J. Alzheimers Dis.* 9(4): 393–398.

Uthus, E.O., Yokoi, K. & Davis, C.D. 2002. Selenium Deficiency in fisher-344 rats decreases plasma and tissue homocysteine concentrations and alters plasma homocysteine and cysteine redox status. *J. Nutr.* 132: 1122–1128.

Van Eersel J., Ke, Y.D., Liu, X., Delerue, F., Kril, J.J., Gotz, J. *et al.* 2010. Sodium selenate mitigates tau pathology, neurode generation, and functional deficits in Alzheimer's disease models. *PNAS* 107(31): 13888–13893.

Selenium in the Environment and Human Health – Bañuelos, Lin & Yin (Eds)
© 2014 Taylor & Francis Group, London, ISBN 978-1-138-00017-9

Study of selenium content in foods in a high HIV prevalence community: A case study in Pala-Bondo district, Kenya

S.B. Otieno
Department of Community Health, Kenyatta University, Kenya

F. Were
School of Medicine, University of Nairobi, Kenya

E. Kabiru
School of Public Health, Kenyatta University, Kenya

K. Waza
Kenneth Kaunda Foundation, Zambia

1 INTRODUCTION

Nutritional deficiencies have long been recognized as an important problem amongst people infected with HIV (Beach *et al.*, 1992). Micronutrient deficiencies have been associated with more rapid HIV—disease progression and higher HIV related mortality (Semba *et al.*, 1993). In some studies, micronutrient supplementation has delayed time to AIDS and improved survival, thus representing a simple and cheap strategy to slow the infection (Jiamton *et al.*, 2003). Trace element selenium (Se) has been proposed to be a key nutrient among persons living with HIV (Foster, 2003). Biochemical Se deficiency has been associated with increased mortality among those infected with HIV (Kupka *et al.*, 2004; Baum *et al.*, 1997; Constans *et al.*, 1995) and an accelerated HIV disease progression through increased viral load (Hurwitz *et al.*, 2007). Selenium as an antioxidant micronutrient is an essential element of selenoproteins, including selenoprotein P, and glutathione peroxidase. In HIV positive patients, lower serum Se has been associated with lower CD4T cells, more advanced AIDS, and higher mortality (Look *et al.*, 1997). Selenium plays an important role in the immune system, both innate and active immunity. It affects antioxidant defense systems, thyroid hormone metabolism and redox control and protein system (Gibson *et al.*, 2005). Selenium is present in soil and enters the food chain through plants (Melse-Boonstra *et al.*, 2007). Dietary intake depends on both availability of Se in the soil and the type of crops grown (Dumont *et al.*, 2006). Selenium intake depends on the Se levels in individual dietary components and their proportion of the diet, therefore analyzing Se contents in different food materials and determining the variation of diet composition among different groups of people in local community will be important.

2 MATERIALS AND METHODS

An investigation of dietary patterns and Se levels in diets of smallholder farmers was carried out in Pala Sub-location, western Kenya. In this study, 17 foods commonly eaten in Pala were sampled. The 0.50 gram sub samples of foods were wrapped into labeled polythene bags, transported to the KEBS laboratories in cool boxes at 0–4°C, and kept at the same temperature until analysis. The sample was analyzed for Se by atomic absorption spectrophotometry (PerkinElmer AA Analyst 300, Germany), as described by William *et al.* (2008) and Ohayo-Mitoko *et al.* (1979). Each food sample of 0.250 grams was mixed with H_2O_2, and perchloric acid, 0.75 ml nitric acid and 2.25 ml hydrochloric acid. The contents were thoroughly mixed and heated in aluminum block at 80°C for one hour. The sample was left to cool, 11.5 ml of distilled water were carefully added and the mixture was thoroughly stirred. A portion of the sample was measured and analyzed by AAS for Se.

3 RESULTS AND DISCUSSION

Selenium levels were determined in typical foods eaten in Pala (Table 1). Selenium concentrations (mg Se/kg) ranged as follows: Achak Achak (*Laurnea cornuta*), 148.5 mg/kg; Spider plant (*Cleome gynandra*), 121.5 mg/kg; Omena

Table 1. Concentrations of Se in typical foods in Pala-Bondo, Kenya.

Food material	Se Concentration (mg/kg)
Black nightshade (*Solanum nigrum*)	<0.02[†]
White sorghum (*Sorghum bicolor*)	<0.02
Green gram (*Vigna radiate*)	<0.02
Finger millet (*Eleucine coracona)*	<0.02
Maize (popcorn) (*Zea mays*)	<0.02
Tilapia (*Lates nilotica*)	<0.02
Nile perch (*O.nilotica*)	<0.02
Fulu (*Hapochromine spp)*	<0.02
Mito (*Crotoralia ochroleuca*)	<0.02
Brown sorghum (*S. bicolor)*	12.27
Maize (yellow) (*Z. mays var. everta*)	15.62
Red sorghum (*S. bicolor*)	19.67
Cowpea (*Vigna unguculata*)	21.97
Omena (*Astrineobola argentea*)	51
Spider plant (*Cleome gynandra*)	121.53
Achak achak (*Laurnea cornuta*)	148.5

[†]Method detection limit: 0.020 mg/kg

(*Rastrineobola argentea*), 51 mg/kg; Cowpeas (*Vignia unguculata*), 21.97 mg/kg, and Nile tilapia (*Oreochromis niloticus*), <MDL (method detection limit: 0.02 mg/kg), and Nile perch (*Lates niloticus*), <MDL. Among the cereal crops, sorghum bicolour spp (red) contained 19.97 mg/kg, while sorghum bicolour spp (white) has a Se concentration of <MDL. Most of these values suggest that traditional vegetables are accumulators of Se, as some earlier studies show that accumulators concentrate Se up to 100 mg/kg in their proteins (Melse-Boonstra *et al.*, 2007), although they are eaten by few of the respondents in this study.

Bioavailability of Se depends on the form of Se ingested (Rayman, 2000). The organic form has higher bioavailability compared to inorganic form, when GSH-px activity is considered as a measure (Clausen *et al.*, 1998). Selenium bioavailability is al so higher from vegetarian diet (with Se in the chemical form of SeMet) compared to animal protein (SeCy) diets (Alaejos *et al.*, 2000). Other factors affecting Se bioavailability include age and gender (Veronique *et al.*, 1991). Most (75.2%) respondents in this study consume fish (*Oreochromis niloticus, Lates niloticus*) and white sorghum (*Sorghum bicolour spp.*) as main foods, which have very little or no detectable Se levels (<20 μg/kg). Although *Cleome gynandra* and *Laurnea cornuta* accumulated high levels of Se, these vegetables have bitter unpleasant taste and hence not popular to local residents. They are mainly consumed by older people in the study region. Additional, these vegetables are generally boiled in water to remove the bitter taste during their cooking process, which will lead to significant Se loss. This study therefore suggests that the dietary intake of Se in this community might be below the US-FDA recommended Se intake of 100 μg/day, which might contribute directly/indirectly to the high HIV prevalence of 14.9% (KACC, 2009) observed in the community. In the study community HIV prevalence was reduced from the peak (30%) in 2001 to 14.9% in 2009 (KACC, 2009), compared with a national rate of 6.9%.

4 CONCLUSIONS

It can be concluded that Se level in the diet in Pala is inadequate and is below daily reference intake. Production and consumption of *C. gynandra* and *L. carnuta* containing relatively high concentrations of Se should be encouraged in the short term, while long term intervention could include fortifying with Se the foods commonly purchased and consumed in the community.

REFERENCES

Alaejos, S.M., Romero, D.F.J. & Romero, D.C. 2000. Selenium and cancer: some nutritional as pects. *Nutrition* 16: 76–83.

Beach, R.S., Mantero-Atienza, E., Shor-Posna, G., Javier, J.J. & Morgan, R. 1992. Specific nutrient abnormalities in asymptomatic HIV-1 infection. *AIDS* 6: 701–708.

Baum, M.K., Campa, A., Miguez-Burbano, M.J., Burbano, X. & Shor-Posner, G. 2001. Role of selenium in HIV/AIDS. In D.L. Hatfield (ed.), *Selenium: Its Molecular Biology and Role in Human Health*, 247–256. Boston: Kluer Academic Publishers.

Baum, M.K., Shor-Posner, G., Lai, S., Zhang, G., Lai. H., Fletcher, M.A. *et al.* 1997. High risk of HIV-related mortality is associated with selenium deficiency. *J. Acquir. Immune Defic. Syndr. Hum. Retrovirol.* 15(5): 370–374.

Constans, J., Pellegrin, J.L., Sergeant, C., Simonoff, M., Pellegrin, I., Fleury, H. *et al.* 1995. Serum selenium predicts outcome in HIV infection. *Journal of Acquired Immune Deficiency Syndrome Hum. Retrovirol.* 10: 392.

Clausen, J. & Nielsen, S.A. 1988. Comparison of whole blood selenium values and erythrocyte glutathione peroxidase activities of normal individuals on supplementation with selenate, selenite, L-selenomethionine, and high selenium yeast. *Biology of Trace Elements Research* 15: 125–38.

Dumont, E., Vanhaecke, E. & Cornenelis, R. 2006. Selenium speciation from food source to metabolites: Critical review. *Anal. Bioanalysis Chemistry* 358: 1304–1323.

Foster, H.D. 2003. Why HIV-1 has diffused so much more rapidly in Sub-Saharan Africa than in North America. *Med. Hypotheses* 60(4): 611–614.

Melse-Boonstra, A., Hogenkamp, P. & Lungu, O.A. 2007. *Mitigating HIV/AIDS in Sub-Sahara Africa through Selenium in Food*. 45–49. Lisaka, Zambia: Golden Valley Agricultural Research Trust.

Gibson, R.S. 2005. *Principles of Nutritional Assessment*. Oxford: Oxford University Press.

KACC (Kenya Aids Control Council). 2009. *Kenya AIDS Indicator Survey 2007 - Final Report*. Nairobi: Kenya Aids Control Council.

Kupka, R., Msamanga, G.I. & Spiegelman, D. 2004. Selenium status is associated with accelerated HIV disease progression among HIV 1 infected pregnant women in Tanzania. *Journal of Nutrition* 134: 2556–2560.

Mitoko-Ohayo, G.J.A. 1995. Concentrations of heavy metals, organochlorine pesticides, organic and microbial pollution in the Nairobi River and its Tributaries, Volume 1: The government of Netherlands, The Royal Netherlands Embassy, Nairobi, Kenya. 19–20.

Semba, R.D. & Grey, G.E. 2001. Human Immunodeficiency Virus Infection in Nutrition and Health in Developing Countries, 237–266. Totowa: Human Press.

William, R.M. & Stephen, G.C. 2008. Inductively-coupled plasma-atomic emissions spectrometric determination of elements in food using microwave assisted digestion. In *Elemental Analysis Manual*: Section 4.4A. on-line at: http://www.fda.gov/Food/FoodScienceResearch/LaboratoryMethods/ucm203652.htm.

Selenium in the Environment and Human Health – Bañuelos, Lin & Yin (Eds)
© 2014 Taylor & Francis Group, London, ISBN 978-1-138-00017-9

Anti-hepatoma effects of selenium-enriched foods on rats

Xiaoqi Lu, Jian Wang & Xuebin Yin
School of Earth and Space Sciences and Advanced Lab for Selenium and Human Health, University of Science and Technology of China, Hefei, Anhui, China

Linxi Yuan
Jiangsu Bio-Engineering Centre on Selenium, Suzhou, Jiangsu, China

1 INTRODUCTION

Liver cancer is the fourth common cancers in China and the second common cause of cancer-related deaths (Chen *et al.*, 2013). Selenium (Se) is an essential trace element for animals and humans and has been reported to improve immune function and ameliorate specific disease conditions. Selenium supplementation with 200 µg Se/d on selected human population in Southeastern United States for 10 years showed that total cancer incidence and mortality significantly deceased (Clark *et al.*, 1996). However, there were few studies on the anti-hepatoma effects of Se-enriched foods, and the differences among them. In the present study, selenite and three kinds of Se-enriched feed materials were chosen to study the intervention of hepatocarcinoma model rats induced by diethylnitrosamine (DEN).

2 MATERIALS AND METHODS

Selenium-enriched corn and broccoli were provided by SeTek Co. and *Cardamine hupingshanesis* (Se-accumulating plant) was collected from Enshi, China. 60 Wistar rats (about 180 g) were randomly divided into 6 groups. The treatments on rats are shown in Table 1. At the end of intervention, rats were terminated by exsanguination under ether anesthesia. Heparinized blood was centrifuged at 1200 rpm to separate serum and kept at −20°C until chemical analysis. Commercial assay kits for determining GSH-Px, SOD, γ-GT and GST were purchased from Jiancheng Biotechnology Research Institute (Nanjing, China). Total Se contents were determined by atomic fluorescence spectrometry (AFS-9230, Tian Co, Beijing).

3 RESULTS AND DISCUSSION

3.1 *Mortality, liver index and nodule number of rats*

The mortality, liver index and nodule numbers of rats fed with different kinds of Se-enriched foods were presented in Table 2. No rat of group 1 died during the experiment, while two died in group 2 and 1 died in each of the other groups, respectively. This observation may indicate that Se-enriched foods may reduce the rat mortality with liver cancer induced by DEN. The different death time of the rats may suggest that Se-enriched foods could slow down the dying process of rats with liver cancer induced by DEN, compared with the selenite treatment.

Table 1. Treatments on rats during intervention.

Treatment group	Treatment	Feed	Se in feed (mg/kg)
1	Pure water (control)	Basal feed	0.07
2	100 mg/L DEN	Basal feed	0.07
3	100 mg/L DEN	Selenite + basal feed	3
4	100 mg/L DEN	Se-enriched corn + basal feed	3
5	100 mg/L DEN	Se-enriched broccoli + basal feed	3
6	100 mg/L DEN	*Cardamine hupingshanesis* + basal feed	3

Table 2. Mortality, liver index, and nodule numbers of rats.

Group	Mortality	Lifetime (week)	Liver index	Nodule number
1	0	-	$2.74 \pm 0.28a^{\dagger}$	0^{a}
2	2	11, 12	10.3 ± 2.16^{b}	52.0 ± 10.7^{b}
3	1	12	10.1 ± 1.38^{b}	46.1 ± 4.96^{bc}
4	1	12	7.77 ± 1.34^{c}	40.8 ± 8.24^{c}
5	1	13	7.85 ± 1.88^{c}	37.9 ± 6.25^{c}
6	1	13	7.94 ± 1.01^{c}	43.0 ± 4.92^{bc}

†Means with different lowercase letters are significantly different ($P < 0.05$).

Liver index of rats fed with DEN in groups 2 to 6 was significantly higher than that of group 1 (the control group), while liver index of rats in groups 4, 5 and 6 were significantly lower than that in groups 2 and 3. These indicate that DEN could cause rat liver enlargement, but Se-enriched corn, Se-enriched broccoli and *Cardamine hupingshanesis* could be more effective in reducing the DEN damage on rats than selenite. Compared with selenite, the nodule number in rats fed with Se-enriched broccoli and *Cardamine hupingshanesis* was significantly decreased, compared with the other treatments.

3.2 *Specific indicators of liver cancer*

Serumγ-GT activities of rats fed with different Se-enriched foods were shown in Figure 1. Usually, serum γ-GT was low in normal rats, but it would significantly increase when the liver was damaged (Yao *et al.*, 2007). In the present study, serum γ-GT contents in groups 4, 5 and 6 were significantly lower than those of group 2, indicating that Se-enriched feed could weaken the damage induced by DEN. Particularly, serum γ-GT contents of group 5 were significantly lower than groups 3, 4 and 6, indicating Se-enriched broccoli more effective on protecting rat liver function.

Serum GST activity of rats fed with different Se-enriched food was shown in Figure 2. GST would be over expressed when liver cancer occurred in rats (Sato, 2005). In the present study, Serum GST in group 3, 4, 5 and 6 was significantly lower than that of group 2, indicating that selenite and Se-enriched food could slow down evolving on liver cancer induced by DEN.

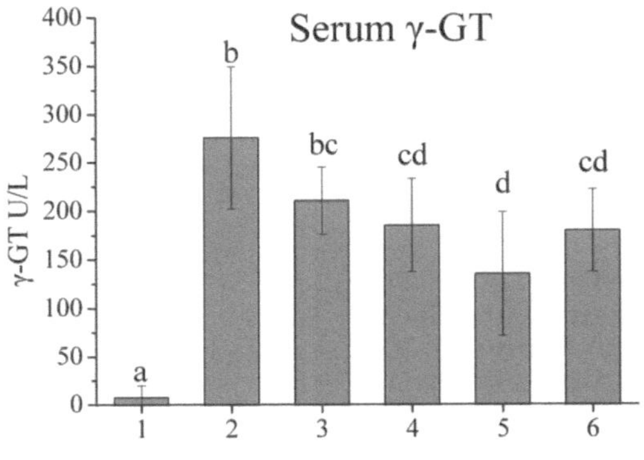

Figure 1. Serum γ-GT activity of rats fed with different Se-enriched foods. Means with different lowercase letters are significantly different ($P < 0.05$). X-axis: Treatment group number.

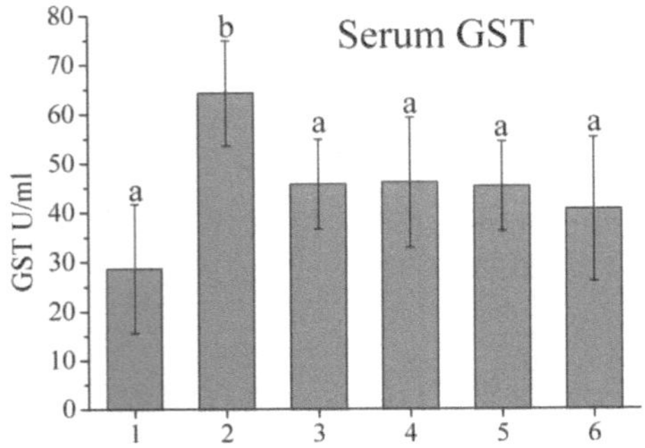

Figure 2. Serum GST activity of rats feed with different Se-enriched food. Means with different lowercase letters are significantly different ($P < 0.05$). X-axis: Treatment group number.

4 CONCLUSIONS

Selenium-enriched feed stuff could reduce the mortality, liver index and nodule number of rats with liver cancer induced by DEN and they could also reduce the serum γ-GT and GST of rats, which were considered as specific indicators of liver cancer. Se-enriched feed has a protective role on the damage of liver induced by chemical carcinogenic agent. Among the selected Se-enriched feeds, Se-enriched broccoli showed the best protective effect. Supplementation of Se-enriched feed, especially Se-enriched broccoli could protect animal and human from liver cancer induced by chemical carcinogenic agents.

ACKNOWLEDGEMENT

This work was supported by Natural Science Foundation of Jiangsu Province for Youth, China (BK 2012202, BK 2012195), Jiangsu Research-Industry Joint Innovation Program (BY 2011186).

REFERENCES

Chen, W.Q., Zhang, R.S. & Zhang, S.W. 2013. Liver cancer incidence and mortality in China, 2009. *Chinese Journal of Cancer Research* 32(4): 162–169.

Clark, L.C., Combs, G.F. Jr., Turnbull, B.W., Slate, E.H., Chalker, D.K., Chow, J. *et al.* 1996. Effects of selenium supplementation for cancer prevention in patients with carcinoma of the skin a randomized controlled trial—A randomized controlled trial. *Journal of the American Medical Association* 276(24): 1957–1963.

Sato, K. 1988. Glutathione S-transferases and hepatocarcinogenesis. *Jpn. J. Cancer Res.* 79: 556–572.

Yao D.F., Dong, Z.Z. & Yao, M. 2007. Specific molecular markers in hepatocellular carcinoma. *Hepatobiliary Pancreat Dis. Int.* 6(3): 241–247.

Selenium in the Environment and Human Health – Bañuelos, Lin & Yin (Eds)
© 2014 Taylor & Francis Group, London, ISBN 978-1-138-00017-9

Selenium accumulation and oxidation resistance in rats fed with typical selenium-enriched plant materials

Xiaoqi Lu, Jian Wang & Xuebin Yin
School of Earth and Space Sciences and Advanced Lab for Selenium and Human Health, University of Science and Technology of China, Hefei, Anhui, China

Linxi Yuan
Jiangsu Bio-Engineering Centre on Selenium, Suzhou, Jiangsu, China

1 INTRODUCTION

Selenium (Se) is an essential trace element for animals and humans and Se-enriched foods is increasingly used as Se supplement for Se-deficiency populations (Rotruck *et al.*, 1973; Thiry *et al.*, 2013). Due to the different Se speciation in different Se-enriched foods, the absorption and accumulation of Se in animal could be different. Se in Se-biofortified corn and broccoli were dominated by SeMet and SeMeCys, respectively. In contrast, SeCys2 was predominantly in *Cardamine hupingshanesis*, a kind of Se hyper-accumulating plant in Enshi, China (Yuan *et al.*, 2013). Thus, in the present study, these specific Se-enriched plant materials were chosen to study Se accumulation and oxidation resistance on rats, which would provide the scientific reference for plant Se supplementation.

2 MATERIALS AND METHODS

Selenium-enriched corn and broccoli were provided by SeTek Co. and *Cardamine hupingshanesis*, a Se-accumulating plant, was collected from Enshi, China. 60 Wistar rats (about 180 g) were randomly assigned to 6 groups. The treatments of rats are showed in Table 1. At the end of intervention, rats were terminated by exsanguination under ether anesthesia. Heparinized blood was centrifuged at 1200 g to separate serum and kept at –20°C until analysis. Commercial assay kits for determining GSH-Px, and SOD were purchased from Jiancheng Biotechnology Research Institute (Nanjing, China). Total Se contents were determined by atomic fluorescence spectrometry (AFS-9230, Tian Co, Beijing).

3 RESULTS AND DISCUSSION

3.1 *Serum Se and erythrocyte Se*

Concentrations of serum Se and erythrocyte Se in rats were shown in Figure 1. Serum Se and

Table 1. The treatments of rat feed in the intervention study.

Treatment group	Treatment	Feed	Se in feed (mg/kg)
1	Pure water (control)	Basal feed	0.07
2	100 mg/L DEN	Basal feed	0.07
3	100 mg/L DEN	Selenite + basal feed	3
4	100 mg/L DEN	Se-enriched corn + basal feed	3
5	100 mg/L DEN	Se-enriched broccoli + basal feed	3
6	100 mg/L DEN	*Cardamine hupingshanesis* + basal feed	3

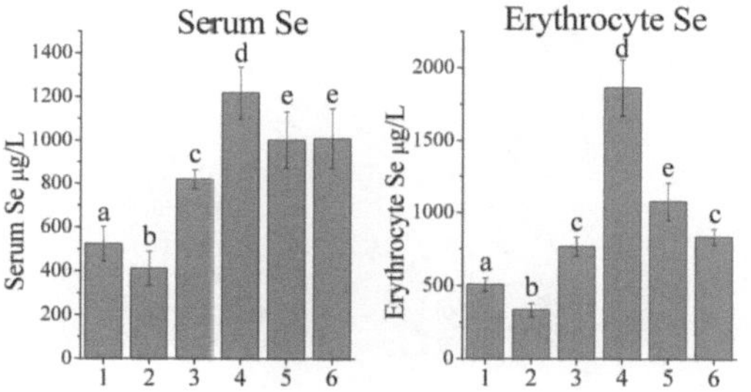

Figure 1. Serum Se and erythrocyte Se of rats fed with different Se-enriched food. Means with different lowercase letters are significantly different ($P < 0.05$). X-axis: Treatment group number.

erythrocyte Se contents of rats fed with selenite and Se-enriched plant materials were significantly higher than those in Group 2. Moreover, serum Se contents of rats fed with Se-enriched plant materials and the erythrocyte Se contents of rats fed with Se-enriched corn and broccoli were higher than those fed selenite, indicating that Se from

Se-enriched plant materials could be more effective to improve Se level in rats.

3.2 *Organ Se accumulation*

The organ Se contents of rats were shown in Figure 2. Concentrations of Se in liver, kidney and heart of rats fed with selenite and three kinds of Se-enriched plant materials were significantly higher than those fed with basal feed, especially Se-enriched corn. The high accumulation of Se in rats fed with Se-enriched corn could be attributed to SeMet as the main Se speciation in Se-enriched corn, which was considered as the main form of Se stored in animals (Thiry *et al.*, 2012). Furthermore, Se contents in liver and heart of rats fed with Se-enriched broccoli was significantly higher than those fed with selenite. However, there was no significantly difference in testis except group 4, which could be due to the function of resisting overmuch Se by blood-testis barrier of rats.

3.3 *Oxidation resistance*

The serum GSH-Px and SOD activities of rats were shown in Figure 3. Serum GSH-Px activities of rats fed with Se-enriched corn and *Cardamine hupingshanesis* were higher than those fed with basal feed and Serum SOD activity of rats fed with three kinds of Se-enriched plant materials was higher than those fed with basal feed. The increase of oxidation resistance after a supplement of Se has been reported elsewhere (Tapiero *et al.*, 2003). Moreover, Serum SOD activity of rats fed with Se-enriched broccoli and *Cardamine hupingshanesis* was significantly higher than those fed selenite,

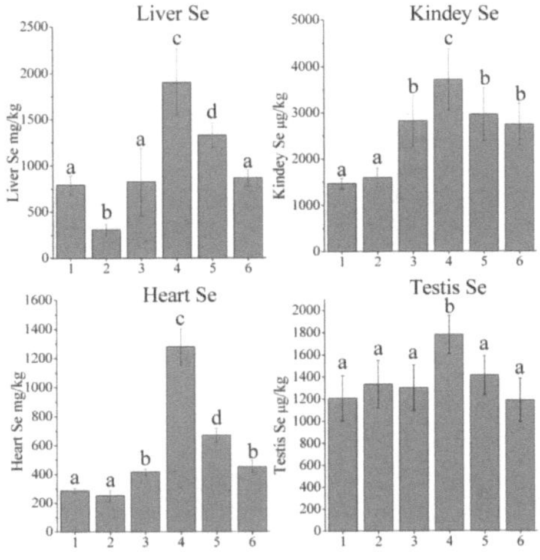

Figure 2. Organ Se of rats fed with different Se-enriched feed. Means with different lowercase letters are significantly different ($P < 0.05$). X-axis: Treatment group number.

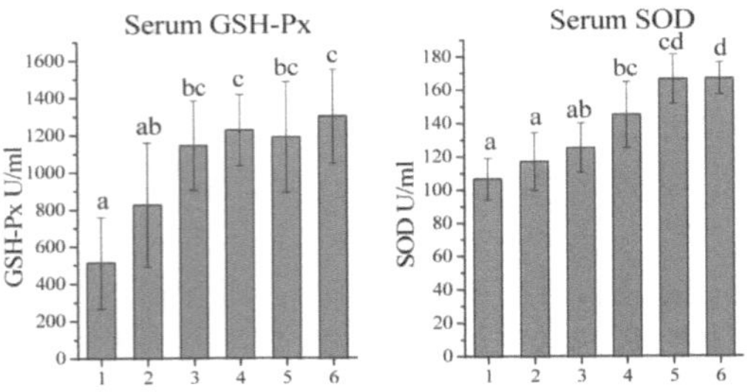

Figure 3. Serum GSH-Px and SOD activity of rats fed different Se-enriched feed. Means with different lowercase letters are significantly different ($P < 0.05$). X-axis: Treatment group number.

which indicated a better oxidation resistance with these two Se-enriched materials than selenite.

4 CONCLUSIONS

Se-enriched plant materials could significantly increase serum Se, erythrocyte Se and organ Se contents of rats, particularly in Se-enriched corn. Se-enriched plant materials could improve the oxidation resistance of rats, and Se-enriched broccoli and *Cardamine hupingshanesis* were better than selenite. In conclusion, Se-enriched plant materials could be a better choice as Se supplementation for Se-deficiency people.

ACKNOWLEDGEMENTS

This work was supported by Natural Science Youth Foundation of Jiangsu Province for Youth, China (BK 2012202, BK 2012195), Jiangsu Research-Industry Joint Innovation Project (BY 2011186).

REFERENCES

Milde, D., Novak, O., Stu, ka V., Vyslou, il K. & Macha, ek J. 2001. Serum levels of selenium, manganese, copper, and iron in colorectal cancer patients. *Biological Trace Element Research* 79(2): 107–114.

Rotruck, J.T. & Ganther, H.E. 1973. Selenium: biochemical role as a component of glutathione peroxidase. *Science* 179: 588–590.

Tapiero, H., Townsend D.M. & Tew, K.D. 2003. The antioxidant role of selenium and seleno-compounds. *Biomedicine & Pharmacotherapy* 57(3–4): 134–144.

Thiry, C., Ruttern, A., De Temmerman, L., Schneider, Y.J. & Pussemier, L. 2012. Current knowledge in species-related bioavailability of selenium in food. *Food Chemistry* 130(4): 767–784.

Thiry, C., Schneider Y.J., Pussemier, L., De Temmerman, L. & Ruttern, A. 2013. Selenium bioaccessibility and bioavailability in Se-enriched food supplements. *Biological Trace Element Research* 152(1): 152–160.

3 *Biological uptake and accumulation of selenium*

Selenium in the Environment and Human Health – Bañuelos, Lin & Yin (Eds)
© 2014 Taylor & Francis Group, London, ISBN 978-1-138-00017-9

Characteristics of selenium in soil and plants of Ping'an county, Qinghai province of China

D.S. Yu, D.L. Liang & J.J. Liu
Northwest A&F University, Yangling, Shaanxi, China

R. Zhang & X.F. Sun
Soil and Fertilizer Institute, Qinghai Academy of Agricultural and Forestry Sciences, Xining, Qinghai, China

1 INSTRUCTION

Soil is the ultimate source of selenium (Se) in the food chain, and plants have a pivotal role in transferring Se from soil into the food chain (Combs, 2001; Hawkesford & Zhao, 2007). The Se content in different kinds of foods is mainly determined by the total Se and available Se in soil (Fordyce, 2005). In China, the range of Se concentration in soil varies between 0.095 μg/kg in low-Se regions, *i.e.*, Bin County in Shaanxi (Wang *et al.*, 2000), and 33.1 mg/kg in high-selenium regions, *i.e.*, Ziyang county in Shaanxi (Wang *et al.*, 2012). The Office of Land and Resources of China found a large area of Se-rich soil in the east of Qinghai province (Xing & Kang, 2010). Ma *et al.* (2012) also found a high selenium area (most of the Se content in soil was greater than 0.44 mg/kg) in Xiaoxia and Hongsuiquan regions in Ping'an County, Qinghai province. However, no research has been conducted for evaluating the geographical distribution of Se in soil and plants, as well as evaluating the Se availability in this high Se area.

In this study, soil and plant samples were collected in Ping'an County in Qinghai province to: (1) determine the concentration of total Se in soil and plant and clarify the selenium status in the area; and (2) better understanding the geographical distribution of selenium both in soil and plants.

2 MATERIALS AND METHODS

Ping'an County was located in the northeast of Tibet plateau, Qinghai province of China. The climate in this region is typically arid; with an annual precipitation of 300–600 mm. Ping'an has a varied topography, and is mostly mountainous with elevation ranging from 2066 to 2300 m.

A total of 573 soil samples were collected from cultivated land in the Xiaoxia, Hongshuiquan, Sanhe and Shagou regions. Meanwhile, the correspond plants grown on these soils were also collected, including cereals, oil crops, vegetables, pasture and fruits (total 43 types). Soil samples were digested with HNO_3-$HClO_4$ (3:2, v/v) and plant samples were digested with HNO_3-$HClO_4$ (4:1, v/v). In addition, the plant available Se in soil was extracted with 1M K_2HPO_4-KH_2PO_4 (pH = 7). All samples were determined by the method using AFS-930. Quality control of Se analysis was assessed using duplicates, method blanks, matrix spikes and certified reference materials. The Kriging method with ArcGIS 10.0 was used to create the geographical distribution map of total Se in soil and plants.

3 RESULTS AND DISCUSSION

3.1 *Selenium in soil*

The concentration of total Se in soil ranged from 0.076 to 1.88 mg/kg, with an average of 0.369 ± 0.236 mg/kg. 86.8% of the soil samples contained over 0.45 mg/kg of Se, which meets the Se-adequate standard (Tan, 1989). In addition, Se content in soil in each village or town is shown in Table 1.

There is a relatively high Se-rich soil distributed in Huang river basin in the cultivated land of Xiaoxia and Hongshuiquan regions, according to the geographical distribution map of total Se in soil (Fig. 1). Moreover, the Se concentration was the highest in irrigated land in river basin, but lowest in high mountain land. Compared with other regions in China, this study region is obviously Se-rich.

Table 1. Soil Se concentrations in different villages and towns in Ping'an County.

Region (n)	Max (μg/kg)	Min	Mean ± SD
Xiaoxia (65)	969.33	146.98	588.61 ± 208.32
Sanhe (103)	794.15	91.11	380.58 ± 203.58
Hongshuiquan (363)	1882.34	75.72	335.68 ± 239.01
Shagou (42)	478.90	169.06	283.54 ± 93.23

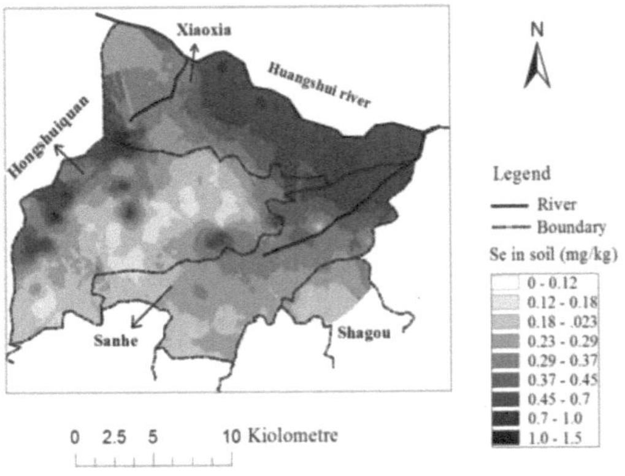

Figure 1. The geographical distribution map of total selenium in soil in Ping'an selenium-rich region.

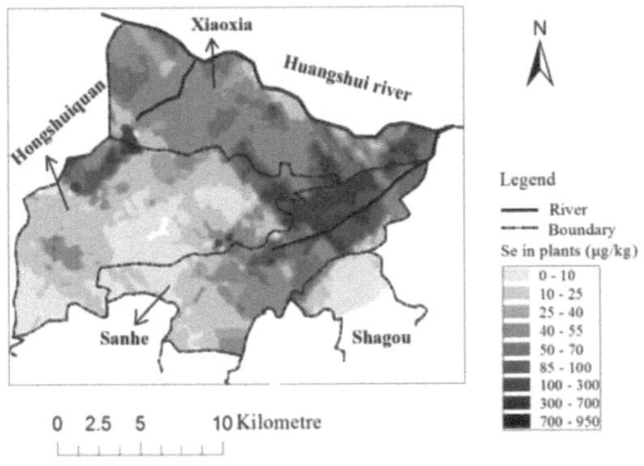

Figure 2. The geographical distribution map of total selenium in plants in Ping'an selenium-rich region.

3.2 *Selenium in plants*

The concentration of total Se in plants ranged from <MDL (0.001 mg/kg) to 903.6 μg/kg with an average of 48.7 ± 67.6 μg/kg. The Se uptake ability among different kinds of plants was significantly different. Se accumulation in plants ranged as follows: pastures > cereal > vegetables > fruits, while Se concentrations in main plants were as follows: 83.82 ± 80.62 μg/kg in wheat, 61.02 ± 53.33 μg/kg in rape, 44.28 ± 35.38 μg/kg in pea, 8.40 ± 4.92 μg/kg in potato, and 183.45 ± 220.78 μg/kg in wheatgrass. Moreover, the geographical distribution of Se in plants was similar with the Se contents in soil (Fig. 2).

Selenium in plants is relatively low compared with other regions of China. Selenium is high in soil but low in plants in this Se-rich region, indicating that total Se content in soil might not reflect the plant available Se level.

3.3 *Available selenium in soil*

The available Se extracted by K_2HPO_4-KH_2PO_4 ranged from <MDL (0.001 mg/kg) to 0.029 mg/kg with an average of 0.011 ± 0.006 mg/kg. This value had significant correlation both with Se content in soil ($r = 0.677, n = 215$) and crops ($r = 0.329, n = 215$). Therefore, the available Se extracted by K_2HPO_4-KH_2PO_4 can be used to evaluate the plant available Se in soil. In addition, available Se in soil accounted for only 2.97% of total soil Se, which is much lower than other regions in China. The Se accumulated by plants mainly depends on the available Se in soil (Wen *et al.*, 2007). Therefore, the low content of available Se in soil may be the main reason of the low Se content in plants in this Se-rich area.

4 CONCLUSIONS

The Se in study area is rich in soil but relatively low in plants, compared with many other regions in China. Moreover, the plant available Se in soil is deficiency, which is the main reason of the low Se content in plants in studied area. Further study needed to be done on increasing the bioavailability of Se in soil in Ping'an region, Qinghai province.

REFERENCES

Combs, G.F. 2001. Selenium in global food systems. *British Journal of Nutrition* 85(5): 517–547.

Fordyce, F.M. 2005. Selenium deficiency and toxicity in the environment. In: O. Selinus (ed.), *Essentials of Medical Geology*, 374–415. London: Academic Press.

Hawkesford, M.J. & Zhao, F.J. 2007. Strategies for increasing the selenium content of wheat. *Journal of Cereal Science* 46(3): 282–292.

Ma, Q., Ji, B.Y., Zhang, Y.F., Reng, W.K., Dai, L. & Huang, B.X. 2012. Geochemical characteristies of selenium in soils and organisms in the east of Qinghai Province. *Advances in Earth Science* 27(10): 1148–1152.

Tan, J.A. 1989. *The Atlas of Endemic Diseases and Their Environments in the People's Republic of China.* Beijing: Science Press.

Wang, J.D., Yu, J.B. & Zhang, X.L. 2000. Geochemical features of elements of selenium etc. in soil of Loess Plateau. *Scientia Geographica Sinica* 20(5): 469–473.

Wang, S.S., Liang, D.L., Wang, D., Wei, W., Fu, D.D. & Lin, Z.Q. 2012. Selenium fractionation and speciation in agriculture soils and accumulation in corn (*Zea mays* L.) under field conditions in Shaanxi Province, China. *Science of the Total Environment* 427–428:159–164.

Wen, T.F. & Li H.P. 2007. Nutrient situation of zinc and selenium in Codetta in northeast farming grazing inter locking distric. *Journal of Jilin Agricultural University* 29(1): 86–90.

Xing, Z. & Kang, W.H. 2010. Selenium-rich soil on 840 km^2 land found in eastern Qinghai. *Qinghai Daily* February 26, 2010.

Selenium in the Environment and Human Health – Bañuelos, Lin & Yin (Eds)
© 2014 Taylor & Francis Group, London, ISBN 978-1-138-00017-9

Selenium accumulation in roots and shoots of different Malaysian rice varieties

N. Alifar, Z.A. Rahman, C.F. Ishak & G. Sharifianpour
Land Management Department, Faculty of Agriculture, University Putra, Malaysia

Y. Awang
Crop Science Department, Faculty of Agriculture, University Putra, Malaysia

1 INTRODUCTION

Selenium (Se) was first identified as an essential nutrient for animals in the late 1950s (Schwarz & Foltz, 1957), although Se had been known as a toxin for more than a century (Whanger, 2002). Low levels of Se can cause an increased risk of cardio-vascular death and cancer in humans (Combs, 2001) and health disorders, such as muscular dystrophy and certain cardiovascular, bone or immune system diseases (Whanger, 2002). A positive correlation has also been reported between Se deficiency and endemic diseases, including liver cancer, breast cancer, fluorosis, goiter and mental retardation (Wang & Gao, 2004). For humans and animals their main dietary source of Se is from the soil through plants, although plants do not require Se to complete their life cycle. Plant foods are the major provider of Se in most countries throughout the world (Longnecker *et al.*, 1991). Rice is the staple food and the main source of nutrient in many Asian countries but it contains low level (1–3 mg/kg) of Se in the polished form (white rice) (Abilgos-Ramos *et al.*, 2007). In order to maintain a healthy quality in life, Se intake should be increased in the daily diet. The purpose of this study was to investigate Se accumulation in different Malaysian rice varieties and identify those which may contribute more Se for human intake.

2 MATERIALS AND METHODS

This study was conducted at faculty of Agriculture, University Putra Malaysia from June to July 2012. The experiment was laid out in a three factorial using Randomized Complete Block Design (RCBD) with three replications. Six rice varieties (MR211, 219, MR220, MR232, MR253, MR263) used in this experiment were germinated for 4–5 days and after seed germination, they were grown in Yoshida nutrient solution with different rates of Se as Na_2SeO_3 (0, 20, and 200 µg Se/L). The nutrient solution was changed every four days and pH was adjusted to 5.5 with NaOH and HCl every day. Seedlings were harvested weekly up to 4 weeks. The dry-matter weights of roots and shoots were determined separately. Samples were wet-ashed, stored, and analyzed later for Se by inductively-coupled plasma spectroscopy. The data obtained were subjected to ANOVA (in RCBD) using the SAS 9.2 version. Pairs of treatment means were compared by applying Duncan's new multiple range test at the $P < 0.05$ level.

3 RESULT AND DISCUSSION

3.1 *Roots and shoots selenium content*

Selenium uptake in both shoot and roots of rice seedling was improved significantly each week at all Se levels. As shown in Tables 1 & 2, there was no significant difference among different rice varieties in the control during four weeks. There were significant differences among rice varieties when 20 and 200 µg Se/L were supplied for 7, 14, 21, and 28 days respectively ($P < 0.05$). By prolonging the treatments, the total Se in shoots and roots of the different varieties, increased (Tables 1 & 2). When 20 and 200 µg Se/L as were supplied, the total Se of roots treated for 28 days was 1.17 to 7 times higher than that of roots treated for 7 days. Hydroponic results reported by Zhang *et al.* (2006) on different rice genotypes showed that total Se values for the high Se genotypes were markedly higher than the low Se varieties.

Among different varieties tested, MR232 showed the strongest capacity to accumulate Se in shoots and roots and MR253 was the weakest. When 20 µg Se/L as Na_2SeO_3 was supplied for 7 or 28 d the total Se content in MR232 was 1.54–3.52 times higher than that of MR253.The total Se of MR232 in root was 3.64–2.09 times higher than that of MR253.

Table 1. Selenium accumulation (μg/plant) in shoots of different rice varieties.

	Control			
Variety	7 day	14 day	21 day	28 day
MR232	0.013a[†]	0.01a	0.016a	0.02a
MR219	0.01a	0.01	0.015a	0.02a
MR211	0.006a	0.013a	0.013a	0.01a
MR263	0.006a	0.01a	0.013a	0.01a
MR220	0.01a	0.01a	0.014a	0.01a
MR253	0.01a	0.013a	0.015a	0.02a
	20 μg/L			
MR232	7.51a	8.82a	12.43b	16.04a
MR219	7.26a	7.43a	17.77a	15.36a
MR211	6.4a	7.81a	11.29b	14.88a
MR263	1.08b	2.42b	3.28c	3.89b
MR220	1.04b	2.2b	2.93c	6.63b
MR253	1.19b	2.6b	3.03c	4.4b
	200 μg/L			
MR232	12.32ab	22.28a	24.42a	29.86a
MR219	13.78a	19.98ab	21.63a	27.69b
MR211	12.53ab	16.27cb	19.56ab	25.96b
MR263	9.47cb	13.7c	15.86b	17.71c
MR220	8.26c	12.57c	15b	16.62c
MR253	9.48cb	12.43c	14.22b	14.25c

[†] Values with the same letter are not significantly different at the $P < 0.05$ level.

Table 2. Selenium accumulation (μg/plant) in roots of different rice varieties.[†]

	Control			
Variety	7 day	14 day	21 day	28 day
MR232	0.013a[†]	0.01a	0.016a	0.02a
MR219	0.01a	0.01	0.015a	0.02a
MR211	0.006a	0.013a	0.013a	0.01a
MR263	0.006a	0.01a	0.013a	0.01a
MR220	0.01a	0.01a	0.014a	0.01a
MR253	0.01a	0.013a	0.015a	0.02a
	20 μg/L			
MR232	7.51a	8.82a	12.43b	16.04a
MR219	7.26a	7.43a	17.77a	15.36a
MR211	6.4a	7.81a	11.29b	14.88a
MR263	1.08b	2.42b	3.28c	3.89b
MR220	1.04b	2.2b	2.93c	6.63b
MR253	1.19b	2.6b	3.03c	4.4b
	200 μg/L			
MR232	12.32ab	22.28a	24.42a	29.86a
MR219	13.78a	19.98ab	21.63a	27.69b
MR211	12.53ab	16.27cb	19.56ab	25.96b
MR263	9.47cb	13.7c	15.86b	17.71c
MR220	8.26c	12.57c	15b	16.62c
MR253	9.48cb	12.43c	14.22b	14.25c

[†] Values with the same letter are not significantly different at the $P < 0.05$ level.

4 CONCLUSIONS

Selenium accumulation in rice is greatly affected by total Se in root and shoot. To produce Se-enriched rice we will need to consider those rice varieties that accumulate higher levels of Se in the seedlings.

REFERENCES

Abilgos-Ramos, R., Murchie, E., Anukul, N., Manaois, R., Julaton, M., Stangoulis, J. *et al.* 2007. Rice biofortification. *Comparative Biochemistry and Physiology A* 146: s247.

Arvy, M.P. 1993. Selenate and selenite uptake and translocation in bean plants (*Phaseolus vulgaris*). *Journal of Experiment Botany* 44: 1083–1087.

Combs, G.F. Jr. 2001. Impact of selenium and cancer prevention findings on the nutrition-health paradigm. *Nutrition and Cancer* 40(1): 6–11.

Cao, Z.H., Wang, X.C., Yao, D.H., Zhang, X.L. & Wong, M.H. 2001. Selenium geochemistry of paddy soils in Yangtze River Delta. *Environment International* 26: 335–339.

De Souza, M.P., Pilon-Smits, E.A.H., Lytle, M.C. Hwang, S., Tai, J., Honma, T.S.U. *et al.* 1998. Rate-limiting steps in Se assimilation and volatilization by Indian mustard. *Plant Physiology* 117: 1487–1494.

Gissel-Nielsen, G., Gupta, U.C., Lamand, M. & Westermarck, T. 1984. Selenium in soils and plants and its importance in livestock and human nutrition. *Advances in Agronomy* 37: 397.

Hu, Q.H., Chen, L.C., Xu, J., Zhang, Y. & Pan. G.X. 2002. Determination of selenium concentration in rice and the effect of foliar application of Se-enriched fertilizer or sodium selenite on the selenium content of rice. *Journal of Science of Food and Agriculture* 82: 869–872.

Longnecker, M.P., Taylor, P.R., Levander, O.A. Howe, M., Veillon, C., McAdam, P.A. *et al.* 1991. Selenium in diet, blood and toenails in relation to human health in a seleniferous area. *The American Journal of Clinical Nutrition* 53(5): 1288–1294.

Schwarz, K. & Foltz, C.M., 1957. Selenium as an integral part of factor-3 against dietary necrotic liver degeneration. *Journal of the American Chemical Society* 79: 3292–3293.

Whanger, P.C. 2002. Selenocompounds in plants and animals and their biological significance. *Journal of the American College of Nutrition* 21: 223–232.

Wang, Z.J. & Gao, Y.X. 2004. Biogeochemical cycling of Se in Chinese environments. *Applied Geochem.* 16: 1345–1351.

Zhang, L., Shi, W., Wang, X. & Zhou, X. 2006. Genotypic difference in selenium accumulation in rice seedlings at early growth stage and analysis of dominant factors influencing selenium content in rice seeds. *Journal of Plant Nutrition* 29: 1601–1618.

Selenium in the Environment and Human Health – Bañuelos, Lin & Yin (Eds)
 ISBN 978-1-138-00017-9

Speciation-dependent uptake of selenium by rice (*Oryza sativa*) from varying substrates

A.K. Nothstein, E. Eiche, M. Stelling, G. Konrad, M. von Brasch,
R. Kosmala & T. Neumann
Institute of Mineralogy and Geochemistry, Karlsruhe Institute of Technology, Karlsruhe, Germany

M. Riemann, R. Brendel & P. Nick
Molecular Cell Biology, Botanical Institute, and Center for Functional Nanostructures, Karlsruhe Institute of Technology, Karlsruhe, Germany

1 INTRODUCTION

The trace element selenium (Se) is known for both its toxicity and essentiality to human and animal nutrition (Navarro-Alarcon & Cabrera-Vique, 2008). Selenium uptake into plants as the first step of its availability in the food chain is governed by many factors, including adsorption/desorption processes at mineral-water interfaces, precipitation, speciation and mobility (Fernández-Martínez & Charlet, 2009). The two main oxidized, inorganic Se species in soil solutions of aerated soils are selenate (SeO_4^{2-}) and selenite (SeO_3^{2-}) (Lemly, 1993).

Though Se has yet to be proven to be essential for higher plants (Läuchli, 1993), uptake into plants is dependent mainly on the concentration of bioavailable Se species (Fernández-Martínez & Charlet, 2009). However, plant physiology also plays a fundamental role in Se assimilation. Uptake of Se is elevated in the form of selenate when plants are sulfur-deficient and in the form of selenite when they are phosphorous-deficient, which gives reason to believe that Se uptake—though species-dependent—is also correlated to regulatory mechanisms of plant specific sulfate or phosphate transport (Li *et al.*, 2010).

Bioavailability of Se is also dependent on its geo-chemical mobility. Though there is substantial amount of knowledge about sorption/desorption-processes onto and from iron oxy-hydroxides (Balistrieri & Chao, 1987) and clay minerals (BarYosef & Meek, 1987) as well as Se uptake by crop plants from natural soils, more needs to be known about the interactions within the Se-soil-water-plant-system to correctly asses and act upon Se deficiency or toxicity in organisms (Schrauzer, 2000).

The overall aim of our research is therefore to combine sorption/desorption findings with uptake of Se into plants in lab scale experiments with defined soil and plant parameters to enable a better understanding of the biogeochemical interactions.

In this part of our research, we focused on the up-take of Se by *Oryza sativa* when Se is added to hydroponic cultures, agar-grown cultures and cultures grown on an artificial soil substrate. Due to the nature of experimental cultures grown in closed plant box systems, these plants are all harvested at seedling growth stages. The objective of these experimental series is to determine differences in Se uptake due to substrate/solution composition and Se speciation in both root and leaf of *Oryza sativa*.

2 MATERIAL AND METHODS

Cultures of rice (*Oryza sativa* L. *ssp. Japonica* cv. *Nihonmasari*) were prepared under clean bench conditions in closed box systems adding Se as $Na_2SeO_4 \times 10H_2O$ or Na_2SeO_3 in concentrations of 0, 5, 10, 25, 50, 100, 250, 500, 1000 and 2500 µg Se/L to either phytoagar or nutrient solution. Plant growth was monitored for a period of 7–14 days after roots were observed to be in contact with the Se-enriched substrate.

In addition to that, an open 3-month culture of the same cultivar was grown on a quartz sand substrate with dosages of 27–55 mL of a customized nutrient solution added every other day. Two weeks before harvesting, the two-daily nutrient solution was supplemented with 2.5 or 5 mg/L Se as Na_2SeO_4.

After a 24-h-freeze-drying procedure, bulk samples of leaves and roots were digested with nitric acid and hydrogen peroxide in closed microwave vessels to avoid volatilization of Se. Digestion quality was monitored with the certified Wheat Flour SRM 1567a standard.

Selenium concentrations in plant digestion samples and agar samples were measured with HG-FIAS (PerkinElmer AAS 400), anions of the nutrient solution were analyzed with IC (ICS 1000, Dionex) and other nutrient elements were measured with ICP-MS (X-Series 2, Thermo Fisher Scientific). The quality of the respective measurement methods was verified by the multi-element standard PromoChem Trace Metals (QCP 050–1 and QCP 050–2).

3 RESULTS AND DISCUSSION

Selenium uptake into leaves for the open 3-month rice plant cultures on a substrate of quartz sand and supplied with nutrient solution was found to be correlated more closely with plant vitality determined through leaf matter mass and leaf color rather than the applied Se concentration. Root uptake was 1.5–2 times higher for 2.5 mg/L of applied Se than for 5 mg/L, suggesting that higher Se concentrations do not automatically mean higher Se uptake into the root.

The plant box study in which half the plants were harvested after one and the other half after two weeks of Se-exposure showed that Se-uptake is not limited to later stages of plant growth, although rice seedlings draw their nutrients from reserves within the seed during the first days of growth. Though uptake during the first week was significantly lower than in the second week, Se—though not enriched—was incorporated into the plant in concentrations equivalent to the Se solution concentration.

Plant box experiments with phytoagar compared to experiments with nutrient solution yielded smaller plants for the same duration of Se uptake. This may be due to the plants being directly exposed to Se from the first day. Furthermore, the reduced nutrient supply from the agar may affect Se uptake as well. Plant roots in the hydroponic cultures with nutrient solution reached the Se containing liquid after 5 days. Though this led to larger plants, these showed chlorosis at elevated Se concentrations whereas plants grown directly in Se containing phytoagar substrate were smaller with stunted root growth but exhibited considerably less to no chlorosis at comparable Se concentrations.

Overall, the plant box cultures displayed a positive correlation between Se uptake and Se accumulation in the plant up to concentrations of about 250 µg/L of Se applied as Na_2SeO_4. Any increase of Se in the agar or nutrient solution beyond that was not mirrored within the Se concentration per plant mass. Selenium concentration in the plant either remained constant or decreased. This trend applied both to concentrations measured in roots and leaves and was valid for both the seedling experiments in phytoagar and nutrient solution, although concentration values for both series were not identical.

Selenium applied as Na_2SeO_3 showed a similar trend, however, the maximum of Se accumulation in the plant for selenite was found at Se concentrations about twice as high as for selenate, though the absolute uptake was only a third of that of selenate. Again, higher Se concentrations in the nutrient solution or phytoagar led to no greater accumulation in the plant, but remained at a constant level or showed a slight decrease.

4 CONCLUSION

When rice plants are treated with Se after a period of Se-free growth, uptake of the element is strongly dependent on the plant's vitality, which can be deduced from plant matter mass, leaf color and plant height. In all experiments we observed that although higher concentrations of Se in the substrate led to greater Se accumulation in the plant, this was only the case up to a certain threshold value of Se supply above which Se uptake remained stable or even decreased. This was observed more prominently in the leaves than in the roots, and is also significantly higher for selenate due to higher bioavailability and also toxicity compared to that of selenite. Plants exhibited stunted root development and reduced leaf length when in contact with Se from the first day of growth. However, chlorosis is more like-ly to occur when plants grow without Se interfer-ence in their early growth stage and are then treated with excessive concentrations of the element.

REFERENCES

Balistrieri, L.S. & Chao, T.T. 1987. Selenium adsorption by goethite. *Soil Science Society of America Journal* 51: 1145–1151.

BarYosef, B. & Meek, D. 1987. Selenium sorption by kaolinite and montmorillonite. *Soil Science* 144: 11–19.

Fernádez-Martínez, A. & Charlet, L. 2009. Selenium environmental cycling and bioavailability: a structural chemist point of view. *Reviews in Environmental Science and Biotechnology* 8: 81–110.

Läuchli, A. 1993. Selenium in plants—uptake, functions, and environmental toxicity. *BotanicaActa* 106: 455–468.

Lemly, A.D. 1997. A teratogenic deformity index for evaluating impacts of Selenium on fish populations. *Ecotoxicological and Environmental Safety* 37: 259–266.

Li, H.F., Lombi, E., Stroud, J.L., McGrath, S.P. & Zhao, F.J. 2010. Selenium speciation in soil and rice: Influence of water management and Se fertilization. *Journal of Agricultural and Food Chemistry* 58: 11837–11843.

Navarro-Alarcon, M. & Cabrera-Vique, C. 2008. Selenium in food and human body: a review. *Science of the Total Environment* 400: 115–141.

Schrauzer, G.N. 2000. Seleniomethionine: a review of its natural significance, metabolism and toxicity. *Journal of Nutrition* 130: 1653–1656.

Selenium in the Environment and Human Health – Bañuelos, Lin & Yin (Eds)
© 2014 Taylor & Francis Group, London, ISBN 978-1-138-00017-9

Distribution of selenium in orange in the citrus orchards under different soil and water conservation conditions

B.J. Zhang
College of Public Health, Nanchang University, Nanchang, Jiangxi, China
Jiangxi Institute of Soil and Water Conservation, Nanchang, Jiangxi, China

1 INTRODUCTION

Selenium (Se), as one of the essential microelements, plays an important physiological role in biological organisms. It is well-known that environmental Se is deficient in a belt stretching from northeast to southwest in China and the levels of Seon the both sides of the Se-deficiency belt are relatively high. Usually, selenium in the human body is derived mainly from the food chain and soil Se is the basic source for food. The Jiangxi Province, located in the southeast of China, belongs to the moderate level Se and its ecological landscapes and red soil contain more Se when compared with the other soil types. Leaching or redistribution of soil Se often occurs in the hilly and rainy area in the process of vigorous development activities. Most fruit trees are planted in disturbed lands and the loss of soil and water is more serious in this province. Korenovska (2003) has determined the Se content in fruits but citrus was not included in his study. Therefore, we selected a representative slope orchard (including eight experimental plots) to determine the Se characteristics in oranges, which would provide a scientific basis for the utilization and protection of soil Se in the hilly red soil.

2 MATERIALS AND METHODS

A total of 16 orange samples were collected from the experimental plots in Jiangxi Province. We collected samples in the uphill and downhill, respectively, from each plot. The field experiment was carried out in October, 2012. After the oranges were taken from the trees, they were placed into a clean plastic bag, labeled, packed, and transported to the laboratory. About 5 g of fresh orange flesh (*i.e.*, excluding orange skin) were weighed and placed into a 50 ml beaker, and we added 5 ml nitric acid and 1 ml perchloric acid into the beaker. After cold digestion for 0.5 h, the beaker was placed on the electric heating plate and we gradually increased the temperature but not exceeding 150°C. More concentrated nitric acid was added if necessary during the course of acid digestion. The digestion of orange samples was stopped when the remaining solution in the beaker was approximately 1 ml, and no residues could be seen clearly. An equal volume of concentrated hydrochloric acid was immediately added to the container in order to induce Se^{6+} to Se^{4+}. Finally, the total Se contents were determined using the method of Hydride Generation Atomic Fluorescence Spectrometry (HG-AFS).

3 RESULTS AND DISCUSSION

Concentrations of Se in the orange samples (wet weight) ranged from 16.62 to 171.03 μg/kg with a mean of 90.06 ± 40.65 μg/kg (n = 16). There were differences of orange Se between uphill and downhill in the experimental plots. Contents of orange Se in the uphill and downhill were 82.47 ± 41.03 μg/kg and 97.66 ± 41.55 μg/kg, respectively, which might be associated with the soil and water loss over a period greater than 10 years and the soil nutrients consequently leaching out. Compared with those experimental plots covered with grasses (dominated or ribbon) in the citrus orchard, interplanting orange trees with soybean or radish could maintain and improve the Se levels in the soil and the oranges. In addition, there were no consistent Se concentration changes among different slope types (including transverse slope and longitudinal slope) in the experimental plots of citrus orchard.

Terracing is one of the most important agricultural measures in the hilly areas. In each experimental plot, a three-level terrace was set and we sampled in the uphill and downhill. Hence, three kinds of treatments in this section included; terrace wall bare, terrace wall grassed and terrace wall grassed plus a ridge in the front and a concavity in the back. Figure 1 shows the concentration of orange Se in the experimental plot greater in the terrace wall grassed 1.74 and 1.53 times greater than in the experimental plot bare in the terrace wall. In theory, the Se levels in plant or edible

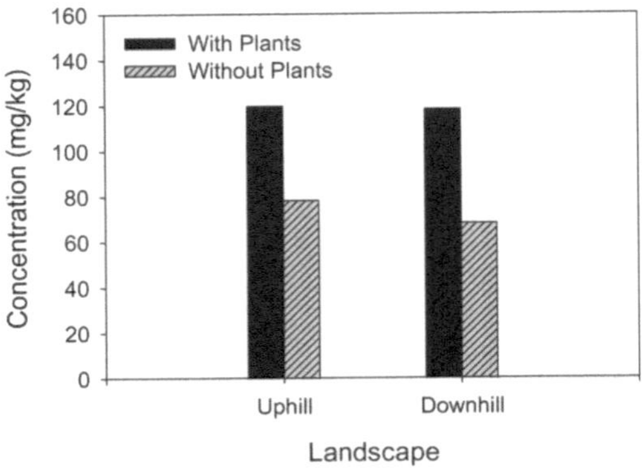

Figure 1. Concentrations of Se in orange under different conservation schemes, comparing effects of terrace wall vegetation with grass on selenium accumulation under different landscape conditions (*i.e.*, uphill versus downhill).

parts would be higher if some soil and water conservation measures has been constructed in the disturbed land which can help improve fruit weight and number (Liu *et al.*, 2012). The observation was that Se content in orange was the lowest from the plot with the terrace wall grassed plus a ridge in the front and concave in the back.

4 CONCLUSIONS

Concentrations of orange Se varied among different soil and water conservation treatments under field conditions. Due to the long-term soil erosion (mainly caused by rain or irrigation surface runoff), Se losses in red soil on an un-vegetated slope can be increasingly significant. As a result, Se accumulation in orange can be reduced over time. This study suggests that increasing ground grass coverage (among orange trees) or inter-cropping could reduce soil Se losses by surface runoff and results in an increase of Se content in fruits. The construction of terraces or especially terrace wall with vegetation, would significantly enhance the orange Se content in the disturbed lands, but the impact of slope types on the orange Se accumulation needs to be examined in future studies.

REFERENCES

Korenovska, M. 2003. Optimization of selenium determination in vegetables, fruits, and dairy products by flow injection hydride generation atomic absorption spectrometry. *Chemical Papers* 57(3): 155–157.

Liu, Qunlong, Ning, Chanjuan, Wang Duo, Wu, Wenjiang & Wu, Guoliang. 2012. Effect of spraying selenium on the pear fruit quality. *Acta Hort.* (ISHS) 940: 545–549.

Selenium in the Environment and Human Health – Bañuelos, Lin & Yin (Eds)
© 2014 Taylor & Francis Group, London, ISBN 978-1-138-00017-9

Different water-and-soil conservation measures in citrus orchard on selenium distribution in red soil

Tang Yin
College of Public Health, Nanchang University, Nanchang, China

1 INTRODUCTION

Selenium (Se) is one of the essential microelements for animals, humans, and agricultural ecosystems. Selenium's biological activities include; anti-oxidation, delaying aging, preventing human from cancer and cardiovascular disease, and enhancing immune function. According to the results from Jiangxi Provincial Institute of Geological Survey, Poyang Lake Basin was one acre of "sufficient Se" and "excessive Se" areas. However, with the regional soil erosion conditions becoming more serious, this process can cause Se loss in the soil, especially in the agricultural soil of the slope. Geographically, DeAn County in the northwest of Jiangxi Province, was a typical hilly area with the Mufu mountains in the northwest and Poyang Lakeshore wavy plains in the southeast. Therefore, in the current study, we examined the Se levels and distribution characteristics of the soil in an experimental ecological area under different water conservationist measures in DeAn County. In Poyang Lake River Basin, we are hoping to improve the impact of soil-and-water conservation and rational use of Se resources in red soil in this area of southern China.

2 MATERIALS AND METHODS

Soil samples of this study were taken from one red soil slope experimental area of the experimental station in DeAn County of Jiangxi Institute of Soil-and-Water Conservation. The sampling areas included bare soil, grass coverage, and tree coverage landscapes. A total of 92 soil samples were collected in an area of 100 m^2 (5×20 m) and all the soil samples were pooled by the sampling depth (0–20 and 20–40 cm). Two kg samples were collected from the pooled soil sample for each sampling depth at each site. All soil samples were sieved with 100 mesh nylon sieves after drying at 65°C and kept in dark. Soil samples (0.2000 g) were placed in 50 ml conical flasks with polypropylene lids containing 6 ml mixtures (HNO_3:$HClO_4$ = 5:1, v/v) at room temperature. Samples were then heated up to 170°C until the solution became colorless or light yellow, and the remaining volume was 1–2 ml. After cooling to room temperature, an equal volume of concentrated HCl was added and the Se(VI) of samples were reduced to Se(IV) in boiling water for 15 min. After cooling to room temperature, the final volume was brought up to 10 ml. The Se content of soil was determined by Hydride Generation Atomic Fluorescence Spectrometry (HG-AFS).

Statistical evaluation of the data was performed using the SPSS 19.0 software package (SPSS Inc., Chicago, IL, USA). Comparisons among the different groups were analyzed using t-test. Descriptive statistics were expressed as the means ± standard deviation and the minimum level of statistical significance was set at $P < 0.05$ for all analyses.

3 RESULTS AND DISCUSSION

The mean soil Se concentrations in the experimental area of soil-and-water conservation was 0.273 ± 0.06 mg/kg (0.136–0.401 mg/kg) which was significantly lower than the average level of soil Se content in Poyang Lake basin (0.308 mg/kg). The result shows that the experimental area with soil and water erosion is more serious than the Lake basin. However, content of Se (0.287 ± 0.057 mg/kg) in surface soil (0–20 cm) was significantly higher than that (0.259 ± 0.06 mg/kg) in the deep soil (20–40 cm). This observation might be related to weakly vertical leaching intensity of Se and the application of fertilizer, including Se.

The Se content of the cross and longitudinal slope planted to fruit trees-crop inter-planting was 0.289 ± 0.057 mg/kg and 0.318 ± 0.027 mg/kg, respectively, and there were no statistically differences between them ($P > 0.05$). However, both of their Se contents were significantly higher than the regional average of soil Se content ($P < 0.05$). It appears that the planting of fruit trees-crop inter-planting was an effective measure for maintaining soil Se.

The Se content in different areas was ranked as follows: bare area (0.202 ± 0.045 mg/kg) < bahia grass coverage area (0.225 ± 0.059 mg/kg) < citrus forest area (0.299 ± 0.012 mg/kg). There were no significant differences between the bare area and grass coverage area. Compared with bare area and grass coverage area, the content of Se of citrus forest area was significantly higher ($P < 0.05$). We concluded that citrus planted forest soil was more effective than grass coverage soil on maintaining soil Se. In addition, the content of selenium in terraced citrus forest (0.316 ± 0.03 mg/kg) was significantly higher ($P < 0.05$) than ordinary citrus forest (0.246 ± 0.054 mg/kg). We concluded that terraces were more effective for preventing soil from erosion and consequently, minimizing the loss of Se in red soil slope than ordinary.

4 CONCLUSIONS

Concentrations of Se in the experimental district of soil erosion in DeAn County were lower than the Se content in Poyang Lake River basin. Moreover, there was an obvious difference in Se content of different soil and water conservation experimental areas. Overall, the Se content of terraced land was highest. Our preliminary results show that the terraced land would be the best method of soil-and-water conservation in this study.

REFERENCES

Girling, C.A. 1984. Selenium in agriculture and the environment. *Agric. Eco. Environment* 11: 37–65.

Rayman, M.P. 2002. The argument for increasing selenium intake. *Proceedings of the Nutrition Society* 61: 203–215.

Selenium in the Environment and Human Health – Bañuelos, Lin & Yin (Eds)
© 2014 Taylor & Francis Group, London, ISBN 978-1-138-00017-9

Effect of soil nano-selenium fertilizers on selenium accumulation and distribution in mulberry trees

Yongjin Deng
Sericulture Research Institute, Anhui Academy of Agricultural Sciences, Hefei, Anhui, China

Limin Yang, Zhilin Wu, Heyu Zhang, Yunhui Xu & Miao Li
Key Laboratory of Agri-Food Safety of Anhui Province, School of Plant Protection, Anhui Agriculture University, Hefei, Anhui, China

Xuebin Yin
Advanced Lab for Selenium and Human Health, Suzhou Institute for Advanced Study, University of Science and Technology of China, Suzhou, Jiangsu, China

1 INTRODUCTION

Selenium (Se) in small amounts is an essential micronutrient for many organisms, including humans and other animals, and a beneficial nutrient for many plants. Selenium plays an important role as an antioxidant and contributes to the antioxidant protection of cells (Hasanuzzaman *et al.*, 2010). Mulberry fruits are fashionable in the world by their high nutrition value and multi-faceted health protection. The biofortification of food crops for human consumption using soil Se fertilizers represents an effective solution to increase a low-Se diet and status (Broadley *et al.*, 2006). The most common way to increase the amount of Se in cultivated crops is to use Se-enriched fertilizers or to spray Se salts on the plants (Pezzarossa *et al.*, 2009). Inorganic or organic supplementation of Se affects in different ways the amount of Se retained in the soil of various agricultural crops (Zhu *et al.*, 2009). The aim of this study was to investigate the effect of root adding of nano selenium-enriched fertilizers on Se accumulation in mulberry tree.

2 MATERIALS AND METHODS

The field experiments were conducted in spring 2013 at the Taoxi Orchard in Hefei, including two mulberry species of *Morus alba* L. (with white color fruits) and *Morus nigra* L. (with black color fruits) (see Fig. 1). The experimental mulberry trees were three years old. The soil nano-Se fertilizer was presented by Suzhou Setek Co., Ltd, and it was applied via soil/root at the application rate of 35 or 50 mg nano-Se fertilizer per mulberry tree. Mulberry trees that were not treated with nano-Se were used as control. Each mulberry variety (i.e., white mulberry or black mulberry) had two Se fertilizer treatments (*i.e.*, 35 mg and 50 mg Se per tree). Each treatment had five replicates. Therefore, a total of 10 trees of each variety were treated with Se fertilizer, along with 10 individual trees of each variety as control.

Mulberry fruits were harvested at commercial maturity (in late May). Plant leaves were also collected from each replicate tree of each treatment and the control. Mulberry fruit and foliar fresh weights were recorded, and their dry weights were determined after oven-dried at 45°C for 48 hours. Oven-dried samples were acid-digested in nitric and perchloric acids, and concentrations of Se in acid digestion solutions were analyzed by atomic absorption spectrophotometry with hydride generation, after Se reduction using hydrochloric acid (Hawrylak-Nowak, 2013). One way ANOVA was used to determine the treatment effects in a completely randomized experimental design. The level of significance was 0.05.

Figure 1. White and black mulberry fruits and trees in the field.

3 RESULTS AND DISCUSSION

The addition of nano selenium-enriched fertilizers in mulberry orchard cv. '*Morus alba* L.' and '*Morus nigra* L.' resulted in an increase of Se concentration (μg/kg fresh weight) both in the leaves and mulberry fruits (Figs. 2 & 3). Concentrations of Se significantly increased by increasing the amount of Se added. The level of Se concentration was much higher in the leaves than in mulberry fruits.

The accumulation of Se in mulberry fruits and leaves might be affected by the age of mulberry tree, the plant growth time period when nano-Se fertilizer was applied, climate conditions, and soil Se bioavailability and distribution. Further research will be conducted to explore the effective approaches for the enhancement of Se accumulation in Mulberry fruits.

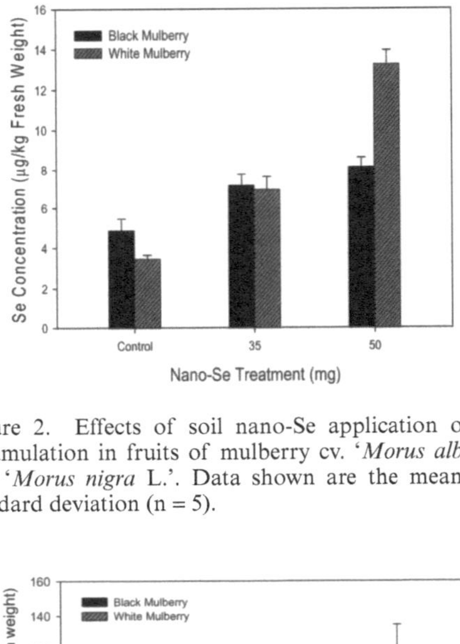

Figure 2. Effects of soil nano-Se application on Se accumulation in fruits of mulberry cv. '*Morus alba* L.' and '*Morus nigra* L.'. Data shown are the mean and standard deviation (n = 5).

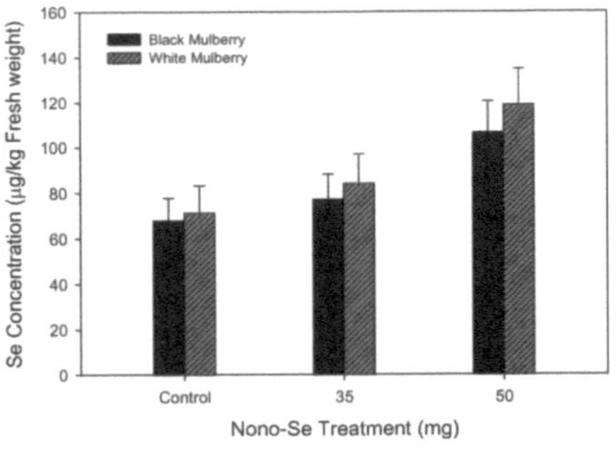

Figure 3. Effects of soil nano-Se application on Se accumulation in leaves of mulberry trees. The two species were '*Morus alba* L.' and '*Morus nigra* L.' Data shown are means and standard deviation (n = 5).

4 CONCLUSIONS

The addition of Se via root Se fertilizer application appears effective in increasing the Se content in mulberry leaf and fruit. Comparing the black mulberry variety to the white mulberry variety, the white significantly increased the Se content in the leaves and fruits, especially in the leaves. These results suggest that root Se adding appeared effective in increasing the Se content in mulberry fruits. Future research needs to focus on the mechanisms associated with the accumulation of Se in the mulberry tree.

REFERENCES

Broadley, M.R., White, P.J., Bryson, R.J., Meacham, M.C., Bowen, H.C., Johnson, S.E. *et al.* 2006. Biofortification of UK food crops with selenium. *Proc. Nutr. Soc.* 65: 169–181.

Hasanuzzaman, M., Hossain, M.A. & Fujita, M. 2010. Selenium in higher plants: physiological role, antioxidant metabolism and abiotic stress tolerance. *J. Plant Sci.* 5: 354–375.

Hawrylak-Nowak, B. 2013. Comparative effects of selenite and selenate on growth and selenium accumulation in lettuce plants under hydroponic conditions. *Plant Growth Regulation* 70(2): 149–157.

Pezzarossa, B., Remorini, D., Piccotino, D., Malagoli, M. & Massai, R. 2009. Effects of selenate addition on selenium accumulation and plant growth in two genotypes of *Prunus* root stocks. *J. Plant Nutr. Soil Sci.* 172: 261–269.

Zhu, Y.G., Pilon-Smits, E.A.H., Zhao, F.J., Williams, P.N. & Meharg, A.A. 2009. Selenium in higher plants: understanding mechanisms for biofortification and phytoremediation. *Trends Plant Sci.* 14: 436–442.

Selenium in the Environment and Human Health – Bañuelos, Lin & Yin (Eds)
© 2014 Taylor & Francis Group, London, ISBN 978-1-138-00017-9

Effects of selenium on crop yield and accumulation of selenium, nutrients, and other elements in paddy rice

X.L. Chen, C.H. Wei, Z.Y. Bao & H. Tian
College of Material Science and Chemical Engineering, China University of Geosciences and The State Key Laboratory of Geological and Environmental Geological, Wuhan China

Y. Guo
Guangzhou Institute of Geological Survey, Guangzhou, China

1 INTRODUCTION

Cadmium (Cd) is an important environmental pollutant. It is ubiquitously distributed in the environment, and is highly toxic to animals and plants. Soil Cd contents range from 0.01 to 30 mg/kg (Walley & Huerta, 2010). Some plant species have the ability of accumulating a large amount of Cd from Cd-polluted soils (Sriprachote *et al.*, 2012). The excessive accumulation of Cd in plants may reduce photosynthesis rates and decrease uptake of plant essential nutrients, inhibit growth and even lead to mortality (Sanitàdi & Gabbrielli, 1999). Therefore, it is important to develop an effective method to reduce Cd accumulation in plants and minimize potential toxic effects of Cd to humans.

Selenium (Se) is an essential element for humans and animals. It plays an important role in enhancing humans' immune function. The Recommended Dietary Allowance (RDA) of Se is 55–70 μg/day. Insufficient daily intake of Se can lead to health problems (Rayman, 2008). About 0.5–1.0 billion people are experiencing Se deficiency issues around the world (Combs, 2001). Previous studies have shown that Se may alleviate the toxicity of metals, such as Cd and Hg, and inhibit plant uptake of heavy metals (Malik *et al.*, 2012; Wei *et al.*, 2011). Some studies have suggested that Se may help plants to accumulate more micronutrients in crops (Wei *et al.*, 2013; Hua *et al.*, 2012) and consequently improve the plant health.

Rice is one of the most popular staple foods in the world, especially in Asia. Studies show that rice from some cropping areas accumulated high levels of Cd, and low amounts of other essential micronutrients. To minimize Cd toxicity and enhance the accumulation of micronutrients in staple foods, it is important to investigate the content and the change of each nutrient element, as well as their interactions in paddy rice. Thus, the objective of this study was to determine effects of different soil Se concentrations on uptake or accumulation of toxic metal Cd and micronutrients Zn, Mn and Ca in paddy rice. Our research findings will provide insight into the implementation of biofortification strategy to improving paddy rice nutrient values (Poletti *et al.*, 2004; Welch, 2002).

2 MATERIALS AND METHODS

Two field experiments were conducted from May to November 2011 at the research field of Wuhan Academy of Agricultural Science & Technology, and repeated again from April to July 2012 in Yangjiang of Guangdong. The two experimental locations had an annual precipitation of 700 mm and 1000 mm, respectively. The experimental design was identical for the two field tests (described below).

Paddy rice (*Oryza sativa* L. wanxian 98) in each plot was sprayed during the heading stage of growth with 10 L Se solution (Na_2SeO_3) at the application rate of 13.4, 26.8, or 40.2 mg Se per plot. Soil not treated with Se was used as control. With three replicates of each treatment, a total of 12 field plots were prepared, and each was 5 m × 5 m with an interrow spacing of 0.5 m. The field experiments were conducted and managed with local normal agronomic practices during the course of each field experiment. At rice maturity, grains were harvested from 3 m^2 area within each plot. The rice yield for the sampling plot was determined.

Paddy rice plants were harvested and washed in deionized water prior to oven-drying at 60°C for two days. Husks were removed by a laboratory-scale dehulling machine. Brown rice was milled and polished, and the bran was collected. In addition to husk, rice grain, and bran, plant root, stem, leaf tissues were separated and collected. Soil samples were also collected, oven-dried along with plant samples, and then ground using a Joyoung mill. Soil and plant samples were acid-digested for chemical analysis, described below.

Concentrations of Cd, Ca, Mg, and Zn were determined by inductively-coupled plasma mass spectrometry (ICP-MS), while concentrations of Se were measured by AFS (PERSEE PF6, Beijing).

3 RESULTS AND DISCUSSIONS

3.1 *Concentrations of Se in plant tissues*

When Se solution was sprayed, the Se concentrations in rice increased significantly with increasing Se treatment level (Fig. 1). From 65 different rice samples, the Se concentration was generally below 0.049 mg/kg (Yong *et al.*, 2008), while the Se content in Se-biofortified rice was 0.04–0.30 mg/kg. The distribution of Se in various organs was investigated, showing a descending order of leaf > stem > root > bran > rice > husk.

3.2 *Effects of Se on Cd and Pb accumulation in paddy rice*

The application of Se solution did not show any consistent effects on Cd accumulation in different rice tissues, including roots and rice grain (data not shown). In comparison to control, the 28.6 mg/plot Se treatment resulted in significant reduction in Cd accumulation in rice tissues. The reduction rates in root, stem, leaf, husk bran and rice were 68%, 74.8%, 50%, 20.8%, 7.8% and 66.7%, respectively. However, when the Se treatment was 40.2 mg/plot, the Cd concentration increased in all different tissues.

The effects of Se application on Pb accumulation in different rice tissues (data not shown) were similar to effects on the accumulation of Cd, however, most of the Pb was distributed in rice roots, leaf, bran, and husk tissues. Rice gain accumulated only very low concentrations of Pb. Overall, the rice Pb concentration did not exceed the national safety standard of 0.2 mg/kg (GB2762-2012).

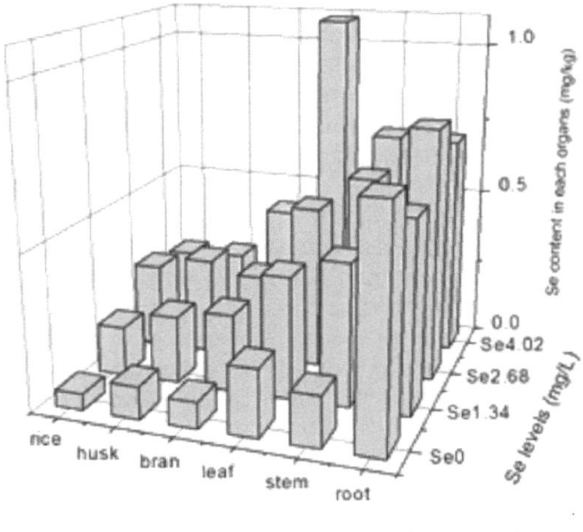

Figure 1. Effect of Se treatments on Se content in each organ.

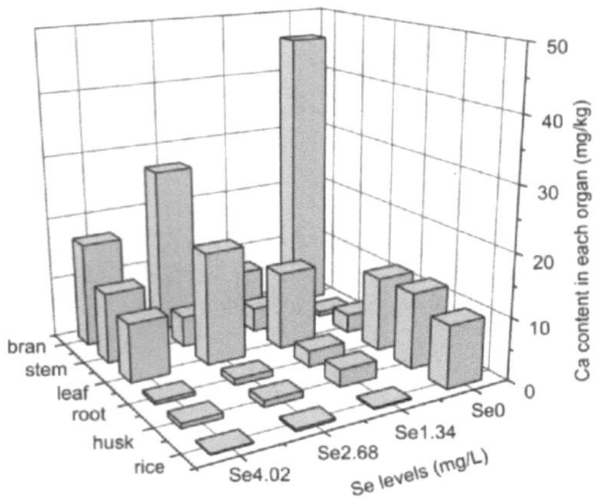

Figure 2. Effects of the Se treatments on Ca content in each organ.

3.3 *Effect of Se on Zn, Mg and Ca in paddy rice*

The addition of Se has different effects on the rice Zn accumulation. Zinc uptake and transport to the edible parts of plants can be increased by foliar fertilizer application (Yong *et al.*, 2008).

When Se solution was sprayed, the concentrations of Ca and Mg in root, husk, bran and rice were decreased although the content of Ca in leaf and stem were increased (Fig. 2). The content of Mg and Ca in shoots was suppressed by increasing Se levels (Wei *et al.*, 2013). Selenium can reduce the content of Mg in tea (Jing *et al.*, 2012). The paddy rice growth and yield is influenced by tissue Mg concentrations. Selenium treatments affected concentrations of Mg in leaf and stem, but the visual growth of paddy rice was not influenced as yield increased.

4 CONCLUSIONS

When applying foliar Se, the distribution of Se in different organs was as follows: leaf > stem > root > bran > rice > husk. The concentration of Ca and Mg were decreased with the increase of Se level in each organ. When Se solution was sprayed, concentrations of Se and Zn in rice were increased by 396% and 146%, respectively. However, the rice Se and Zn concentrations still cannot meet the micronutrient requirements for the population primarily feeding on the rice. When Se was supplied via foliar-spray, the rice yield was generally improved.

REFERENCES

Hua, Z., Bin, X.F., Ming, J.Z., Atindra, S., Bo, M., Heng, Y. *et al.* 2012. Selenium n soil inhibits mercury uptake and translocation in rice. *Environ. Sci. Technol.* 46: 10040–10046.

Jing, Y.Y., Lin, M.C., Ye, Y.X., Ming, Z.J. & Jia, Z.W. 2012. Effect of applying biogas fertilizer on tea farm soil and changes of magnesium, zinc and selenium in tea. *Guizhou Science* 30(3): 68–71.

Malik, J.A., Goel, S., Kaur, N., Sharma, S., Singh, I. & Nayyar, H. 2012. Selenium antagonizes the toxic effects of arsenic on mungbean (*Phaseolus aureus* Roxb.) plants by restricting its uptake and enhancing the antioxidative and detoxification mechanisms. *Environ. Exp. Bot.* 77: 242–248.

Poletti, S., Gruissem, W. & Sautter, C. 2004. The nutritional fortification of cereals. *Curr. Opin. Biotechnol.* 15(2): 162–165.

Rayman, M.P. 2008. Food-chain selenium and human health: emphasis on intake. *Br. J. Nutr.* 100: 254–268.

Sanitàdi, T.L. & Gabbrielli, R. 1999. Response to cadmium in higher plants. *Environ. Exp. Bot.* 41: 105–130.

Sriprachote, A., Kanyawongha, P., Ochiai, K. & Matoh, T. 2012. Current situation of cadmium-polluted paddy soil, rice and soybean in the Mae Sot District, Tak province, T hailand. *Soil Sci. Plant Nutr.* 58: 349–359.

Walley, J.W. & Huerta, A.J. 2010. Exposure to environmentally relevant levels of cadmium primarily impacts transpiration in field grown soybean. *J. Plant Nutr.* 33(10): 1519–1530.

Wei, R.F., Yang, C.W., Xin, S.T., Rong, S.T. & Chang, F.W. 2011. Detoxification of antimony by selenium and their interaction in paddy rice under hydroponic conditions. *Microchem. J.* 97: 57–61.

Wei, R.F., Yang, C.W., Xin, S.T., Zhen, Y.D. & Guo, Z.S. 2013. A Dual Role of Se on Cd toxicity: Evidences from the uptake of Cd and some essential elements and the growth responses in paddy rice. *Biol. Trace Elem. Res.* 151(1): 113–121.

Welch, R.M. 2002. Breeding strategies for biofortified staple plant foods to reduce micronutrient malnutrition globally. *J. Nutr.* 132(3): 495–499.

Yong, F., Lin, W., Hong, Z.X., Yan, Z.L., Xin, X.A. & Hui, Q.H. 2008. Effect of foliar application of zinc, selenium, and iron fertilizers on nutrients concentration and yield of rice grain in China. *J. Agri. Food. Chem.* 56: 2079–2084.

Selenium in the Environment and Human Health – Bañuelos, Lin & Yin (Eds)
© 2014 Taylor & Francis Group, London, ISBN 978-1-138-00017-9

Effects of soil selenate and selenite on selenium accumulation and distribution in Pak choi

L. Guo, D.L. Liang, N. Man & J.Y. Xie
College of Resources and Environment, Northwest A&F University Yangling, Shaanxi, China

1 INTRODUCTION

Selenium (Se) is an essential element for humans and animals (Lenz & Lens, 2009). At present, there are a large number of people in the world who have inadequate intake of Se (Williams *et al.*, 2009). Thus, increasing Se content in plants could have a positive impact on reducing Se deficiency in different parts of the world (Djanaguiraman *et al.*, 2005).

Selenium accumulation and transport in plants is dependent on Se forms in soil, plant species and growth stages. Plants absorb Se from soil primarily as selenate and selenite (Zhao *et al.*, 2005; Xue *et al.*, 2011). Selenium levels vary greatly in different plants, as well as in the same plants but at different growth stages (Sharma *et al.*, 2010; Keskinen *et al.*, 2010). Therefore, the aim of the present study was to investigate the differences of selenium uptake pattern and bioavailability on Pak choi (*Brassica chinensis* L.) when amended selenate and selenite. Meanwhile, the dynamic changes of soil available Se was analyzed to explore the reason. In addition, the optimum maturity stage of Pak choi for harvest was also recommended.

2 MATERIALS AND METHODS

Pot experiments were conducted to cultivate Pak choi (*Brassica rapa* subsp. *chinensis* L.). There were seven Se treatments to be added to soils: 0.5, 1.0, and 2.5 mg Se/kg (dry soil weight) as selenate, and 1.0, 2.5 and 5.0 mg Se/kg as selenite. The soil without Se was designated as the control (CK). Each treatment had 4 replicates. From a total of 196 identical experiment pots, 28 (4 pots for each treatment) were randomly selected and harvested after 2, 3, 4, 5, 6, and 7 weeks, respectively.

After each harvest, plants were separated into root and shoots to measure their biomass and Se concentration. Soil available Se was determined for each treatment. Concentrations of Se were determined by hydride generation atomic fluorescence spectrophotometry. Statistical analysis was conducted with SPSS 16.0.

3 RESULTS AND DISCUSSION

3.1 *Effect of selenate and selenite on Se uptake pattern in Pak choi across time*

No significant differences in the biomass of pakchoi were observed between selenate and selenite treatments (data not presented). Significant difference was observed between selenite and selenate treatments for shoot Se concentration during the Pak choi growing period (Fig. 1).

Concentrations of Se in Pak choi decreased rapidly across growing period for selenate treatments, but slowly increased for selenite treatments, especially during the first 5 weeks period. Selenium concentration increased significantly as the amount of Se added to the soil increased. The Se concentration was the highest in the young leaves and roots, and lowest in the shoots and roots of mature plants for soils amended with selenate. In contrast, the Se concentration was highest both in shoots and roots of mature plants for soils amended with selenite. In the selenate treatments, the pakchoi shoot Se concantration was significantly higher than the roots, but the opposite results were found in the selenite treatments. In the whole growth period of pakchoi, Se concentration in selenate treatment was 2–60 times of that in selenite treatments. Therefore, the Se uptake ability of Pak choi for selenate and selenite was such different.

3.2 *Effect of selenate and selenite on soil available Se with the growth of the Pak choi*

Soil available Se can be directly absorbed by plants Se (Lee *et al.*, 2011), which changed cross Pak Choi growing period for selenate and selenite were showed in Figure 2. The content of soil available Se decreased rapidly with the growth of Pak choi in the selenate treatments (Fig. 2a), especially during the first 5 weeks when Pak choi grown quickly and took up more Se from soil (Fig. 1a). In contrast, soil available Se content slightly decreased in the selenite treatments during Pak choi whole growing period (Fig. 2b) due to less Se uptake took place (Fig. 1b).

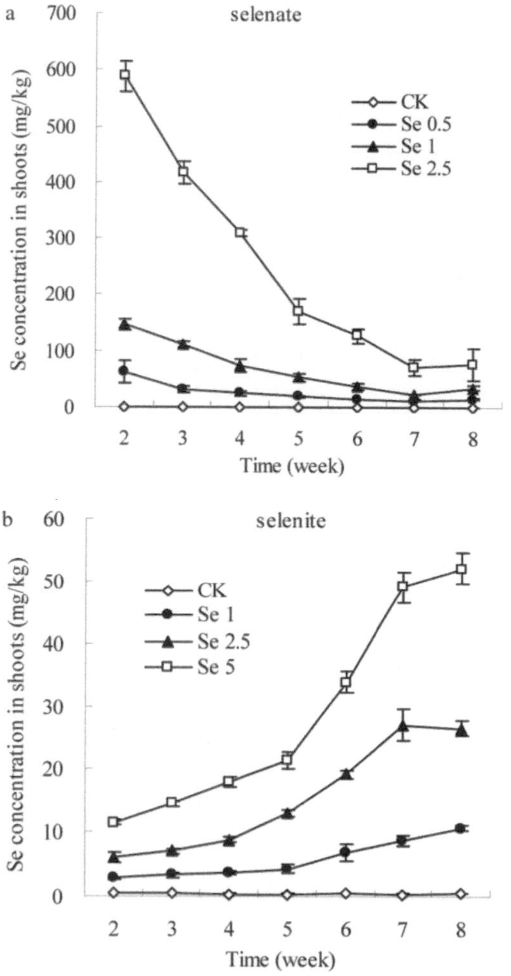

Figure 1. Effects of selenate and selenite on Se concentration of Pak choi shoots during the growing period. Top: selenite treatment; Bottom: selenite treatment.

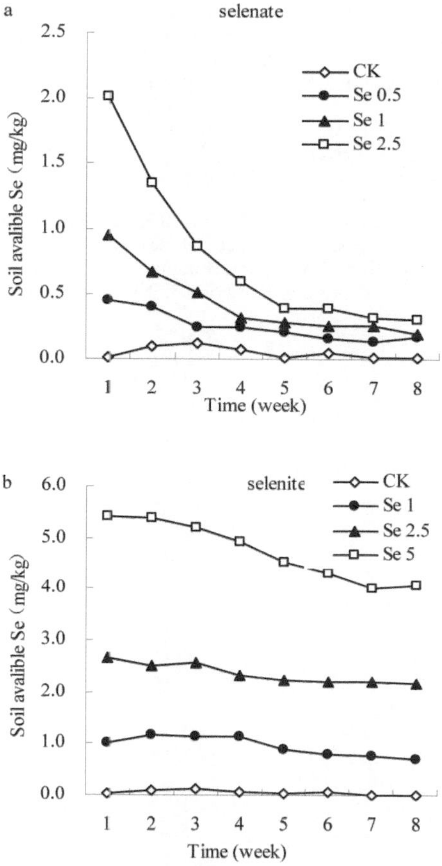

Figure 2. Changes of selenate and selenite on soil available selenium with the growth of the pak choi. Top: selenite treatment; Bottom: selenite treatment.

3.3 *Relationship of soil available Se and Se concentrations of the Pak choi*

The tendency of soil available Se change matched well to the variation of Se concentration in Pak choi across the whole growth period. In this regard, a significant correlation was found between soil available Se and Se concentration either in shoots or in roots of Pak choi for selenate treatments ($P < 0.05$). In contrast, no significant correlation was observed between soil available Se and plant Se concentration either in roots and shoots of Pak choi for selenite treatment. Based on the results, the ability of soil to provide Se to Pak choi will be dependent more on the form of Se in soil.

4 CONCLUSIONS

The differences between the selenium uptake pattern of selenate and selenite on Pak choi was the comprehensive function of the ability of soil providing Se and crop uptake Se ability. So it was necessary to defined suitable reaping time in order to produced Se enrichment plants.

REFERENCES

Djanaguiraman, M., Durga Devi, D., Shanker, A.K., Annie Sheeba, J. & Bangarusamy, U. 2005. Selenium-an antioxidative protectant in soybean during senescence. *Plant Soil* 272(1–2): 77–86.

Keskinen, R., Turakainen, M. & Hartikainen, H. 2010. Plant availability of soil selenate additions and selenium distribution within wheat and ryegrass. *Plant Soil* 333(1–2): 301–313.

Lee, S., Woodard, H.J. & Doolittle, J.J. 2011. Selenium uptake response among selected wheat (*Triticum awstivum*) varieties and relationship with soil selenium fractions. *Soil Science and Plant Nutrition* 57(6): 823–832.

Lenz, M. & Lens, P.N.L. 2009. The essential toxin: The changing perception of selenium in environmental sciences. *Science of the Total Environment* 407(12): 3620–3633.

Sharma, S., Bansal, A., Dhillon, S.K. & Dhillon, K.S. 2010. Comparative effects of selenate and selenite on growth and biochemical composition of rapeseed (*Brassica napus* L.). *Plant Soil* 329(1–2): 339–348.

Williams, P.N., Lombi, E., Sun, G.X., Scheckel, K., Zhu, Y.G., Feng, X. *et al.* 2009. Selenium characterization in the global rice supply chain. *Environmental Science and Technology* 43(15): 6024–6030.

Xue, R.L., Liang, D.L., Wang, S.S., Fu, D.D., Duan, M.L. & Xu, M. 2011. Valence state changes and bioavailability of selenium in soil treated with selenite and selenate. *Environmental Science* 32(6): 1726–1733.

Zhao, C.Y., Ren, J.H., Xue, C.Z. & Lin, R.D. 2005. Study on the relationship between soil selenium and plant selenium uptake. *Plant and Soil* 277(1–2): 197–206.

Selenium in the Environment and Human Health – Bañuelos, Lin & Yin (Eds)
 ISBN 978-1-138-00017-9

Selenium distributions in *Cordyceps militaris* from different areas in China

Qin Li & Chi Zhang
School of Biological Science and Technology, Hubei Institute for Nationalities, Hubei, China
Key Laboratory of Biologic Resources Protection and Utilization of Hubei Province, Enshi, Hubei, China

1 INTRODUCTION

Organic Se has been the study hotpot in recent years, since Se is an essential micronutrient elements for human and animals (Rotruck *et al.*, 1973; Berry *et al.*, 1991; Kvicala, 1999). Recent studies indicated that fungi (*e.g.*, *Cordyceps militaris*) are important carriers to transform inorganic Se into organic Se compounds because of their rapid propagating, fast growing, and strong adaptation (Chen *et al.*, 2009). *Cordyceps militaris* were found to have similar active constituents and pharmacological effects as those of *Cordyceps sinensis* (Wei *et al.*, 2004; Wu *et al.*, 2011). Although Se-enriched *C. militaris* was identified as a popular hearth medicinal species (Hu *et al.*, 2006), there were few studies or attentions on Se distributions in the fungal tissues. In the present study, the distribution of Se-protein and Se-polysaccharide in *Cordyceps militaris* was determined with the samples collected from different climate regions. This study provided experimental evidence showing biotransformation of organic Se compounds in *C. militaris*.

2 MATERIALS AND METHODS

The fruiting bodies of *Cordyceps militaris* were collected from Yunnan and Hubei, respectively. The protein content was determined by Coomassie brilliant blue (G-250) staining technology (Li *et al.*, 2000). The content of polysaccharide was determined by the phenol-sulfuric acid method (Zhang *et al.*, 1987). The soluble protein contents were divided into four classes based on their saturation ranges by the $(NH_4)_2SO_4$ salting-out method: A, 0–30%; B, 30–50%; C, 50–75%; and D, 75–95%. The Se concentrations in the samples were determined by atomic fluorescence spectro-photometer.

3 RESULTS AND DISCUSSION

3.1 *Contents of proteins and polysaccharides*

The contents of soluble proteins were included in Table 1. Overall, the contents of four groups/classes of soluble protein in *C. militaris* tissues collected from two different climate regions were similar, showing a descending order of C > D > B > A. The contents of polysaccharide in *C. militaris* tissues collected from Hubei and Yunnan were 130.3 mg/g and 122.1 mg/g, respectively.

Table 1. The contents of soluble protein of different saturation ranges in fruiting bodies of *C. militaris* collected from Hubei and Yunnan provinces.

	Soluble proteins (mg/g)			
Location	A	B	C	D
Hubei	0.549	1.178	4.298	1.418
Yunnan	0.265	1.049	4.666	1.464

3.2 *Concentrations of total Se in soluble proteins and polysaccharides*

The concentration of total Se in *C. militaris* was 210 μg/g from Hubei and 57.5 μg/g from Yunnan. The Se distribution among different classes of soluble Se protein was shown in Figure 1. For *Cordyceps militaris* from Hubei, the amount of Se associated with soluble proteins was less than 5% of the total Se. However, for *Cordyceps militaris* from Yunnan, the amount of Se associated with soluble proteins was much higher than those of *Cordyceps militaris* from Hubei. Amongst the soluble protein associated Se, the Se associated with A and B classes accounted for 17%, followed by 14% in the D class and about 10% in the C class. The polysaccharide-Se accounted for 11.1% of the total Se accumulated in *Cordyceps militaris* from Hubei and 26.7% from Yunnan.

There was no statistically significant ($P > 0.05$) correlation between the protein content and the soluble protein Se concentration (data not shown). Although the concentration of total Se in *C. militaris* was higher in Hubei, the proportions of protein-Se and polysaccharide-Se were lower than those in Yunnan. According to previous studies by Tie *et al.* (2006) and Zhong *et al.* (2008), organic Se in the Se-enriched *Cordyceps militaris*

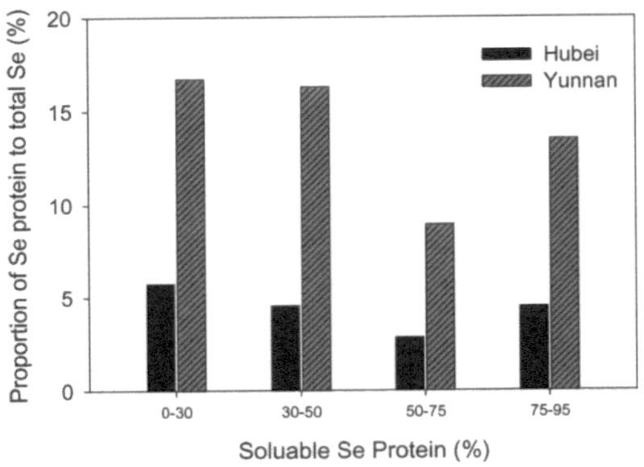

Figure 1. The proportion of soluble Se protein to the total Se accumulated in *C. militaris* collected from two different climate regions (*i.e.*, Hubei and Yunnan, China).

accounted for 69–85% of the total Se accumulated in the tissues. It was speculated that the transformation of organic Se in *Cordyceps militaris* could be related to the growth medium, the amount of Se applied in the medium, as well as different environmental conditions. Therefore, our future study will focus on finding or developing effective growth medium and optimizing the Se content in the growth medium.

4 CONCLUSIONS

Cordyceps militaris grown in Hubei and Yunnan provinces showed different levels of Se accumulation and different amounts of Se associated with soluble protein and polysaccharide in their fruiting bodies. The contents of Se bound with soluble protein and polysaccharide in *Cordyceps militaris* varied from 57.5 μg/g in Yunnan to 210 μg/g in Hubei. In Se-biofortified *Cordyceps militaris* tissues, higher proportions of protein Se and polysaccharide Se can be obtained, showing effective biotransformation of organic Se.

REFERENCES

Berry, M.J., Banu, L. & Larsen, P.R. 1991. Type iodothyronine deiodinase is a selenocysteine—containing enzyme. *Nature* 349(6308): 438–440.

Chen, Z.J. & Yuan, S.L. 2009. Research on selenium enrichment and product development of edible and medicinal fungi. *Edible Fungi of China* 28(6): 9–11. (in Chinese).

Hu, K., Zhong, M., Liu, Y.D. & Jiang, M.L. 2006. Effect of Selenium on the nutritional and active components of *Cordyceps militaris*. *Studies of Trace Elements and Health* 23(1): 6–8. (in Chinese).

Kvicala, J. 1999. Selenium and organism. *Cas Lek Cesk* 138(4): 99–106.

Li, H.S., Sun, Q., Zhao, S.J. & Zhang, W.H. 2000. *Principles and Techniques of Plant Physiological Biochemical Experiment*. 184–185. Beijing: Higher Education Press. (in Chinese).

Rotruck, J.T., Pope, A.L., Ganther, H.E., Swanson, A.B., Hafeman, D.G. & Hoekstra, W.G. 1973. Selenium: biochemical role as a component of glutathione peroxidase. *Science* 179(4073): 588–590.

Tie, M., Zang, S.L., Fang, Y.Z., Sun, T.B. & Li, H.W. 2006. Applying study of SEC-ICP-MS in speciation analysis of selenium-containing proteins in selenized *Cordyceps militaris* Link. *Chemical Journal of Chinese Universities* (7): 1232–1236. (in Chinese).

Wei, H.P., Xiao, B. & Hu, K.Z. 2004. Medicinal value of *Cordyceps militaris*. *Journal of Chinese Medicinal Materials* 27(3): 215–217. (in Chinese).

Wu, Y. & Chen, W.D. 2011. The research overview of pharmacological effects of *Cordyceps militaris*. *China Pharmacist* 14(5): 732–734. (in Chinese).

Zhang, W.J. 1987. *The Biochemical Research Techniques of Compound Polysaccharide*. 33–40. Shanghai: Shanghai Science and Technology Press. (in Chinese).

Zhong, M. & Wang, L.H. 2008. Analysis on speciation of selenium and Se-protein in *Cordyceps militaris*. *Guang Dong Wei Liang Yuan Su Ke Xue* 15(3): 35–40. (in Chinese).

Selenium in the Environment and Human Health – Bañuelos, Lin & Yin (Eds)
© 2014 Taylor & Francis Group, London, ISBN 978-1-138-00017-9

Drumming up selenium and sulphur in Africa: Improving nutrition with *Moringa oleifera*

G.H. Lyons & C. Gondwe
School of Agriculture and Wine, University of Adelaide, Glen Osmond, SA, Australia

G.S. Bañuelos & M.C. Zambrano
USDA-ARS, Water Management Unit, Parlier, CA, USA

A. Haug & O.A. Christophersen
Department of Animal and Aquacultural Sciences, The Norwegian University of Life Sciences, Norway

1 THE RELEVANCE OF GOLDSCHMIDT'S CONCEPTS FOR AGRICULTURE, NUTRITION AND MEDICINE

Victor Moritz Goldschmidt (1888–1947), the father of modern geochemistry, introduced the term *biophile* for elements found at high absolute or relative concentrations in living organisms. These include nitrogen, sulphur, phosphorus, potassium, selenium, iodine, zinc, boron and bromine.

Goldschmidt was also interested in how different elements are transported and become fractionated relative to each other because of geological processes (*e.g.*, volcanism, weathering and erosion), both inside the earth and at its surface. For this he used the term *geochemical cycles* to describe how the transport processes are often cyclic and form closed loops. Combining his concept of geochemical cycles with biophile elements is useful for understanding problems of mineral nutrient depletion in soils due to environmental degradation processes such as deforestation and erosion.

The deep roots of trees that grow on highly weathered soil can scavenge nutrient elements from deep soil layers where chemical weathering is not complete. Consequently, biophile elements that have been deposited from rainwater (*i.e.*, sulphur, selenium, iodine and bromine) are recycled and prevented from being removed by leaching processes because of their active uptake into tree roots and mycorrhiza. But when the forest is cut down or burnt, the vertical transport of nutrient elements upwards from deep soil layers ceases and the system becomes more open than before, with enhancement of the rate of nutrient element removal by leaching and topsoil erosion. The only viable solution for the future is to try to replace these open agricultural ecosystems with new ones that in a geochemical sense are similar to the natural forest ecosystems that they have replaced (Christophersen *et al.*, 2012).

2 MEDICAL CONSEQUENCES OF BIOPHILE ELEMENT DEPLETION

Selenium and sulphur are strongly biophile elements. Selenium (as selenate) and sulphur (as sulphate) are very leachable under humid climatic conditions, and can also be lost to the atmosphere in the form of SeO_2 and SO_2, respectively, as a consequence of anthropogenic fires, especially on the savannah (Christophersen *et al.*, 2012). Sulphur is necessary, similarly to nitrogen, for protein synthesis in plants, with the synthesis of more sulphur-rich protein molecules being even more strongly inhibited than total protein synthesis when the soil is sulphur-deficient. In the more humid parts of sub-Saharan Africa there are large areas where the human diet is deficient in sulphur amino acids, which is in part explained by a low protein intake, but which may also be due to sulphur deficiency in the soil (Christophersen *et al.*, 2012; Gondwe *et al.*, unpubl.). For example, a survey of Zambian maize grain conducted in 2012 revealed a median sulphur concentration of only 1030 mg/kg and high nitrogen: sulphur ratios of 13–15 (Gondwe *et al.*, unpubl.). Critical deficiency level for sulphur and nitrogen: sulphur ratios in maize grain are around 1700 mg/kg and 8–9, respectively (Reuter & Robinson, 1997). Similarly, available selenium is very low in many soils in Zambia, Malawi, Rwanda, Burundi, Zaire and other sub-Saharan African countries, with levels of less than 20 μg/kg selenium common in maize grain (Chilimba *et al.*, 2011; Gondwe *et al.*, unpubl.; Hurst *et al.*, 2013; Vanderpas *et al.*, 1990).

Protein deficiency is likely to be associated with deficiencies in vitamins (*e.g.*, vitamin A, vitamin B12) and minerals (*e.g.*, zinc, iron, selenium and iodine). Medical consequences include brain damage in fetuses and children (due to iodine and zinc deficiencies), and kwashiorkor, general impairment of immunological defence *i.a.* as a consequence of reduced growth rate in leukocytes, and more rapid growth of tubercle bacilli. Importantly, more rapid replication and population dispersal of the HIV virus occurs (*i.a.* because glutathione and selenium depletion leads to enhanced activation of oxidatively activated transcription factors such as NF-*kappa*B and Sp1 that enhance the transcription of the HIV provirus gene in infected leukocytes) (Christophersen *et al.*, 2012). A marked decline in human selenium associated with progression of HIV disease was observed in rural Malawi (van Lettow *et al.*, 2004).

3 DRUMMING UP A VILLAGE-LEVEL SOLUTION: THE PROMISE OF THE DRUMSTICK TREE (MORINGA OLEIFERA) FOR AFRICA

The "drumstick tree" or "miracle tree" (*Moringa oleifera*) is already well known for its high nutritional value. It grows well in tropical and subtropical regions, even on poor soils, and produces abundant leaves high in protein (with a favourable amino acid balance), vitamins, minerals and a range of beneficial phytochemicals, including anticancer glucosinolates (Anwar *et al.*, 2007; Mbikay 2012). Moringa leaves had the highest beta-carotene level (427 mg/kg) of all leaf samples collected by our group in South Pacific countries and Northern Australia in 2012 (Lyons *et al.*, unpubl.), and bioconversion of beta-carotene in Moringa leaves to vitamin A is efficient (Ejoh *et al.*, 2010). Moringa leaves are low in antinutrients such as phytate, hence the bioavailability of nutrients and micronutrients in the leaves is high (Anwar *et al.*, 2007; Mbikay 2012).

Furthermore, Moringa is drought-tolerant and its new leaves appear at the end of the dry season when other greens are scarce (Anwar *et al.*, 2007). Moringa leaves can be produced intensively in a small garden, with just two or three trees being sufficient for a family.

Perhaps the most exciting feature of this plant, in view of the widespread deficiency of plant-available selenium and sulphur in sub-Saharan Africa (and the relevance of these nutrients to HIV disease), is its exceptional ability to take up and accumulate both of these elements, even when grown on soils where they are poorly available to most other plants. We have found that Moringa leaves commonly accumulate around 11 times the concentration of selenium and three times the concentration of sulphur compared to a range of other plants growing on the same soil (Lyons *et al.*, unpubl.). The main forms of selenium in Moringa leaves are presently being determined.

4 CONCLUSIONS

The drought-tolerant, nutritious drumstick tree (*Moringa oleifera*) has exceptional ability to accumulate sulphur and selenium in its leaves when grown on soils which are low in plant-available forms of these minerals. Widespread planting, and consumption of this species by humans and animals, are recommended, especially in tropical and sub-tropical regions of sub-Saharan Africa where HIV disease and TB are prevalent.

REFERENCES

Anwar, F., Latif, S., Ashraf, M. & Gilani, A.H. 2007. *Moringa oleifera*: a food plant with multiple medicinal uses. *Phytotherapy Research* 21: 17–25.

Chilimba, A.D., Young, S.D., Black, C.R., Rogerson, K.B., Ander, E.L., Watts, M.J. *et al.* 2011. Maize grain and soil surveys reveal suboptimal dietary selenium intake is widespread in Malawi. *Scientific Reports* 1: 72. doi: 10.1038/srep00072.

Christophersen, O.A., Lyons, G., Haug, A. & Steinnes, E. 2012. Selenium. In B.J. Alloway (ed) *Heavy Metals in Soils: Trace Elements and Metalloids in Soils and their Bioavailability* 3rd edition. Dordrecht: Springer.

Ejoh, R.A., Dever, J.T., Mills, J.P. & Tanumihardjo, S.A. 2010. Small quantities of carotenoid-rich tropical green leafy vegetables indigenous to Africa maintain vitamin A status in Mongolian gerbils (*Meriones unguiculatus*). *British Journal of Nutrition* 103: 1594–1601.

Hurst, R., Siyame, E.W., Young, S.D., Chilimba, A.D., Joy, E.J., Black, C.R. *et al.* 2013. Soil-type influences human selenium status and underlies widespread selenium deficiency risks in Malawi. *Scientific Reports* 3: 1425. doi: 10.1038/srep01425.

Mbikay, M. 2012. Therapeutic potential of *Moringa oleifera* leaves in chronic hyperglycemia and dyslipidemia: a review. *Frontiers in Pharmacology* 3(24): 1–12.

Reuter, D.J. & Robinson, J.B. 1997. *Plant analysis: an interpretation manual.* 2nd edition. Melbourne: CSIRO.

Vanderpas, J.B., Contempre, B., Duale, N.L., Goossens, W., Bebe, N., Thorpe, R. *et al.* 1990. Iodine and selenium deficiency associated with cretinism in northern Zaire. *American Journal of Clinical Nutrition* 52: 1087–1093.

Van Lettow, M., Harries, A.D., Kumwenda, J.J., Zijlstra, E.E., Clark, T.D., Taha, T.E. *et al.* 2004. Micronutrient malnutrition and wasting in adults with pulmonary tuberculosis with and without HIV co-infection in Malawi. *BMC Infectious Diseases* 4: 61 doi: 10.1186/1471-2334-4-61.

Selenium in the Environment and Human Health – Bañuelos, Lin & Yin (Eds)
© 2014 Taylor & Francis Group, London, ISBN 978-1-138-00017-9

Evaluation of selenium on germplasm and selenium content in wheat seedlings

L.R.G. Guilherme, G.A. Souza & J.G. Carvalho
Federal University of Lavras, Lavras, MG, Brazil

J.C. Albrecht
Embrapa CPAC, Planaltina, DF, Brazil

M. Rutzke & L. Li
USDA-ARS, Cornell University, Ithaca, NY, USA

1 INTRODUCTION

Micronutrient malnutrition is a well-known human health problem affecting more than 3 billion people worldwide (WHO, 2002). Deficiencies in selenium (Se) are common and estimated to affect approximately 15% of the world population (White & Broadley, 2009). Selenium is not only essential for human nutrition, but also exert profound effects on plant growth and development (Graham *et al.*, 2001). Biofortification of crops with enhanced mineral content provides a complementary strategy in conjunction with dietary diversification, mineral supplementation, and food fortification to address the worldwide micronutrient deficient problem (Welch & Graham, 2002; White & Broadley, 2009). We studied 20 wheat accessions from the wheat breeding program at Embrapa Cerrados CPAC, Brazil to evaluate the effects of adding Se on wheat seedling growth and Se content. In addition, the effects of Se treatments on the accumulation of other nutrients in these wheat germplasm were also examined to provide a general view of mineral interactions in wheat plants. Such information is relevant for selecting wheat accessions with an ability to accumulate minerals and hence, be most useful for crop biofortification.

2 MATERIALS AND METHODS

Details of this study are described in Souza *et al.* (2013). Briefly, we selected 15 varieties and 5 cultivars of bread wheat (*Triticum aestivum* L.) from the Embrapa breeding program. In this selection, we used the lines with high grain production and good quality to bread industry. Seeds of wheat accessions were surface-sterilized in 0.5% NaOCl for 15 min, rinsed in distilled milli-Q water, and germinated in a chamber at 20°C for 3 days on moistened filter paper. The uniform 3-day old young seedlings were transplanted into 2.2 L pots containing a modified Johnson's nutrient solution for wheat (Hart *et al.*, 1998). Seedlings were grown with aeration in a greenhouse at 20 ± 2°C under 16-h day length. Four days after transplant, these plants were either grown in the nutrient solution (control) or exposed to the nutrient solution with addition of 10 μM Na_2SeO_4 or 10 μM Na_2SeO_3. The solutions were changed twice each week. Following 10 days of treatments the plants were harvested individually and rinsed in distilled water. Shoots and roots were separated and oven-dried at 65°C for 72 h and the total minerals concentrations in the samples were determined using inductively-coupled plasma mass spectrometry.

3 RESULTS AND DISCUSSIONS

When these wheat lines were grown in the nutrient solution containing selenate, the plant growth was unaffected in some lines, and enhanced or suppressed in other lines (Fig. 1A). However, when the wheat varieties and cultivars were exposed to selenite, a general inhibition of plant growth was observed in nearly all lines. An average of 36% suppression of growth was noticed. These results indicate that wheat lines exhibited genetic variation for shoot growth in response to Se treatments. Each line responded differently to Se treatments, indicating a different capacity in uptake, assimilation, and use of these nutrients within the plant. A similar response pattern as shoot growth was observed for root growth of these wheat lines when they were exposed to various Se treatments ($P < 0.05$) (Fig. 1B). SeO_4 treatment had minimum effects on root growth in some lines and increased the root biomass in half of the other lines. A general reduction of root growth was observed in SeO_3-treated

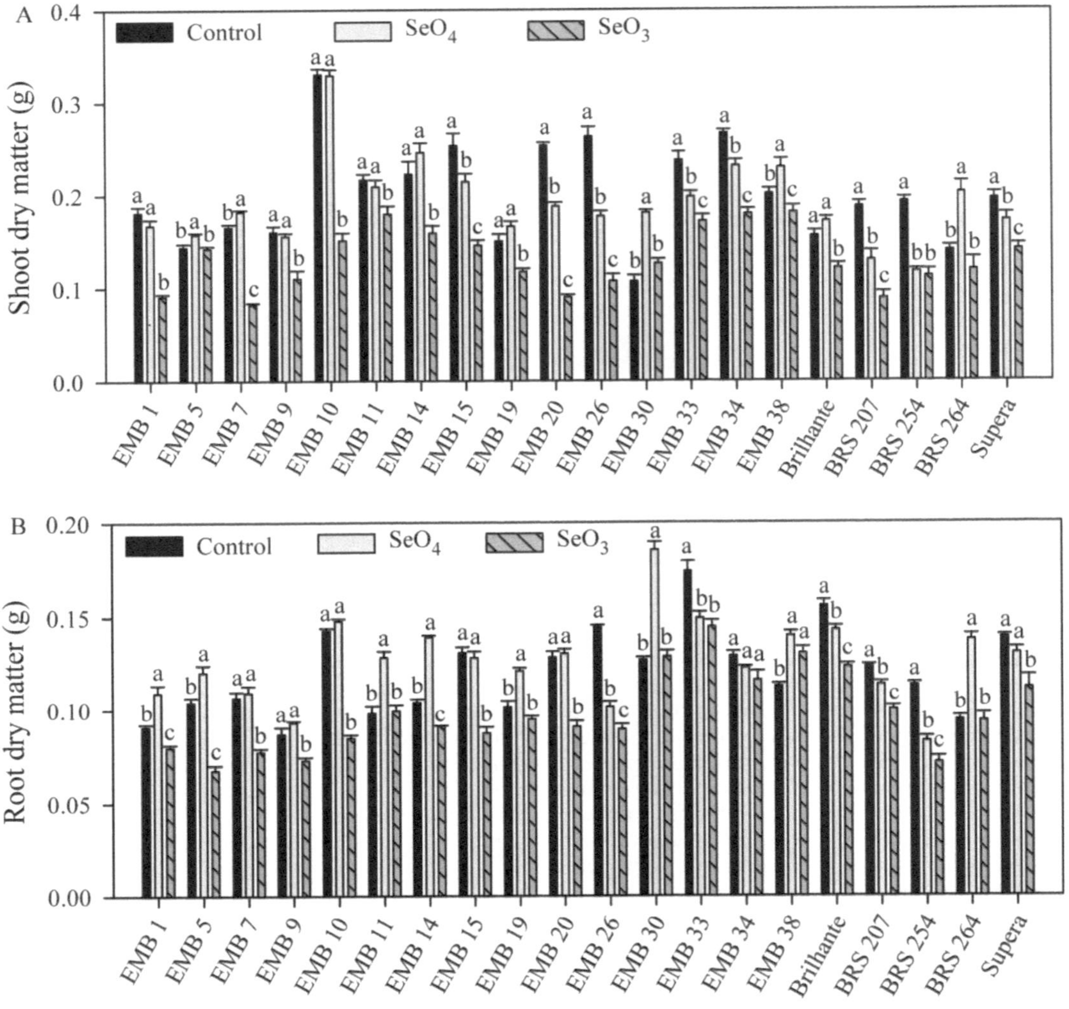

Figure 1. Effect of Se treatments in shoot (A, top) and root (B, bottom) dry matter production of wheat varieties and cultivars. Values shown are mean ± SE (n = 4). Different letters above the columns indicate significant difference between the treatments at $P < 0.05$.

plants, which indicates that selenite is more toxic to plant growth than selenate. Selenium was undetected in seedlings of the wheat lines that grew in the control nutrient solution. When the seedlings of these wheat lines were exposed to SeO_4 and SeO_3, the total Se levels showed 4- and 2-fold difference, respectively, between lines containing high and low level of Se (Fig. 2).

The SeO_4-treated wheat seedlings contained 3- to 7-fold more total Se than the SeO_3-treated ones, which is consistent with previous reports that plants accumulate more Se when treated with SeO_4 than SeO_3 (Li *et al.*, 2010).

We also investigated the effect of Se supplement on the accumulation of micronutrients—iron (Fe) and zinc (Zn). We measured their respective mineral content in response to Se treatments and examined their interactions. Iron content was stimulated by SeO_4 supplement in over half of the wheat lines and unaffected by SeO_3 treatment in majority of the lines. For Zn, SeO_4 caused a slight increase of Zn content in wheat seedlings in half of the wheat lines, while SeO_3 suppressed Zn levels in majority of the lines. Further information concerning such interactions can be found in Souza *et al.* (2013).

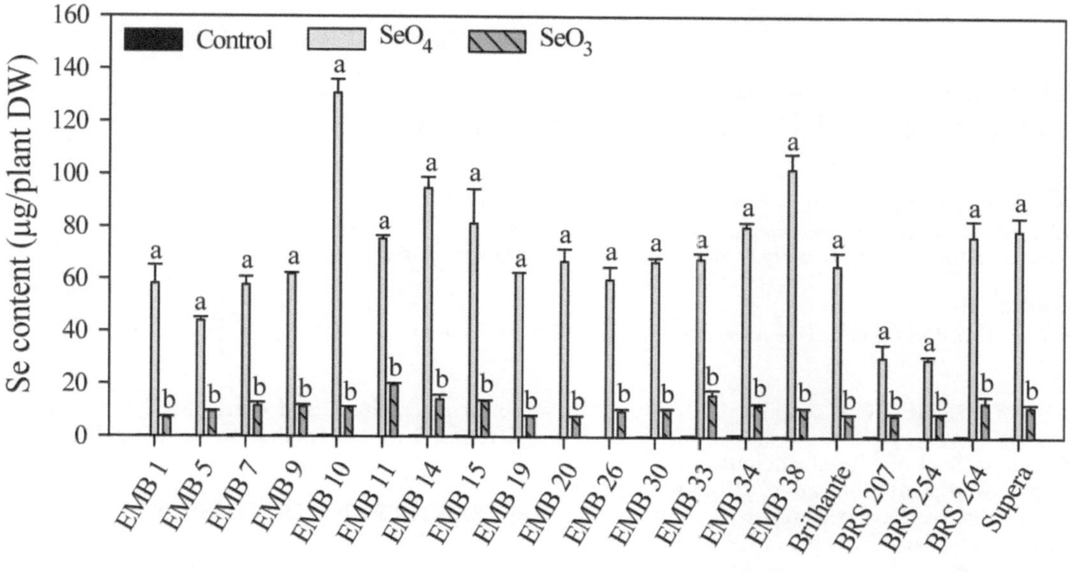

Figure 2. Selenium content in seedlings of wheat varieties and cultivars. Values shown are mean ± SE (n = 4). Different letters above the columns indicate significant difference between the treatments at $P < 0.05$.

4 CONCLUSIONS

Based on nutrient accumulation, plant growth, and mineral interaction, a number of accessions such as EMB 10, EMB 14, EMB 30 and BRS 264 appeared to be the better lines for breeding wheat cultivars with the addition of selenate, while EMB 10, EMB 11, EMB 33 and BRS 264 accumulate essential micronutrients in edible grains with the addition of selenite.

REFERENCES

Graham, R.D., Welch, R.M., & Bouis, H.E. 2001. Adressing micronutrient malnutrition through enhancing the nutritional quality of staple foods: Principles, perspectives and knowledge gaps. *Advances in Agronomy* 70: 77–142.

Hart, J.J., Welch, R.M., Norvell, W.A., Sullivan, L.A., & Kochian, L.V. 1998. Characterization of Cadmium Binding, Uptake, and Translocation in Intact Seedlings of Bread and Durum Wheat Cultivars. *Plant Physiology* 116: 1413–1420.

Li, H.-F., McGrath, S.P., & Zhao, F.-J. 2008. Selenium uptake, translocation and speciation in wheat supplied with selenate or selenite. *The New Phytologist* 178(1): 92–102.

Ramos, S.J., Rutzke, M., Hayes, R.J., Faquin, V., Guilherme, L.R.G., & Li, L. 2010. Selenium accumulation in lettuce germplasm. *Planta* 233(4): 649–660.

Souza, G.A., De Carvalho, J.G., Rutzke, M., Albrecht, J.C., Guilherme, L.R.G., & Li, L. 2013. Evaluation of germplasm effect on Fe, Zn and Se content in wheat seedlings. *Plant Science*. doi:10.1016/j.plantsci.2013.05.015

Welch, R.M., & Graham, R.D. 2002. Breeding crops for enhanced micronutrient content, 205–214.

White, P.J., & Broadley, M.R. 2009. Biofortification of crops with seven mineral elements often lacking in human diets-iron, zinc, copper, calcium, magnesium, selenium and iodine. *The New Phytologist*, *182*(1), 49–84.

WHO. 2002. *The World Health Report: Reducing Risks, Promoting Healthy Life*, 230. Geneva: WHO.

Selenium in the Environment and Human Health – Bañuelos, Lin & Yin (Eds)
© 2014 Taylor & Francis Group, London, ISBN 978-1-138-00017-9

Evaluation of wheat tolerance to high selenium fertilization and its associated SNP markers in common wheat

An Wang, Tao Li, Fayu Sun, Guofang Lan & Shiliang Gu
Jiangsu Provincial Key Laboratory of Crop Genetics and Physiology, Key Laboratory of Plant Functional Genomics of Ministry of Education, Yangzhou University, Yangzhou, China

Guihua Bai
USDA-ARS, Hard Winter Wheat Genetics Research Unit, Manhattan, KS, USA

1 INTRODUCTION

Selenium (Se) is the beneficial nutrition element in animals and plant (Sun *et al.*, 2010). More than 40 human diseases have been proven to be associated with Se deficiency (Zeng *et al.*, 2012). Thus, appropriate intake of Se can be beneficial for populations living in Se-deficient regions (Zhang *et al.*, 2006). Wheat is one of the staple crops that have a strong capacity to accumulate Se in grains (Lyons *et al.*, 2005). However, wheat is also a Se-sensitive crop and the growth of wheat may be suppressed with excessive application of Se. Only those genotypes tolerant to Se fertilization may be ideal for a Se-biofortification strategy. The objectives of this paper were to investigate the genotypic differences in Se-tolerance in wheat, and to associate Single Nucleotide Polymorphism (SNP) markers with Se-tolerance, which have been seldom addressed.

2 MATERIALS AND METHODS

A set of 120 wheat cultivars (or inbred lines) from domestic and overseas were planted in plastic pots filled with 12 kg of soil, with four pots per cultivar and six plants per pot. Two pots (replicates) were for control (CK) and the other two pots for selenate application (Trt). A 120 mg of selenate in a volume of 1000 mL double deionized water (ddH$_2$O) was applied per pot (Trt) at the initial stage of stem elongation. Wheat tolerance to Se was evaluated (30 days after the Se treatment) by measuring five agronomic traits including; 1000-seed weight (after harvest), plant height (before harvest), chlorophyll measurements (SPAD values), number of productive tillers and number of wheat leaves per plant, and number of functional leaves. SNP markers were genotyped using Wheat_9000SNP_Infinium method in Kansas State University (USA). Association of SNP markers with traits and estimation of their explained phenotypic variation were performed using Mixed Linear Model (MLM) in TASSEL 3.0 pipeline (Bradbury *et al.*, 2007).

3 RESULTS AND DISCUSSION

3.1 *Variation of wheat tolerance to high Se-fertilization*

The plant height of all the cultivars was reduced by different degrees after Se-fertilization by an average decrease of 15.94 cM. For SPAD values, half of the plants decreased and the other half increased but there was no significant change between control and the treatment. Selenium application increased 1000-seed weight in 3/4 of the cultivars by 2.46 g on an average. The number of leaves per plant, the number of productive tillers and the number of functional wheat leaves in Se-fertilized samples were lower than those in their controls by 3.94, 0.83, and 3.5, respectively. All of the above observations showed that the dosage of selenate fertilizers used in this experiment was excessive to most of the cultivars, and the responses of wheat to excessive Se treatment was genotype-dependent. Only those cultivars that demonstrated increased or unchanged values or insignificant reduction in these traits after Se fertilization were considered to have tolerance to excessive Se-fertilization.

Since these traits gave an ambiguous description of response in some genotypes to evaluate their tolerance in a panoramic and unbiased view, Principal Component Analysis (PCA) was used to extract the information embedded in the six agronomic traits. The differences between the treatment and the control (i.e., treatment-control) were used as the variables for tolerance evaluation. The first three Principal Components (PC) contributed 0.49, 0.22 and 0.16 of phenotypic variation, respectively, with an accumulative contribution of 87% of observed variation in selenium tolerance. The number of productive tillers, the number of

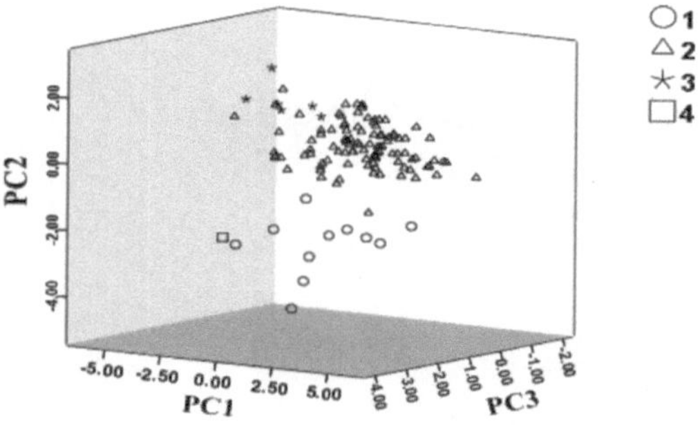

Figure 1. Cluster analysis based on three principal components (PC1, PC2, and PC3) in 120 wheat cultivars. Cultivars having similar Se tolerance were shown as the same symbols.

wheat leaves per plant and the number of functional leaves accounted for the 1st PC (PC1), and plant height and 1000-weight for the 2nd PC (PC2) and SPAD values for the 3rd PC (PC3). The cultivars were clustered into four groups based on the three PCs (Fig. 1). Group 1 contained 11 varieties including cvs. Corrine and Aurora, *etc.*, representing the high Se-sensitive type. Group 2 had 102 varieties including Ning7840, Anhui11, Haiyanzhong and Wheaton, *etc.*, representing the moderate tolerance to moderate sensitive group. Six cultivars were included in Group 3, representing the high tolerance catergory, and Group 4 contains only 1 variety, Chokwang, a Se-sensitive group.

3.2 *SNP markers associated with selenium tolerance and their allelic variation*

Thirty three SNP markers were significantly associated with PC1 ($P < 0.01$), and explained 6.2% (SNP5925) –10.5% (SNP421) of phenotypic variation, with an average of 7.3%. Up to 223 markers were associated with PC2, and explained 6.58% (SNP5843) –19.8% (SNP3527) of phenotypic variation, with an average contribution of 12.14%. Sixteen markers were associated with PC3, with an average contribution rate of 7.79%.

Among the associated SNP markers, the allele SNP5827_CT had the largest positive effect on Se tolerance in PC1, followed by the allele SNP8487_AG, and cv. Yangmai2 carried those alleles. In PC2, SNP585_CC and SNP5940_GG were the favorable alleles. Cv. Yangmai13 carried the favorable allele SNP8498_GG in PC3.

4 CONCLUSIONS

This study demonstrated that wheat tolerance to selenate addition at 120 mg/L were genotype-dependent. Those genotypes that were sensitive to Se were not good candidates for Se biofortification. The number of productive tillers, the number of wheat leaves per plant, and the number of functional leaves and plant height were important agronomic traits for evaluating wheat tolerance to Se fertilization. The favorable alleles at the SNP loci associated with Se tolerance identified in this study will be valuable in Se biofortification programs in high-Se regions or environments.

ACKNOWLEDGEMENT

This work was supported by grants from the Priority Academic Program Development of Jiangsu Higher Education Institutions, and by the Innovative Research Team of Universities in Jiangsu Province. We also thank Dr. Eduard Akhunov for SNP genotyping.

REFERENCES

Bradbury, P.J., Zhang, Z., Kroon, D.E., Casstevens, T.M., Ramdoss, Y. & Buckler, E.S. 2007. TASSEL: software for association mapping of complex traits in diverse samples. *Bioinformatics* 23: 2633–2635.

Lyons, G.H., Genc, Y., Stangoulis, J.C., Palmer, L.T. & Graham, R.D. 2005. Selenium distribution in wheat grain, and the effect of postharvest processing on wheat selenium content. *Biol Trace Elem Res* 103: 155–168.

Sun, G.X., Liu, X., Williams, P.N. & Zhu, Y.G. 2010. Distribution and translocation of selenium from soil to grain and its speciation in paddy rice (*Oryza sativa* L.). *Environmental Science & Technology* 44: 6706–6711.

Zeng, M.S., Li, X., Liu, Y., Zhao, H., Zhou, J.C., Li, K. *et al.* 2012. A high-selenium diet induces insulin resistance in gestating rats and their offspring. *Free Radic. Biol. Med.* 52: 1335–1342.

Zhang, L.H., Shi, W.M. & Wang, X.C. 2006. Difference in selenium accumulation in shoots of two rice cultivars. *Pedosphere* 16: 646–653.

Selenium in the Environment and Human Health – Bañuelos, Lin & Yin (Eds)
© 2014 Taylor & Francis Group, London, ISBN 978-1-138-00017-9

Determination of total selenium and identification of selenium compounds using ICP-MS in Australian wheat flours

W. Tjong, P.D. Morrison & D.M. Small
RMIT University, Melbourne, Victoria, Australia

1 INTRODUCTION

Recognition of selenium (Se) as an essential trace element has attracted increased interest. It was initially feared for its toxicity that had caused the death of thousands of sheep and cattle. Selenium is now recognized as an essential trace element playing a role in immune function, viral suppression, delaying the aging process and further confirmed as a chemopreventative agent (Pedrero & Madrid, 2009). Diet is the major source of Se intake for human and recent studies have indicated that the primary food sources of Se are cereals, meat and fish, while dairy products and eggs contribute a small proportion of total intakes in most countries (Navarro-Alarcon & Cabrera-Vique, 2008). Selenium content in cereal foods depends on the amount of Se available in soils where the plants were grown, and therefore this can vary geographically. Crops grown on high Se soil will yield grains with greater Se concentration.

The current recommended dietary intake for Se in Australia is 70 and 60 µg/day for men and women respectively (NHMRC, 2006). There is increasing evidence that, globally, many populations do not currently have sufficient Se in their diet to support maximal expression of selenoproteins. Furthermore, current recommended values for intake of this micronutrient do not take into account that Se is present in food systems in various chemical forms, including both organic and inorganic species. The essential and toxic nature of Se depends on both the concentration and chemical forms in which this element is present in the sample (Thiry *et al.*, 2012).

In this context, the aim of the current work has been to evaluate, adapt and validate procedures to determine the total Se content and the primary species of Se present in commercial wheat flours.

2 MATERIALS AND METHODS

2.1 *Instrumentation*

Digestion of solid samples was carried out by microwave digestion (Mars X, CEM Corp, USA) with closed vessels (Omni™ Teflon). Enzymatic hydrolysis was carried out at 37°C in a digital control immersion thermostat (Ratek TH6C, Ratek Instruments, Australia) and extracts were centrifuged (BOECO C-28 A, Hamburg, Germany).

Speciation was performed by HPLC-ICP-MS using an Agilent 1100 HPLC and a Hamilton PRP-X100 column (250 mm × 4.1 mm × 10 µm) (Hamilton, USA). An Agilent model 7700x ICP-MS was used as detector with the sample introduction system consisting of a MicroMist nebuliser with a Scott double-pass quartz spray chamber cooled to 2°C. Plasma conditions and acquisition parameters are presented in Table 1.

2.2 *Reagents and standards*

All reagents were of at least analytical grade and ultrapure water (18.2MΩ cm, Merck Millipore, MA, USA) was used throughout the work. Sodium selenate, sodium selenite, Se-(methyl) selenocyteine, selenomethionine, selenocystine, protease (type XIV), lipase (type VII), Trizma® hydrochloride, ammonium citrate dibasic, nitric acid 70% purified by redistillation, hydrogen peroxide ≥30% TraceSELECT® and methanol TraceSELECT® were obtained from Sigma-Aldrich Pty. Ltd. (NSW, Australia). Selenium ICP standard was obtained

Table 1. ICP-MS operating conditions and data acquisition parameters.

Plasma parameters	
RF power (W)	1550
Plasma gas flow rate (L/min)	15
Auxiliary gas flow rate (L/min)	1
Carrier gas (L/min)	1.03
Data acquisition parameters (total Se)	
Monitored isotope	78
Points per peak	3
Acquisition time per point (s)	0.99
Replicates	3
Sweeps/replicate	40
Data acquisition parameters (Se speciation)	
Monitored isotope	78
Integration time/mass (s)	0.75

from AccuStandard (USA) and wheat flour reference material (NIST 1567a, USA) was used for total Se analysis by ICP-MS. Se-working standards were prepared fresh prior to use. Cellulose acetate syringe filters (Labquip Technologies, Australia) with a pore size of 0.22 μm were used for extract filtration.

2.3 *Samples*

Commercial wheat bread pre-mix and biofortified pre-mix were purchased from local supermarkets in Melbourne. These were ground to smaller particle size and homogenized prior to storage in air-tight containers.

2.4 *Quantification of total selenium*

Representative samples (0.1 g) were weighed into Teflon lined heavy duty vessels after which 1 mL nitric acid, 1 mL of hydrogen peroxide and 8 mL of ultrapure water were added. Upon sealing the vessels, samples were digested in a microwave using a preset program as detailed in Table 2. Subsequently, extracts were made up to 25 mL using ultrapure water with 1% methanol, filtered and analysed by ICP-MS.

2.5 *Quantification of selenium species*

Enzymatic hydrolysis was conducted by adding 5mL of 30 mM TRIS-HCl buffer solution (pH 7.5) containing 40 mg of protease and 20 mg of lipase to 0.4 g of sample. After 18 h of incubation at 37°C, samples were centrifuged at 2100 *g* for 20 minutes. The supernatants were filtered and analyzed by HPLC coupled to ICP-MS. The mobile phase was 5 mM ammonium citrate in 2% methanol with the chromatographic conditions as shown in Table 3.

Table 2. Microwave settings used in digestion of wheat flour for total Se analysis.

Stage	1	2	3
Power (W)	1200	1200	1200
Temperature (°C)	100	150	175
Heating time (min)	5	5	3
Holding time (min)	5	5	15

Table 3. Chromatographic conditions for Se speciation by anion exchange HPLC.

Operating conditions	
Flow rate (mL/min)	1
Injection volume (μL)	100
pH of mobile phase	5.58
Column temperature (°C)	25

3 RESULTS AND DISCUSSION

3.1 *Total Se in wheat flours*

In the current study, validation experiments were carried out to confirm the adequacy of microwave digestion and ICP-MS in determining total Se in representative food samples. A Standard Reference Material (SRM) wheat flour from NIST was used to achieve this and the results confirmed the certified value (Table 4).

3.2 *Selenium speciation by HPLC-ICP-MS*

The chromatographic conditions as shown in Table 3 are similar to those applied previously in the study described by Huerta *et al.* (2003) for Se speciation in wheat flour. These were evaluated and Figure 1 shows the chromatogram obtained by HPLC-ICP-MS for the mixture of standard compounds of the selenium species. It is evident that a complete separation and elution of all five Se compounds can be achieved within 10 minutes.

The typical chromatogram for the bread pre-mix wheat flour Figure 2(a) showed a major peak which was identified as SeMet based upon the retention time. This is also the predominant species in the extracts of biofortified pre-mix (b). In the latter sample, a smaller peak was also detected with a retention time corresponding to that of the $SeCys_2$ standard. This indicates the presence of detectable levels of another Se compound in addition to SeMet in wheat flours that have been biofortified.

Table 4. Total Se in wheat flour.

Sample	Certified value (μg/g)	Content found (μg/g)
SRM	1.1 ± 0.2	1.13 ± 0.02
Bread pre-mix		0.075 ± 0.007
Biofortified pre-mix		1.66 ± 0.06

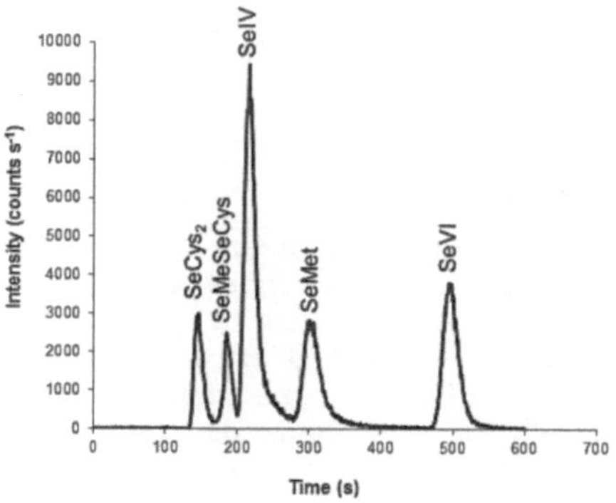

Figure 1. Chromatogram (counts $^{78}Se^{s-1}$) of a mixture of standards: selenocystine, Se-(methyl)selenocysteine, selenite, selenomethionine and selenate.

(a)

(b)

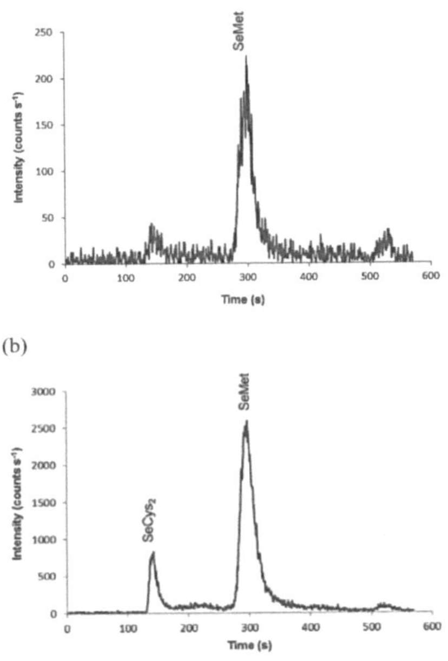

Figure 2. Chromatograms (counts $^{78}Se^{s-1}$) of (a) bread pre-mix and (b) biofortified pre-mix.

Table 5. Concentrations (µg/g) of different Se chemical species in wheat flour extracts.

Se species	Bread pre-mix	Biofortified pre-mix
$SeCys_2$†	0.001 ± 0.003	0.19 ± 0.02
SeMeSeCys†	<MDL	0.021 ± 0.001
Selenite*	<MDL	0.021 ± 0.002
SeMet*	0.089 ± 0.003	1.10 ± 0.03
Selenate*	<MDL	<MDL
Sum of species*	0.090	1.33
Recovery (%)	<MDL	80

MDL: Method Detection Limit.

3.3 *Evaluation of Se extraction recoveries*

Enzymatic extraction procedures are often the method of choice to extract Se species from food matrices with minimum risk of species transformation, while allowing high extraction efficiencies (Thiry *et al.*, 2012). The predominant extractable Se species in the wheat flours was SeMet (Table 5), which confirms the results from other studies (Hart *et al.*, 2011; Huerta *et al.*, 2003; Warburton & George-Infante, 2007).

In the current study, total Se was determined for the samples in order to evaluate the yield of enzymatic hydrolysis. The results show that at least 80% recovery has been achieved with the extraction procedures using protease and lipase.

4 CONCLUSIONS

The current results demonstrate the validity of microwave digestion for the analysis of total Se content of formulated pre-mixes based upon Australian wheat flours. The biofortified material contained a level of total Se more than twenty times higher than the normal pre-mix. The separation of the Se species could be achieved using the HPLC-ICP-MS system in less than ten minutes and when applied to the analysis of enzymatic hydrolysates of the pre-mix samples gave relatively repeatable results. Further work is continuing into the hydrolysis of the samples in relation to the recoveries of total Se of the samples. In addition, the procedures will form the basis of further studies of the effect of processing on the stability of the various species of selected cereal grain foods.

REFERENCES

Hart, D.J., Fairweather-Tait, S.J., Broadley, M.R. Dickinson, S.J., Foot, I., Knott, P. *et al.* 2011. Selenium concentration and speciation in biofortified flour and bread: Retention of selenium during grain biofortification, processing and production of Se-enriched food. *Food Chemistry* 126(4): 1771–1778.

Huerta, V.D., Reyes, L.H., Marchante-Gayon, J.M., Sanchez, M.L.F. & Sanz-Medel, A. 2003. Total determination and quantitative speciation analysis of selenium in yeast and wheat flour by isotope dilution analysis ICP-MS. *Journal of Analytical Atomic Spectrometry* 18(10): 1243–1247.

Navarro-Alarcon, M. & Cabrera-Vique, C. 2008. Selenium in food and the human body: A review. *Science of the Total Environment* 400(1): 115–141.

NHMRC. 2006. Nutrient reference values for Australia and New Zealand. Available on line: http://www.nhmrc.gov.au/guidelines/publications/n35-n36-n37.

Pedrero, Z. & Madrid, Y. 2009. Novel approaches for selenium speciation in foodstuffs and biological specimens: A review. *Analytica Chimica Acta* 634(2): 135–152.

Thiry, C., Ruttens, A., De Temmerman, L., Schneider, Y-J. & Pussemier, L. 2012. Current knowledge in species-related bioavailability of selenium in food. *Food Chemistry* 130(4): 767–784.

Warburton, E. & Goenaga-Infante, H. 2007. Methane mixed plasma—improved sensitivity of inductively coupled plasma mass spectrometry detection for selenium speciation analysis of wheat-based food. *Journal of Analytical Atomic Spectrometry* 22(4): 370–376.

Selenium in the Environment and Human Health – Bañuelos, Lin & Yin (Eds)
© 2014 Taylor & Francis Group, London, ISBN 978-1-138-00017-9

The effects of selenium on yield and active ingredients of medicinal Chrysanthemum flower at different growth stages

X.Q. Yao, C. Si & J.Z. Chu
College of Life Sciences, Hebei University, Baoding, China

1 INTRODUCTION

Selenium (Se) is an essential trace element for many organisms (Fairweather-Tait *et al.*, 2011; Chilimba *et al.*, 2012). Se deficiency in soils is overall distributed around the world (Cartes *et al.*, 2011). Clinical medical trials have shown that about 40 kinds of diseases threatening human health are related to Se deficiency in human body (Schrauzer and Meginness, 1978). Thus, it is very important for improving Se deficiency status in human by increasing Se content in diets. Researches have identified that appropriated Se application could not only increase Se content in plants but also promote growth and development of plants (Hu *et al.*, 2010; Chu *et al.*, 2013). However, there are few studies on medicinal plants and Se.

Medicinal chrysanthemum (*Chrysanthemum morifolium Ramat*) flowers are commonly used for tea and traditional Chinese medicine in China. The present investigation studied primarily the effects of Se application time on yield and active ingredients of flowers, and provided a basis for rational application of Se fertilizer and production of Se-rich chrysanthemum tea.

2 MATERIAL AND METHODS

2.1 *Plant material and experimental design*

The research was conducted with pot experiments in Hebei University, Baoding, China. Huai *chrysanthemum morifolium* Ramat seedlings originated from Jiaozuo Chinese herbal medicine planting base, Henan province, China. One seedling was transplanted into each of the plastic pots. The substrate was sieved topsoil from farmland. N (0.4 g/kg), P (0.2 g P_2O_5/kg) and K (0.3 g K_2O/kg) were added to the soil as base fertilizers before transplanting.

Se was supplied as sodium selenite (Na_2SeO_3). The Se solutions (30 mg/L) were sprayed at seedling, vigorous growth, flower bud differentiation, bud and flower stages of chrysanthemum. Spraying volume per pot was 100 mL. Each treatment had four replications.

2.2 *Yield and Se content of flowers*

Fresh flowers were harvested and dried to a constant mass. Flower yield was expressed as g Dry Weight (DW) per plant. Concentrations of Se were determined by Atomic Flu-orescence Spectrometry (AFS-830a, Beijing Jitian, Beijing, China) and was expressed as mg per kg.

2.3 *Penylalanine ammonia lyase enzyme*

Penylalanine Ammonia Lyase (PAL, EC 4.3.1.5) enzyme activity was measured according to Liu and Li (2007). The PAL activity was expressed as μmol t-cinnamic acid/h/g FW.

2.4 *Total flavone and chlorogenic acid content*

Total flavone content was determined according to He and Liu (2007) with minor modifications. The flavone content was expressed as g rutin 100/g DW. Chlorogenic acid content was measured according to Zhang *et al.* (2006). The results were expressed as% DW.

3 RESULTS AND DISCUSSION

3.1 *The effects of Se on flower dry weight and Se content*

Se application significantly increased flower dry weight at vigorous growth, flower bud differentiation and bud stages (Fig. 1). However, Se supplies at seedling and flower stages did not affect flower biomass accumulation, indicating that the effects of Se on flower biomass depended on application stage. The results were similar to Li *et al.* (2010) who reported that Se was beneficial for biomass accumulation of chrysanthemum flowers.

Se application significantly increased Se content in flowers at different growth stages (Fig. 1). This result is consistent with the results of previous studies (Li *et al.*, 2010; Tian *et al.*, 2008).

3.2 *The effects of Se on PAL activity in flowers*

PAL is key enzyme of secondary metabolites that catalyzes the deamination of L-phenylalanine to

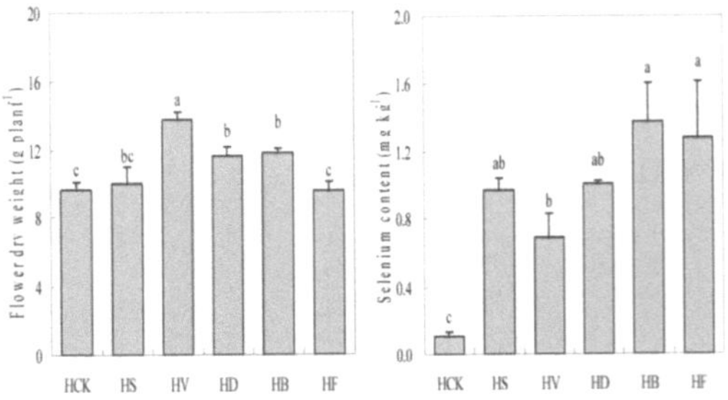

Figure 1. Effects of Se application on dry weight and Se content of flower at different development stages. The bars with different letters are significantly different from each other ($P < 0.05$). Values are means of four replicates ± SE. HCK, the control; HS, Se supplied at seedling stage; HV, Se supplied at vigorous growth stage; HD, Se supplied at flower bud differentiation stage; HB, Se supplied at bud stage; HF, Se supplied at flower stage.

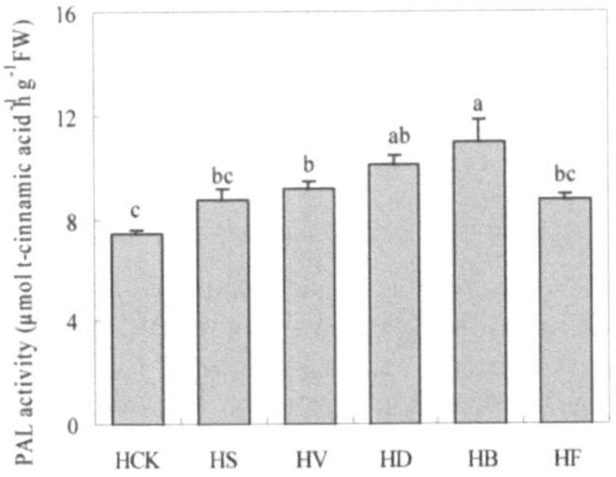

Figure 2. Effects of Se application on PAL activity in flower at different development stages.

trans-cinnamic acid. The carbon skeleton of cinnamic acid serves as a major substrate for many pathways of phenolic biosynthesis in plants (Zucker, 1969). In this study, applied Se significantly increased PAL activity at vigorous growth, flower bud differentiation and bud stages, indicating that Se affected the processes of secondary metabolism at the three growth stages.

3.3 *The effects of Se on flavone and chlorogenic acid content in flower*

Flavone and chlorogenic acid are the main active ingredients in herb plants. Li *et al.* (2010) and Tian *et al.* (2008) reported that the appropriate amount of Se increased total flavone content in flower of chrysanthemum and tartary buckwheat, respectively. However, our results indicated that the application of Se did not significantly affect flavone content in flowers (Fig. 3) but it did significantly increase chlorogenic acid at flower bud differentiation and bud stages.

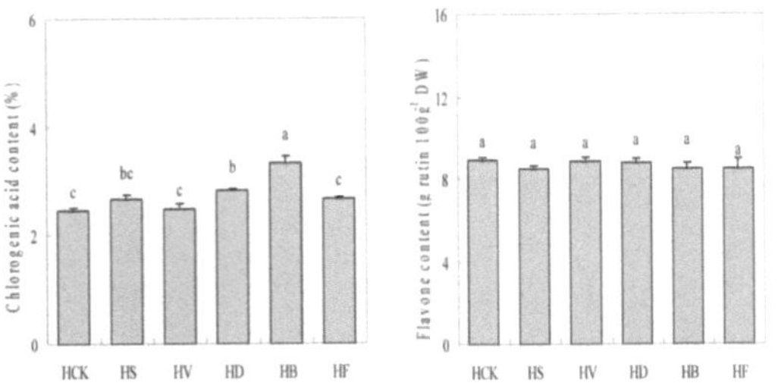

Figure 3. Effects of Se application on flavone and chlorogenic acid content in flower at different development stages.

4 CONCLUSION

At vigorous growth, flower bud differentiation and bud stages, the application of Se significantly increased flower biomass, Se content and PAL activity in flowers. However, Se treatment at seedling and flower stages only increased Se content in flowers. In addition, Se treatments at flower bud differentiation and bud stages also significantly increased chlorogenic acid content in flowers. The results indicated that effects of Se on flower biomass accumulation and active ingredients in flowers depend on the time of Se application.

REFERENCES

Cartes, P., Gianfreda, L., Paredes, C. & Mora, M.L. 2011. Selenium uptake and its antioxidant role in ryegrass cultivars as affected by selenite seed pelletization. *Journal of Soil Science Plant Nutrition* 11: 1–14.

Chilimba, A.D.C., Young, S., Black, C.R., Meacham, M.C., Lammel, J. & Broadley M.R. 2012. Agronomic biofortification of maize with selenium (Se) in Malawi. *Field Crops Research* 125: 118–128.

Chinese Pharmacopoeia Commission. 1985. *Chinese Pharmacopoeia* (I). Beijing: Chemical Industry Press.

Chu, J., Yao, X., Yue, Z., Li, J. & Zhao, J. 2013. The effects of selenium on physiological traits, grain selenium content and yield of winter wheat at different development stages. *Biological Trace Element Research* 151: 434–440.

Fairweather-Tait, S.J., Bao, Y., Broadley, M.R., Collings, R., Ford, D., Hesketh, J.E. *et al.* 2011. Selenium in human health and disease. *Antioxidant & Redox Signaling* 14: 1337–1383.

He, S.M. & Liu, J.L. 2007. Study on the Determination Method of Flavone Content in Tea. *Chinese Journal of Analytical Chemistry* 35: 1365–1368.

Hu, H.F., Hu, C.X., Jie, X.L., Liu, S.L., Guo, X., Hua, D.L. et al. 2010. Effects of selenium on herbage yield, selenium nutrition and quality of alfalfa. *Journal of Food Agriculture & Environment* 8: 792–795.

Li, Y.M., Sun, Y.X. & Li, D.H. 2010. Effects of applying selenium on contents of total flavonoid, chlorogenic acid and selenium in flower of *Chrysanthemum morifolium. Soil* 42: 618–623 (in Chinese).

Liu, P. & Li, M.J. 2007. *Experimental Technology in Plant Physiology.* Beijing, China: Science Press.
Schrauzer, G.N. & Meginness, J.E. 1978. Observations on human selenium supplementation. *Trace Substances and Environmental Health* 13: 64–82.
Tian, X.Y. & Wang, Z.Y. 2008. Effects of selenium application on content, distribution and accumulation of selenium, flavonoids and rutin in tartary buckwheat. *Plant Nutrition and Fertilizer Science* 14: 721–727 (in Chinese).
Zucker, M. 1969. Induction of phenylalanine ammonia-lyase in xanthium leaf disks. Photosynthetic requirement and effect of day length. *Plant Physiology* 44: 912–922.

Selenium in the Environment and Human Health – Bañuelos, Lin & Yin (Eds)
© 2014 Taylor & Francis Group, London, ISBN 978-1-138-00017-9

Protective effect of selenium in oilseed rape plants subjected to cadmium and lead exposure

Zhilin Wu & Miao Li
Key Laboratory of Agri-Food Safety of Anhui Province, School of Plant Protection, Anhui Agriculture University, Hefei, Anhui, China

Linxi Yuan
Jiangsu Bio-Engineering Research Centre on Selenium, Suzhou, Jiangsu, China

Xuebin Yin & Ying Liu
Advanced Lab for Selenium and Human Health, Suzhou Institute for Advanced Study, University of Science and Technology of China, Suzhou, Jiangsu, China

1 INTRODUCTION

The presence of cadmium (Cd) and lead (Pb) in the metal-contaminated soils is toxic for the crops causing their impaired growth and development. Cadmium is emitted from natural and anthropogenic sources, which can induce stress symptoms, including the appearance of chlorosis and growth reduction (Pedrero *et al.*, 2008). Lead is one of the most common and dangerous contaminant whose accumulation in plant tissues results in numerous disturbances of physiological processes (Mroczek-Zdyrska & Wójcik, 2012). Selenium has been shown to counteract various abiotic stresses induced in plants such as heavy metals (Feng *et al.*, 2013). However, the associated mechanisms are rather complicated and still remain to be fully elucidated. The experiments were performed on rape (*Brassica napus* L.), which is very widely cultivated throughout the world for the production of animal feed and vegetable oil for human consumption (Filek *et al.*, 2008, 2010). In the present study, the protective effect of Se against Cd and Pb in oilseed rape plants was investigated. The results could provide evidence that Se supplementation helps plants to minimize heavy metal toxicity.

2 MATERIALS AND METHODS

Oilseed rape (*Brassica napus* L.) was cultivated in certain soil culture from February 1 to April 2, 2012 in the greenhouse (25 ± 5°C). Selenium was added to the soil as sodium selenite, Cd as cadmium chloride, and Pb as acetate. These plants were raised under normal fertilization management in the presence of (a) Se alone (0, 1, 5, 10, 15, 20 mg/kg), (b) Cd (5 mg/kg) combining with Se (0, 1, 5, 10, 15, 20 mg/kg) correspondingly, (c) Pb (500 mg/kg) combining with Se (0, 1, 5, 10, 15, 20 mg/kg) correspondingly. After 60 days, concentrations of Se, Cd, and Pb in plant shoot and root samples were detected by atomic fluorescence spectrometer (AFS-9230) and atomic absorption spectroscopy (AA-700) with graphite furnace. The activity of glutathione peroxidase (GPx) in fresh plant tissues were measured by following the product kit instruction (Nanjing Jiancheng Biological Engineering Institute, Nanjing, China).

3 RESULTS AND DISCUSSION

Selenium could significantly diminish the Cd and Pb contents in the oilseed rape (Fig. 1). The optimal Se dosage against Cd and Pb was 15 mg/kg. Cd content in the upper shoot was reduced regularly with the increasing of Se (<15 mg/kg). It could reduce Cd in the plant to 46.7% at most. There were no significant differences between lead treatment group whether Se was added as 1 mg/kg or not. Se could reduce Pb in the plant to 33.9% at most. But the suppression was not so well as if that Se concentration was higher than 15 mg/kg no matter for Cd and Pb.

The concentrations of Cd and Pb in the roots significantly decreased with increasing the Se treatment level, suggesting that Se inhibited Cd and Pb accumulation in roots and further reduced heavy metal toxic effects in shoots (Fig. 2). Compared with the control, at the Se treatment of 15 mg/kg, the root Cd concentration was reduced by 84.3%, while the root Pb concentration was reduced by 66.3% at the Se treatment of 20 mg/kg. The reduction of heavy metal concentration in roots was much greater than in shoots.

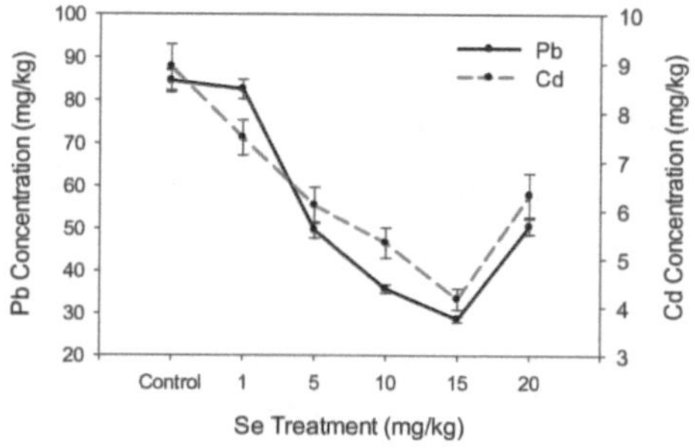

Figure 1. Concentrations of Pb and Cd in shoots of oilseed rape treated with soil selenium.

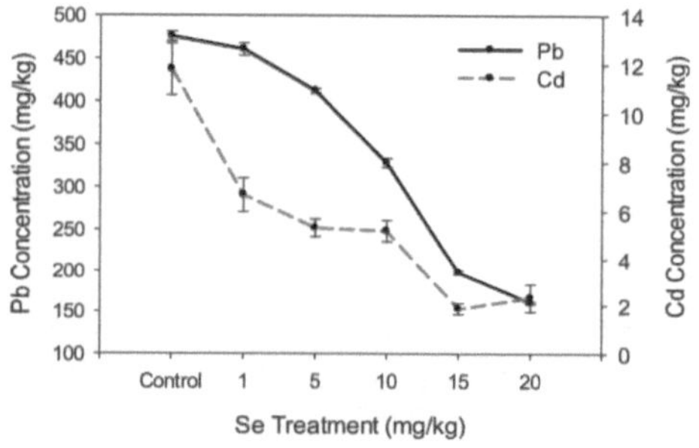

Figure 2. Concentrations of Pb and Cd in roots of oilseed rape treated with soil selenium.

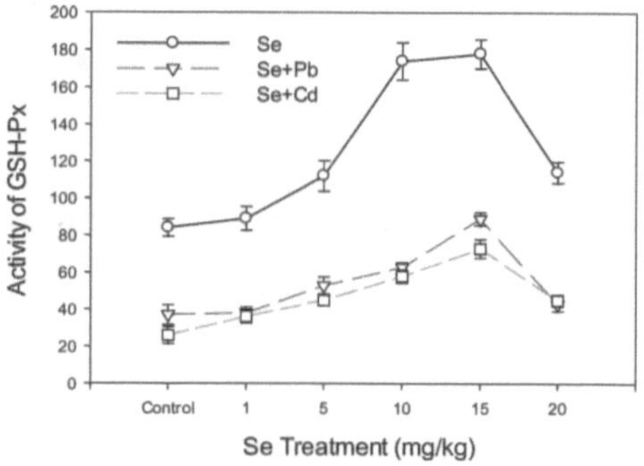

Figure 3. The activity of Glutathione peroxidase (GSH-Px) in shoots of oilseed rape treated with soil Se, Se+Cd, or Se+Pb.

The increased concentration of Se in media from 1 to 20 mg/kg was associated with a significant increase in the activity of GSH-Px in the upper shoot (Fig. 3). Selenium achieved its best effect when added at 15 mg/kg, but its efficacy decreased at concentrations greater than 15 mg/kg. The same result was observed in the plant with Pb or Cd. Selenium addition reduced Cd and Pb negative effects on GSH-Px activity at a concentration of 15 mg/kg. Selenium-induced increases in GSH-Px activity added Cd and Pb was about 64.4% and 56%, respectively, in the shoot. The GSH-Px activity consistently increased with the addition of Se at 15 mg/kg. However, higher Se concentrations (>15 mg/kg) reduced GSH-Px activity in both Pb and Cd-treated plants.

4 CONCLUSIONS

This preliminary study demonstrated that a Se concentration of 15 mg/kg had a protective effect in oilseed rape plants subjected to Cd and Pb exposure. The effect was related to a decreased lipid peroxidation and enhanced glutathione peroxidase (GPX) activity in Se-treated plants. Selenium could enhance the activity of GSH-Px and reduce the toxicity of heavy metals in the plants. The concentration of Pb and Cd was clearly depressed in the roots than that in the upper parts, which may indicate that Se exerted an antagonistic effect in the roots. Based on previous studies, we hypothesized that there was a Se-Pb or Se-Cd macromolecular compound forming in the plants, and consequently they were expelled from plant body by metabolic process. Generally speaking, two possible mechanisms for the action of Se were considered: (1) removal of heavy metals from metabolically active cellular sites, and (2) reduction of oxygen radicals.

ACKNOWLEDGEMENTS

This work was partially supported by the Natural Science Youth Foundation of Jiangsu Province for Youth, China (BK2012195, BK2012202), Jiangsu Research-Industry Joint Innovation Project (BY2011186).

REFERENCES

Feng, R.W., Wei, C.Y. & Tu, S.X. 2013. The roles of selenium in protecting plants against abiotic stresses. *Environ. Exper. Bot.* 87: 58–68.

Filek, M., Keskinen, R., Hartikainen, H., Szarejko, I., Janiak, A., Miszalski, Z. *et al.* 2008. The protective role of selenium in rape seedlings subjected to cadmium stress. *J. Plant Physiol.* 165: 833–844.

Filek, M., Kościelniak, J., Łabanowska, M., Bednarska, E. & Bidzińska, E. 2010. Selenium-induced protection of photosynthesis activity in rape (*Brassica napus*) seedlings subjected to cadmium stress: Fluorescence and EPR measurements. *Photosynth Res.* 105: 27–37.

Mroczek-Zdyrska, M. & Wójcik, M. 2012. The influence of selenium on root growth and oxidative stress induced by lead in *Vicia faba* L. minor plants. *Biol. Trace Elem. Res.* 147: 320–328.

Pedrero, Z., Madrid, Y., Hartikainen, H. & Cámara, C. 2008. Protective effect of selenium in Broccoli (*Brassica oleracea*) plants subjected to cadmium exposure. *J. Agri. Food Chem.* 56: 266–271.

Selenium in the Environment and Human Health – Bañuelos, Lin & Yin (Eds)
 ISBN 978-1-138-00017-9

Selenium-resistant microorganisms from rhizosphere of novel selenium hyperaccumulator *Cardamine hupingshanesis* (Brassicaceae) in Enshi, China

Xinzhao Tong
School of Nanoscience and Advanced Lab for Selenium and Human Health, University of Science and Technology of China, Suzhou, Jiangsu, China

Lei Luo
Jiangsu Bio-Engineering Research Centre on Selenium, Suzhou, Jiangsu, China

Linxi Yuan
Advanced Lab for Selenium and Human Health, University of Science and Technology of China, Suzhou, Jiangsu, China

Xuebin Yin
Suzhou Setek Co., Ltd., Suzhou, Jiangsu, China

1 INTRODUCTION

Cardamine hupingshanesis (Brassicaceae) from the Se-mine drainage area in Enshi, China was identified as a new Se secondary accumulator, and it could accumulate up to 99% of total Se as $SeCys_2$ without showing phytotoxic symptoms (Yuan *et al.*, 2013). Previous studies revealed that rhizospheric microorganisms likely played an important role in the Se hyperaccumulation for Se hyperaccumulating plants (Garbisu *et al.*, 1996; Kessi *et al.*, 1999; Klonowska *et al.*, 2005). In the present study, Se-resistant bacteria were isolated from the rhizosphere of *C. hupingshanesis* (Brassicaceae), and their metabolic processes on selenite were studied to understand the Se-hyperaccumulation mechanism for *C. hupingshanesis.*

2 MATERIALS AND METHODS

According to Yuan *et al.* (2013), plants with the highest Se contents in roots were uprooted *in situ*, and the roots were rinsed in DI water after removing visible soil particles. The obtained water filtered via 1 µm filter was cultured in TSA medium with 200 µg Se /L. Five Se-tolerant strains were isolated and they were identified by 16SrRNA analysis as *Microbacterium oxydans, Bacillus aerophilus, Bacillus thuringiensis, Brevibacterium frigoritolerans,* and *Wauterisiella falsenii*. The first strain, *Microbacterium oxydans*, was selected to culture in TSB medium with 1.5 mg Se/L as selenite for 48 h at 30°C. The cultured medium was centrifuged with 5000 rpm for 30 min. The depositions and the supernatant were separately collected for total Se and Se speciation analysis by AFS and LC-UV-AFS (Titan Co, Beijing), respectively. The supernatants were passed through a 0.22 µm filter before analysis. The reducing sugar content in the culture medium was determined by UV (Shimadzu Co, Japan).

3 RESULTS & DISCUSSION

During 48 h culture period, the variations of reducing sugar concentrations in culture medium and total Se content in medium supernatant and bacterial deposit were shown in Figure 1.

For reducing sugar, the concentrations were approximately 2.2 mg/mL during first 4 h,

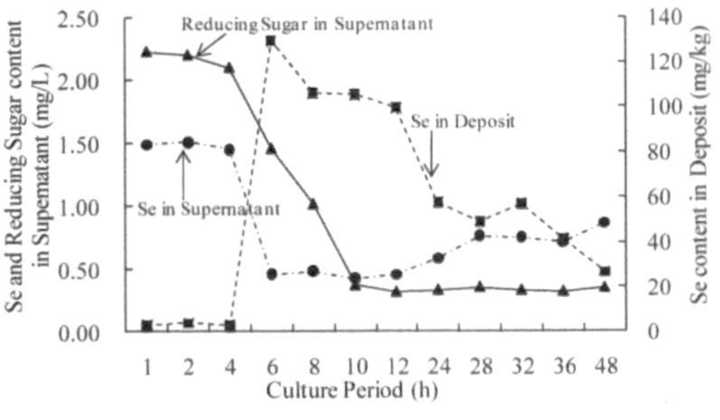

Figure 1. Variations of reducing sugar content in medium and total Se content in supernatant and bacterial deposit during culture period (48 h).

Table 1. Concentrations of Se^{4+}, SeCys2, SeMeCys and SeMet in supernatants and bacterial deposits.

Time	Se in supernatant (mg/L)				Se in deposit (mg/kg)			
	Se(IV)	$Secys_2$	SeMeCys	SeMet	Se(IV)	$Secys_2$	SeMeCys	SeMet
1 h	1.54	<DL†	<DL	<DL	–‡	–	–	–
2 h	1.57	<DL	<DL	0.02	–	–	–	–
4 h	1.52	<DL	<DL	<DL	<DL	<DL	1.77	<DL
6 h	0.03	<DL	<DL	<DL	0.37	14.09	<DL	<DL
8 h	0.08	<DL	<DL	<DL	<DL	22.20	<DL	<DL
10 h	0.06	<DL	<DL	<DL	<DL	12.26	3.54	<DL
12 h	0.07	<DL	<DL	<DL	<DL	22.05	<DL	<DL
24 h	0.07	<DL	<DL	<DL	0.19	7.06	<DL	<DL
28 h	0.11	<DL	<DL	<DL	0.10	12.19	<DL	<DL
32 h	0.08	<DL	<DL	<DL	0.20	11.92	<DL	<DL
36 h	0.07	<DL	<DL	<DL	0.23	7.42	<DL	<DL
48 h	0.07	<DL	<DL	<DL	<DL	2.64	<DL	<DL

†: below the detection limit; ‡: no sample.

decreased to 0.50 mg/mL at 10 h, and remained relatively stable thereafter. The total Se contents in medium supernatants decreased from almost 1.5 mg/L during first 4 h to 0.5 mg/L at 6 h, and then gradually increased to about 1 mg/L at 48 h. In contrast, the deposit contained no Se during first 4 h, then sharply increased to 130 mg Se/kg, and gradually decreased to 25 mg/kg at 48 h. Based on Figure 1, the selected Se-resistant rhizospheric bacterium, *Microbacterium oxydans,* needs 6 h to achieve logarithmic phase and its growth cycle is 24–28 h.

As shown in Table 1, Se(IV) was the predominant Se species in medium supernatants during the 48 h culture period. However, an abundance of $SeCys_2$ was detected in bacterial deposit, accounting for more than 80% of total Se. This observation indicates that indicating the selected Se-resistant rhizospheric bacterium, *Microbacterium oxydans,* can transform Se(IV) into SeCys2.

ACKNOWLEDGEMENTS

This study was financially supported by Natural Science Youth Foundation of Jiangsu Province for Youth, China (BK 2012202, BK2012195), and Jiangsu Research-Industry Joint Innovation Project (BY2011186).

REFERENCES

Garbisu, C., Ishii, T., Leighton, T. & Buchanan, B.B. 1996. Bacterial reduction of selenite to elemental selenium. *Chemical Geology* 132:199–204.

Kessi, J., Ramuz, M., Wehrli, E., Spycher, M. & Bachofen, R. 1999. Reduction of selenite and detoxification of elemental selenium by the phototrophic bacterium *Rhodospirillum rubrum. Applied and Environmental Microbiology* 65(11): 4734–4740.

Klonowska, A., Heulin, T. & Vermeglio, A. 2005. Selenite and tellurite reduction by *Shewanella oneidensis. Applied and Environmental Microbiology* 71: 5607–5609.

Shamberger, R.J. 1983. *The Biochemistry of Selenium* New York: Plenum Press.

Van Fleet-Stalder, V., Chasteen, T.G.., Pickering, I.J., George, G.N. & Prince, R.C. 2000. Fate of selenate and selenite metabolized by *Rhodobacter sphaeroides. Applied Environmental Microbiology* 66: 4849–4853.

Vallini, G., Di Gregorio, S. & Lampis, S. 2011. Rhizosphere—induced selenium precipitation for possible applications in phytoremediation of Se-polluted effluents. In G.S. Banuelos, Z.-Q. Lin & X.B. Yin (eds), *Selenium: Deficiency, Toxicity, and Biofortification for Human Health.* Hefei: University of Science and Technology of China Press.

Yuan, L.X., Zhu, Y.Y., Lin, Z.-Q., Banuelos, G., Li, W. & Yin, X. 2013. A novel selenocyctine-accumulating plant in selenium-mine drainage areas in Enshi, China. *PLOS One*, PONE-D-13-02208.

4 *Selenium biofortification*

Selenium in the Environment and Human Health – Bañuelos, Lin & Yin (Eds)
© 2014 Taylor & Francis Group, London, ISBN 978-1-138-00017-9

Effects of nationwide addition of selenium to fertilizers on foods and animal and human health in Finland

G. Alfthan
Department of Chronic Disease Prevention, National Institute for Health and Welfare, Helsinki, Finland

1 INTRODUCTION

Selenium (Se) is unevenly distributed in soils worldwide. For climatic and geochemical reasons, Finland is a low Se region. Signs of human selenium deficiency have not been reported from Finland, but epidemiological studies had suggested an increased risk for cardiovascular diseases and cancer at very low Se intakes such as seen in Finland (Salonen *et al.*, 1982, 1984). Although livestock had been treated with selenite since 1969, Se deficiency diseases were still relatively common. To improve the quality of Finnish foods and animal health and to increase the Se intake of the population, an official decision was made in 1984 to supplement compound fertilizers with selenium in the chemical form of sodium selenite (Na_2SeO_4). Practically all fertilizers used in Finland have contained selenium since 1985. Currently all crop fertilizers contain 15 mg Se/kg (Eurola *et al.*, 2011). The effects of the supplementation of Se in commercial fertilizers on soils, feeds, basic foodstuffs, dietary Se intake, human tissues and animal and human health will be discussed.

2 MATERIALS AND METHODS

In a national monitoring program, sampling of cereals, basic foodstuffs, feeds, fertilizers, soils and human tissues has been carried out annually since 1985. Four research organizations, including the National Institute for Health and Welfare, the MTT Agri-food Research Finland, Department of Food and Environmental Sciences, University of Helsinki, and the Finnish Food Safety Authority Evira, carried out data and sample collection and the analyses. The Se supplementation programme was coordinated by the Finnish Ministry of Agriculture and Forestry. Sampling of cereals and foods has been done four times per year to form large composite specimen. Samples of human blood have been taken annually from the same urban (n = 30) and rural (n = 40) adults. The systematic error in Se chemical analyses has been verified annually by a quality control assessment scheme involving seven participating laboratories using three different analytical techniques.

3 RESULTS AND DISCUSSION

3.1 *Foods and animal products*

The Se concentration of spring cereals in-creased on average 15-fold compared with the level before the Se fertilization practice. The mean Se concentrations in beef and pork increased six- and twofold, respectively, and in milk threefold, compared with levels before Se fertilization. Organically grown foods of plant origin have generally a Se concentration comparable to pre-fertilizing levels. Milk from organically fed cows has, however, a Se concentration only 50% less than the usual milk (Eurola *et al.*, 2011).

3.2 *Intake and human status*

The average dietary human intake increased from 0.04 mg Se/day to a present plateau of 0.08 mg Se/day, which is well above the current nutrition recommendations. Foods of animal origin contributed about 70% of the total daily intake of Se. The mean human plasma Se concentration increased by 60% to a general level of 1.40 μmol/L. The human Se status can be considered optimal. The absence of Se deficiency disease and a reference population have made conclusions on human health difficult. However, the rates of cardiovascular diseases and cancers have remained similar during the pre- and post-Se supplementation indicating medical and life-style factors to be much stronger determinants than Se (Alfthan *et al.*, 2010). Evident signs of Se transport from the fertilized agricultural fields into natural waters have not been found (Wang *et al.*, 1994, 1995).

4 CONCLUSIONS

Due to the relatively uniform geochemical conditions in Finland, the nationwide application of soil fertilizers containing sodium selenate has proved to be an effective, safe, and controlled way of bringing the Se intake of the whole population to the recommended level. An alternative way of increasing the Se in foods and the Se intake of people could be to supplement animal feeds with organic Se, since most of our food Se is of animal origin.

This would also secure the wellbeing of animals (Aro, 1996).

REFERENCES

Alfthan, G., Aspila, P., Ekholm, P., Eurola, M., Hartikainen, H., Hero, H. et al. 2010. Nationwide supplementation of sodium selenate to commercial fertilizers: History and 25-year results from the Finnish selenium monitoring program. In B. Thompson & L. Amoroso (eds.), *Combating Micronutrient Deficiencies: Food-Based Approaches:* 312–337. Rome: FAO/ CAB International.

Aro, A. 1996. Various forms and methods of selenium supplementation. In J.T. Kumpulainen & J.T. Salonen (eds.), *Natural Antioxidants and Food Quality in Atherosclerosis and Cancer Prevention*: 168–171, Cambridge, UK.: Royal Society of Chemistry.

Eurola, M., Alfthan, G., Ekholm, P., Root, T., Suoniitty, T., Venäläinen, E.R. & Ylivainio, K. 2011. Seleenityöryhmän raportti (Report of Selenium Working Group, in Finnish).

Salonen, J.T., Alfthan, G., Huttunen, J.K, Pikkarainen, J. & Puska, P. 1982. Association between cardiovascular death and myocardial infarction and serum selenium in a matched-pair longitudinal study. *Lancet* 2: 175–179.

Salonen, J.T., Alfthan, G., Huttunen, J.K. & Puska, P. 1984. Association between serum selenium and the risk of cancer. *Amercan Journal of Epidemiology* 120: 342–349.

Wang, D., Alfthan, G., Aro, A., Lahermo, P., Väänänen, P. 1994. The impact of selenium fertilisation on the distribution of selenium in rivers in Finland. *Agriculture, Ecosystems and Environment* 50:133–49.

Wang, D., Alfthan, G., Aro, A., Mäkelä, A., Knuuttila, S. & Hammar, T. 1995. The biogeochemistry of selenium in Finnish lake ecosystems. *Agriculture, Ecosystems and Environment* 54:137–148.

Selenium in the Environment and Human Health – Bañuelos, Lin & Yin (Eds)
© 2014 Taylor & Francis Group, London, ISBN 978-1-138-00017-9

High-selenium upland rice: Agronomic biofortification strategies to improve human nutrition

A.R. Reis & L.R.G. Guilherme
Federal University of Lavras, Lavras-MG, Brazil

M.F. Moraes
Federal University of Mato Grosso, Barra do Garças-MT, Brazil

S.J. Ramos
Vale Institute of Technology, Belo Horizonte-MG, Brazil

1 INSTRUCTION

Cereal production has maintained the same growth rate as that of human population. It is anticipated that world demand for food will double in the period 1999–2030, and up to three and half times in developing countries (Daily *et al.*, 1998; Garvin *et al.*, 2006). On the other hand, malnutrition has increased, reaching almost half of mankind, particularly pregnant women, infants and children (Welch, 2001; Graham *et al.*, 2007). This development is partly due to the focus of plant breeding, which aims on improving gains productivity. These gains are inversely related to the mineral content of the grains (Garvin *et al.*, 2006).

Malnutrition is a consequence of large intake of cereals (*e.g.*, rice, maize, wheat) poor in vitamins and minerals (Graham *et al.*, 2007). Deficiencies of iron, selenium (Se) and zinc are the major concern in relation to human health today, especially in developing countries. Among the strategies applied for reducing the prevalence of Se deficiency problem in human populations, enrichment (biofortification) of food crops with Se through agricultural approaches is a widely applied strategy. Agronomic biofortification (*e.g.*, fertilizer application) represents a complementary and cost-effective agricultural approach to the problem (White & Broadley, 2009).

This study is aimed to evaluate Brazilian upland rice genotypic variation on the accumulation of Se in edible parts, and to establish an agronomic biofortification strategy to enhance Se content in rice grains for better human nutrition.

2 MATERIALS AND METHODS

Experiment 1: A pot experiment was carried in a greenhouse at University of Sao Paulo, Brazil. One month before sowing, recommended amounts of both lime and nutrients were applied using pots with 3 dm^3 of an acid Oxisol soil. In all pots, 0.1 mg dm^3 of Se was applied as sodium selenate. Thirty five rice upland cultivars (Table 1) were selected according to the following characteristics: old and new cultivars, duration of the cycle, high and low content of minerals in grains, yield, type, and color of grain (red, black or white).

After maturation, plants were separated into root, stem, leaf, husk and grain. Total Se content in the samples was determined in a PerkinElmer Analyst 800 atomic absorption spectrophotometer (PerkinElmer Inc., San Jose, USA) with electrothermal atomization by (pyrolytic) graphite furnace. Certified reference material BCR402 (white clover) were used for certification of the quality of analysis.

Experiment 2: The experiment was carried out in a greenhouse at University of Sao Paulo, Brazil. After evaluation of genotypic variation on Se accumulation, 8 genotypes of rice were selected based on Se accumulation efficiency. The experiment was conducted in completely randomized design in a 8 × 2 factorial scheme, with three replicates by means of eight upland rice genotypes (*e.g.*, CNA 10929, BRS Carisma, Piaui, BRS Relampago, IAC 202, Dourado Precoce, IAC 1246 and IAC 4), with or without Se (0.50 and 0.00 mg/kg, respectively) applied as sodium selenate. Total Se content was analyzed as described in Experiment 1. Data presented are mean ± standard deviations.

3 RESULTS AND DISCUSSIONS

Rice is one of major dietary source of Se. In addition, combining technologies such as breeding crops that acquire and/or use Se more efficiently and agronomic biofortification is one important strategy to improve human health.

Selenium content was identified in grains of 35 efficient upland rice varieties under greenhouse conditions and genotypic variation that ranged from 15 to 122 μg/kg DW (see Table 1).

Table 1. Upland rice genotypic variation on Se concentration and accumulation in grains.

		Se content (DW)	
	Rice genotypes	In grains (μg/kg)	Accumulation (ng/pot)
1	IAC 4440	57	1370
2	PB 11	35	479
3	PB 05	17	368
4	BRSMG Relâmpago	16	178
5	BRSMG Caravera	15	236
6	BRSMG Curinga	74	1274
7	Pérola	64	100
8	Bonança	42	542
9	Jaguari	48	82
10	Primavera	53	286
11	IAC 201	61	592
12	IAC 600	47	76
13	Guarani	56	669
14	Caiapó	63	595
15	Bico ganga	27	225
16	IAC 435	63	116
17	BRS Talento	55	714
18	Canastra	73	526
19	PB 01	17	264
20	Arroz preto	47	422
21	IAC 165	58	409
22	BRSMG Conai	40	528
23	Carajás	56	698
24	IAC 47	44	297
25	IAC 202	63	707
26	Pratão	39	29
27	Beira campo	36	487
28	Maravilha	42	270
29	IAC 25	48	450
30	Batatais	49	49
31	Dourado precoce	85	216
32	IAC 4	122	286
33	Cateto	40	99
34	IAC 1246	115	941
35	Cateto seda	71	484

After screening the rice genotypes based on Se accumulation in rice grains, 8 cultivars were selected for agronomic biofortification studies (Table 2). Upland rice genotypic variation for Se content in grains ranged from 16.3 to 91 μg/kg DW in the absence of Se fertilizer (Table 2). In this study, Se application significantly increased the Se content in grains of all varieties (Table 2), showing that agronomic biofortification of rice genotypes with Se can be an important strategy for improving Se human nutrition. The use of Se fertilizers offers an effective means of increasing Se in upland rice grains. Therefore, new approaches involving genetic breeding to enhance Se accumulation potential are required for further studies.

Table 2. Agronomic biofortication of upland rice genotypes with Se.

Treatments	Grain Se (μg/kg) DW	Productivity (g/pot)
BRSMG Relampago –Se	16.3 ± 3.4	27.4 ± 2.1
BRSMG Relampago +Se	6764.6 ± 317.4	27.9 ± 1.1
Piaui –Se	16.7 ± 3.6	10.4 ± 0.6
Piaui +Se	6344.2 ± 487.5	10.9 ± 3.9
IAC 202 –Se	70.2 ± 7.4	17.5 ± 4.3
IAC 202 +Se	9237.3 ± 305.5	23.6 ± 0.7
Dourado Precoce –Se	83.1 ± 2.1	10.1 ± 3.1
Dourado Precoce +Se	8889.3 ± 191.8	13.9 ± 3.9
IAC 4 –Se	89.4 ± 7.9	17.5 ± 4.3
IAC 4 +Se	8515.1 ± 356.7	17.1 ± 1.7
IAC 1246 –Se	91.1 ± 6.5	15.9 ± 4.7
IAC 1246 +Se	8446.7 ± 79.9	13.3 ± 3.1
Carisma –Se	88.26 ± 7.31	18.9 ± 2.1
Carisma +Se	7735.53 ± 655.1	19.2 ± 4.6
CNA 10929 –Se	85.8 ± 5.5	4.2 ± 1.3
CNA 10929 +Se	8715.7 ± 145.9	1.8 ± 1.1

4 CONCLUSIONS

Significant varietal differences in the Se concentration in Brazilian upland rice grains were observed, ranging from 15 to 122 μg/kg DW. Selenium application significantly increased the Se concentration in rice grains, showing that agronomic biofortification can be an important effort to improve Se human nutrition.

REFERENCES

Daily, G., Dasgupta, P., Bolin, B., Cross, P., Guerny, J., Ehrlich, P. *et al.* 1998. Global food supply: food production, population growth, and the environment. *Science* 281(5381): 1291–1292.

Garvin, D.F., Welch, R.M. & Finley, J.W. 2006. Historical shifts in the seed mineral micronutrient concentrations of US hard red winter wheat germplasm. *Journal of the Science of Food and Agriculture* 86(13): 2213–2220.

Graham, R.D., Welch, R.M., Saunders, D.A., Ortiz-Monasterio, I., Bouis, H.E., Bonierbale, M. *et al.* 2007. Nutritious subsistence food systems. *Advances in Agronomy* 92: 1–74.

Welch, R.M. 2001. Micronutrients, agriculture and nutrition: linkages for improved health and well being. In: K. Singh, S. Mori, R.M. Welch (eds). *Perspectives on the Micronutrient Nutrition of Crops*. 247–289, Jodhpur: Scientific Publishers.

White, P.J. & Broadley, M.R. 2009. Biofortification of crops with seven mineral elements often lacking in human diets—iron, zinc, copper, calcium, magnesium, selenium and iodine. *New Phytologist* 182: 49–84.

Selenium in the Environment and Human Health – Bañuelos, Lin & Yin (Eds)
© 2014 Taylor & Francis Group, London, ISBN 978-1-138-00017-9

Biofortification of field crops with selenium in Mediterranean conditions

S.P. McGrath
Rothamsted Research, Harpenden, Hertfordshire, UK

M.J. Poblaciones & S.M. Rodrigo
Department of Agronomy and Forest Environment Engineering, University of Extremadura, Badajoz, Spain

1 INTRODUCTION

Diets that are based on mainly vegetarian staple crops may be low in selenium (Se), and provide food intakes less than the recommended amounts for humans. Previous effort of Se fortification of field crops to improve human dietary intakes have focused on high latitudes (Broadley *et al.*, 2010) and little research has been done to date in the drier regions of southern Europe (Rodrigo *et al.*, 2013).

It is therefore important to test additions of Se under relatively dry Mediterranean conditions and to consider the types of crops and products produced in these areas.

This research examined the efficacy of biofortification of barley, wheat, chickpea and peas treated with selenite or selenate applications in the field. The concentrations and in some cases the chemical speciation of Se in the harvested grain and seed was measured, and in some of the processed products.

2 MATERIALS AND METHODS

The field crops were grown in Badajoz in S.W. Spain in replicated trials conducted during 2011–2012 (Poblaciones *et al.*, 2013; Rodrigo *et al.*, 2013). Selenium was applied as either disodium selenite (SeIV) or disodium selenate (SeVI) in doses between 0 and 80 g Se/ha. Crops were harvested at maturity, yields of grain or seed recorded, and samples ground and digested for total Se analysis using an Agilent 7500ce ICP-MS (Agilent Technologies, Palo Alto, CA, USA), or following proteolytic digestion with protease and lipase according to the methods of Huerta et al. (2003) for speciation analysis using an Agilent 1050 HPLC pump and a Hamilton PRP-X100 250 × 4 mm column (Cole-Parmer, London, UK) linked to the ICP-MS.

3 RESULTS AND DISCUSSION

3.1 *Soil Se status*

All crops were grown on Xerofluvent soils under rain fed Mediterranean conditions. Soil pHs ranged from 6.9 to 7.4 (average 7.2) in which inorganic Se species are likely to be relatively available (Chilimba *et al.*, 2011). However, soil total Se ranged from 110–147 (mean 131) µg/kg and KH_2PO_4-extractable Se from 4.3 to 7 (mean 5.4) µg/kg, which are likely to produce crops that do not provide sufficient Se intake in the human diet (Stroud *et al.*, 2010).

3.2 *Crop trials*

The relative uptakes of Se (µg/kg in harvested part divided by the g/ha added to soil) of barley, durum wheat, bread wheat, chickpea and pea were assessed in crops harvested in both 2011 and 2012 (except pea). Relative uptakes were compared by using the slopes of the linear regression of the concentrations of Se in the harvested parts of crops and in pasta made from the durum wheat, against the amounts of Se IV or VI added in the field. Relative uptake of SeIV ranged from 6–38, and SeVI from 30–148. Averaged across all the crops/products, SeVI was around 5 times more effective in boosting crop Se than SeIV (Table 1). Where the same crops were grown in 2011 and 2012, the uptake was 1.4 to 3.9 times higher in 2011 (average 2.3). Weather data show that 2011 had more annual rainfall (589 mm) than 2012 (338 mm). In 2011, there was a 40-day drought period during flowering and early seed formation of these crops and this gave lower yields and higher protein contents in 2011 (Poblaciones, pers. comm.). Protein was inversely related to yield (data not shown), possibly due to a dilution effect. Taking 2011 for inter-crop comparison, as all of the crops were grown, the order of efficiency of SeIV additions was pea > durum wheat > chickpea > bread wheat > barley (Table 1).

Table 1. Linear regressions of the relationships between Se concentrations in harvested parts of crops versus the amounts of SeIV or SeVI applied to soils (x; in g/ha) in Badajoz, and the relative bioavailability factor (slope with Se VI applied/ slope with Se IV applied).

	Se IV		Se VI		
Crop	Regression	R^2	Regression	R^2	Factor
Barley 2011	10.4x + 113	0.995	55.3x + 93	0.997	5.3
Barley 2012	6.1x + 58.1	0.951	32.9x – 42.8	0.996	5.4
Durum wheat 2011	38.1x – 3.6	0.985	137.6x + 8.3	0.995	3.6
Durum wheat 2012	11.1x + 55.3	0.995	35.4x + 49.6	0.995	3.2
Durum pasta 2012	7.34x + 13.6	0.938	30.2x – 22.5	0.989	4.1
Bread wheat 2011	30.5x – 105	0.866	79.0x + 7.7	0.993	2.6
Bread wheat 2012	8.45x + 70.4	0.925	37.5x – 12.2	0.987	4.4
Chickpea 2011	25.4x + 102	0.984	125.9x + 416	0.977	5.0
Chickpea 2012	19.2x + 268	0.919	87.3x + 15.1	0.989	4.5
Pea 2011	19.3x + 337	0.910	147.6x + 337	0.970	7.6

Speciation measurements in durum wheat and chickpea seed harvested in 2012 showed that applying SeIV or VI increased the concentrations of Se-methionine in all crops compared to untreated controls, and decreased the concentrations of inorganic Se (data not shown). With increasing amounts of Se added, Se-methionine reached >90% of total Se in the seed. Pasta cooked from the durum wheat contained slightly less total Se (Table 1) and similar speciation as the milled wheat due to Se in bran and germ removed by milling.

3.3 *Comparison with UK bread wheat data*

More is known about biofortification of crops in more northerly latitudes, and it is of interest to compare the effect of biofortification of Mediterranean crops with these. To do this, we selected data for two breadmaking wheat varieties grown with a similar range of SeVI additions at four sites in E. England (Broadley *et al.*, 2010).

Total Se in the soils was higher than in Spain, from 245–590 μg/kg (mean 429) but KH_2PO_4-extractable Se was from 4.9 to 9.6 (mean 6.9) μg/kg, i.e. similar to the Spanish soils and insufficient to provide adequate Se intake in the human diet. Soil pHs ranged from 6.8 to 8.0 (average 7.5). Rainfall and yields of wheat were all higher in England, but the relative uptakes were 16–26 (mean 20). This is approximately half to a quarter of the values for Spanish wheat grown in 2012 and 2011 respectively (Table 1). Again, although some Se was lost during milling, the bread cooked from the English wheat retained the majority of the Se in the form of Se-methionine (Hart *et al.*, 2011), as with the Spanish pasta.

4 CONCLUSIONS

A range of grain and pulse crops grown in Spain can be effectively biofortified with relatively low rates of Se addition in fertilizer form. SeVI fertilizer is around 5 times more effective than selenite in boosting grain/seed concentrations. There was an inverse relationship between yield (due to climatic factors) and Se concentrations in four crop species. Comparison of the relative uptake in Spain and England by breadmaking wheats showed apparently higher efficiency under the Mediterranean climate in Spain. This could be due to higher yields in England effectively diluting the concentrations in grain/seed. Relatively small losses of Se occur during milling of grains, but the majority of the Se in grain/seed is in the form of Se-methionine, which is highly bioavailable.

REFERENCES

Broadley, M.R., Alcock, J., Alford, J., Cartwright, P., Foot, I., Fairweather-Tait, S.J. *et al.* 2010. Selenium biofortification of high-yielding winter wheat (*Triticum aestivum* L.) by liquid or granular Se fertilisation. *Plant Soil* 332: 5–18.

Chilimba, A.D.C., Young, S.D., Black, C.R., Rogerson, K.B., Ander, E.L., Watts, M.J. *et al.* 2011. Maize grain and soil surveys reveal suboptimal dietary selenium intake is widespread in Malawi. *Sci. Rep.* 1: 72.

Hart, D.J., Fairweather-Tait, S.J., Broadley, M.R., Dickinson, S.J., Foot, I., Knott, P. *et al.* 2011. Selenium concentration and speciation in biofortified flour and bread: Retention of selenium during grain biofortification, processing and production of Se-enriched food. *Food Chemistry* 126: 1771–1778.

Huerta, V.D., Reyes, L.H., Marchante-Gayon, J.M., Sanchez, M.L.F. & Sanz-Medel, A. 2003. Total determination and quantitative speciation analysis of selenium in yeast and wheat flour by isotope dilution analysis ICP-MS. *Journal of Analytical Atomic Spectrometry*, 18: 1243–1247.

Poblaciones, M.J., Rodrigo, S.M. & Santamaría, O. 2013. Evaluation of the potential of peas (*Pisum sativum* L.) to be used in selenium biofortification programs under Mediterranean conditions. *Biol. Trace Elem. Res.* 151: 132–137.

Rodrigo, S., Santamaría, O., López-Bellido, F.J. & Poblaciones, M.J. 2013. Agronomic selenium biofortification of two-rowed barley under Mediterranean conditions. *Plant Soil Environ.* 59(3): 115–120.

Stroud, J.L., Broadley, M.R., Foot, I., Fairweather-Tait, S.J., Hart, D.J., Hurst, R. *et al.* 2010. Soil factors affecting selenium concentration in wheat grain and the fate and speciation of Se fertilisers applied to soil. *Plant Soil* 332: 19–30.

Selenium in the Environment and Human Health – Bañuelos, Lin & Yin (Eds)
© 2014 Taylor & Francis Group, London, ISBN 978-1-138-00017-9

Uptake of selenate versus selenite in *Zea mays*: Biofortification of crops and forage

M. Longchamp
UPMC, UMR 7618 BioEMCo, Paris Cedex 05, France
Present address: LGCgE, Lille1 Univ, Villeneuve d'Ascq Cedex, France

M. Castrec-Rouelle
UPMC, UMR 7618 BioEMCo, Paris Cedex 05, France

1 INTRODUCTION

Selenium (Se) is an essential micronutrient for health. It enters food systems from soils, water and via plants that are the primary source of Se in the diet. To enhance Se contents in the diet, agronomic fortification of crops with soluble selenite or selenate salts added to mineral fertilizers has become a widespread agricultural practice.

Although maize ranks among the major cereal crops and could represent a significant part of the daily Se intake for some populations, there are only few results published about the bioaccumulation of Se in this plant (Wang *et al.*, 2012; Qin *et al.*, 2013). There is almost nothing published about the uptake and quantities of bioavailable Se in grains and forage, even though some recent publications discuss concentrations of Se in the whole corn plant and within their tissues (Chilimba *et al.*, 2012). Therefore, our study presents original data about the dissimilar responses in accumulation and rate of metabolization of Se in organs (*e.g.*, seed, leaves, steam and root) of Zea mays when selenite or selenate was added to solution culture.

2 MATERIALS AND METHODS

Three weeks after germination maize seedlings were transferred in 20 L containers filled with a modified Hoagland solution supplemented with Se (1 mg Se /L) in two treatments (five corn seedling per treatment): selenate treatment (Se^{VI}-T) and selenite treatment (Se^{IV}-T). Contrary to previous published results on maize grown in hydroponic system (Longchamp *et al.*, 2013), experiments were carried out until reproductive stage. At maturity, plants were harvested and separated into grains, leaves, stems and roots. Each sample was freeze-dried, grounded and the tissues were digested in a mixture of HNO_3 (70%) and H_2O_2 (30%). Total Se concentrations were determined by inductively coupled plasma mass spectroscopy (CRC-ICP-MS, Thermo X Series II). Selenate and selenite contents in the tissues were evaluated after protease hydrolysis; and the pool of various Se-organic species was estimated as the difference between total Se and sum of selenite and selenate.

3 RESULTS AND DISCUSSION

3.1 *Influence of selenate and selenite on maize seed production*

Grain dry weights decreased compared to the control plants, *i.e.*, 60% and 80% in the Se^{VI}-T and Se^{IV}-T plants, respectively. However, the grain number produced by each plant, decreased significantly (70%) with selenite supplied. Biomass allocation was affected by Se; final grain dry weights to shoot dry weights ratios were lower for both Se treatments than for the C-T. In fact, grain biomass for the C-T represented 33% of aerial biomass, but decreased down to 21% in presence of Se in nutrient solution.

As already observed in white lupine or sunflower, (Ximenez-Embrun *et al.*, 2004), in a selenite treatment, the development of all tissues was affected: dry forage biomass as well as quantity (number) and quality (dry weight) of grains decreased. On the other side, selenate does not seem to influence the quantity of grains but seems to inhibit their normal filling.

3.2 *Uptake and accumulation of selenium*

The supply of selenite versus selenate had a significant effect on the distribution of Se in the whole plants and between the tissues of maize. For the whole plant, selenium was 68% more accumulated for the Se^{IV}-T than for the Se^{VI}-T plants respectively with Se concentrations of 210 µg/g (156; 225) and 125 µg/g (103; 126). Values are medians (Q1; Q3). For the Se^{VI}-T, tops (above-ground plant material) Se amount represented more than 90% of total plant Se while almost 50% of total plant Se was located in leaves (Fig. 1). The Se allocation among the 3 tissues for the Se^{IV}-T differed dramatically: tops Se amount was low (about 40% of total plant Se); whereas almost 60% of total plant Se was in roots. Irrespective of Se-form in nutrient solution, grain Se content represented 15% of the total plant Se (Fig. 1). However, grain Se concentrations were 1.7 times higher, *i.e.*, 73%, for the Se^{IV}-T than for the Se^{VI}-T plants (Fig. 2).

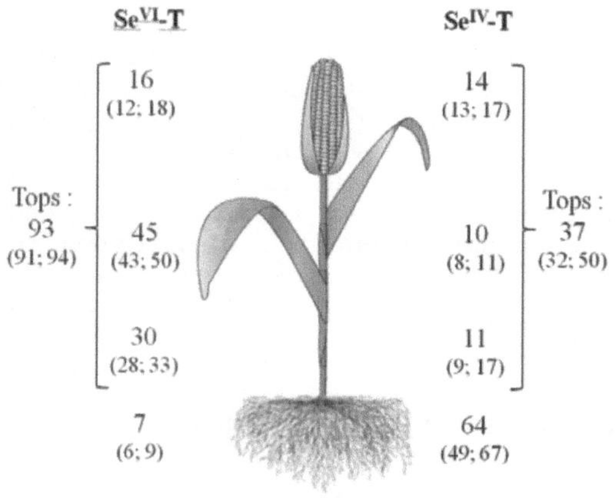

Figure 1. Schematic representation of Zea mays plant showing the Se amount (in %) in roots, stems, leaves, grains treated with SeVI-T (left) or SeIV-T (right). Values are medians (Q1; Q3); n = 5.

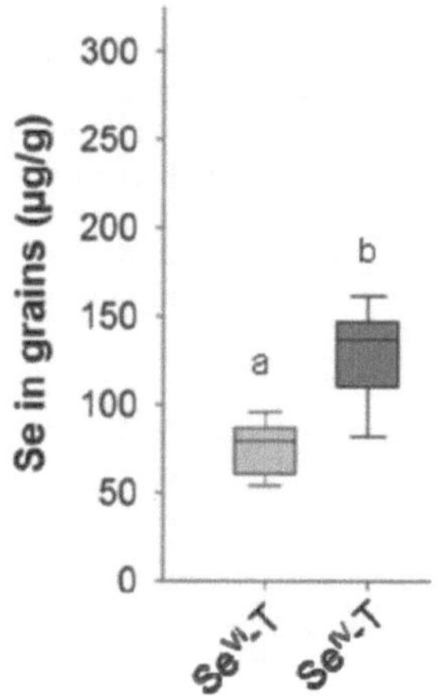

Figure 2. Concentrations of Se (µg/g DW) in grains. Se^{VI}-T (in gray) and Se^{IV}-T (in dark gray). Different letters mean statistically significant difference ($P < 0.05$) by Mann and Whitney test.

3.3 *Enrichment of selenium in Zea mays*

For selenite treatment organoselenium production is greater than for selenate, although most of the Se is taken up by Zea mays as selenate and translocated and accumulated in tops of plants, especially in leaves: Selenate is less metabolized than selenite in whole plants. Selenium in grains of both treatments is exclusively organic compounds. To improve the nutritional intake of Se in the diet, the quality and quantity of bioavailable Se are almost as important as the Se daily ingested but the bioavailable Se amounts to humans is evaluated at 90% for the organoselenium compounds and 50% for selenite or selenate (Thomson *et al.*, 2004). The current results show that supplying selenite is the best treatment to enrich grains in bioavailable Se to animals and humans. Thus, since selenate treatment does not affect shoot biomass, it is the best supplement to enrichment forage (based all aerial parts of the plants—stems + leaves + grains-) in bioavailable Se to animals.

4 CONCLUSIONS

The Se bioavailability depends on the chemical Se-forms, with organic forms being more bioavailable than the inorganic forms. We concluded that Se bioavailablelity for humans is greater with selenite treatment but selenate treatment is the better Se supplement to enrich forage material.

REFERENCES

Chilimba, A.D.C., Young, S.D., Black, C.R., Meacham, M.C., Lammel, J. & Broadley, M.R. 2012. Agronomic biofortification of maize with selenium (Se) in Malawi. *Field Crop Research* 125: 118–128.

Longchamp, M., Angeli, N. & Castrec-Rouelle, M. 2013. Selenium uptake in *Zea mays* supplied with selenate or selenite under hydroponic conditions. *Plant and Soil* 362 (1): 107–117.

Qin, H., Zhu, J., Liang, L., Wang, M. & Su, H. 2013. The bioavailability of selenium and risk assessment for human selenium poisoning in high-Se areas, China. *Environment International* 52: 66–74.

Thomson, C. 2004. Assessment of requirements for selenium and adequacy of selenium status. *European Journal of Selenium in Food* 58: 391–402.

Wang, S., Liang, D., Wang, D., Wei, W., Fu, F. & Lin, Z.-Q. 2012. Selenium fractionation and speciation in agriculture soils and accumulation in corn (*Zea mays* L.) under filed conditions in Shaanxi Province, China. *Science of the Total Environment* 427–428: 159–164.

Ximénez-Embún, P., Alonso, I., Madrid-Albarrán, Y. & Cámara, C. 2004. Establishment of selenium uptake and species distribution in lupine, Indian mustard and sunflower plants. *Journal of Agricultural and Food Chemistry* 52: 832–838.

Selenium in the Environment and Human Health – Bañuelos, Lin & Yin (Eds)
© 2014 Taylor & Francis Group, London, ISBN 978-1-138-00017-9

Selenium accumulation in selenium-biofortified tea from Xuancheng, Anhui, China

Juanjuan Kong, Gongfang Xi, Xin Hong & Shoubiao Zhou
Anhui Normal University, Wuhu, Anhui, China

Qiankun Zhou
University of Science and Technology of China, Suzhou, Jiangsu, China

1 INTRODUCTION

Selenium (Se) is an essential trace element for living beings. Due to the extensive distribution of Se deficiency in certain countries like China, a Se supplement maybe beneficial to humans (Hu *et al.*, 2003). Selenium-enriched food is known as a potentially healthy product for Se supplementation (Underwood *et al.*, 1997). Applying Se fertilizer on crops can improve the Se contents of crops without showing adverse effects on plant growth (Yin *et al.*, 1991). Tea is the Chinese daily main drink and it may be a feasible strategy to supplement Se via drinking teas, if Se-enriched teas could be produced (Xu *et al.*, 2003). Thus, in the present study, a solid Se fertilizer was applied on different teas grown in Xuancheng, Anhui to produce Se-enriched tea.

2 MATERIALS AND METHODS

Different tea cultivars, including *Yongxihuoqing, Aiminyangjian* and *Tingxilanxiang*, were selected to be grown with added Se fertilizers in Xuancheng, Anhui. The Se fertilizer (in powder) was provided by Setek Co. (Suzhou, China), and was added in December one time and in March twice (beginning and mid of the month) at a rate of 60 g/hm^2 to the surface of underlying soils of tea plants. During late spring, new leaves, old leaves, and underlying soils and water of tea plants were collected. The plant samples were dried at 52°C and then ground into powders. The soil samples were air-dried, ground, and sieved through No. 100-mesh. No sample preparation was for water samples, and water samples were directly measured. Concentrations of total Se in all samples were determined by Atomic Fluorescence Spectrometer (AFS-9230, Titan Co, Beijing).

3 RESULTS AND DISCUSSION

Selenium contents of new leaves, old leaves, underlying soils and water were shown in Table 1. The highest Se content (508 μg/kg) was observed in new leaves of Aiminyangjian Tea, which was 2–3 times greater than those of other Se-biofortified teas. Overall, the Se contents in leaves could be improved by 20–60 times via Se-biofortification strategy of adding Se fertilizer compared with the control tea (without the Se treatment). To evaluate the Se accumulation ability in different tea plants, enrichment factors were considered. Overall, the Se bioconcentration factors in all tea samples were lower than 1, indicating tea plants were clearly not a Se-accumulating species. However, the bioconcentration factors of all Se-biofortified tea trees were clearly greater than that of the control

Table 1. Selenium contents in new leaves, old leaves, underlying soils of different teas, and calculated enrichment factors.

Selenium concentrations				
Sample	New leaves (μg/kg)	Old leaves (μg/kg)	Underlying soil (μg/kg)	Enrichment factor† (%)
Yongxihuoqing tea	254	165	643.5	39.5
Aiminyangjian tea	508	304	712.5	71.3
Tingxilanxiang tea	173	97	201.9	85.7
Ordinary tea‡	9.48	7.49	158.5	6.0

(*i.e.*, without application of Se fertilizer), especially Aiminyangjian tea and Tingxilanxiang tea. Previous studies revealed that the Se content in tea drink could reach 10% to 20% of total Se measured in leaves (Hu *et al.*, 1999). Thus, 17.3 to 111.6 μg Se/kg could be released into the tea drinks in the present study, and absorbed by tea drinkers.

4 CONCLUSIONS

In conclusion, Se fertilizer can be used to improve the Se content in new leaves of tea trees by 20–60 times, especially in *Aiminyangjian* Tea with 508 μg Se/kg. Although tea plants were not Se-accumulating species, this study indicates that adding Se fertilizer to the tested tea plants is a potential biofortification strategy to produce Se-enriched tea.

REFERENCES

Hu, Q.H., Pan, G.X. & Lv, X.Q. 1999. Study on Extracting Rate of Selenium and Property of Selenium of Se-enriched Tea. *Scientia Agricultura Sinica* 32(5): 69–72.

Hu, Q., Xu, J. & Pang, G. 2003. Effect of selenium on the yield and quality of green tea leaves harvested in early spring. *Journal of Agricultural and Food Chemistry*, 51(11): 3379–3381.

Underwood, E.J. 1997. *Trace Element in Human and Animals Nutrition*. 38. New York, San Francisco, London: *Academic Press.*

Xu, J., Zhu, S., Yang, F., Cheng, L., Hu, Y., Pan, G. *et al.* 2003. The influence of selenium on the antioxidant activity of green tea. *Journal of the Science of Food and Agriculture*, 83(5): 451–455.

Yin, H.Y., Hao, Y.H. & Yu, H.B. 1991. Effects on selenium content and yield of millet grain of low selenium areas. *Journal of Chinese Soil Science* 22(1): 15.

Selenium in the Environment and Human Health – Bañuelos, Lin & Yin (Eds)
© 2014 Taylor & Francis Group, London, ISBN 978-1-138-00017-9

Selenium accumulation in Chinese milk vetch (*Astragalus sinicus* L.) from southern Anhui, China

Qiankun Zhou & Shoubiao Zhou
College of Life Science, Anhui Normal University, Wuhu, Anhui, China

Xuebin Yin & Linxi Yuan
Advanced Lab for Selenium and Human Health, Suzhou Institute for Advanced Study, University of Science and Technology of China, Suzhou, Jiangsu, China
Jiangsu Bio-Engineering Research Centre on Selenium, Suhzou, Jiangsu, China

1 INTRODUCTION

Chinese milk vetch (*Astragalus sinicus* L.) is known as one of the native selenium (Se) accumulators (Davis, 1972). The plant is widely grown as a green manure crop in southern China to help produce Se-enriched grains, milk, pork, tea and etc. (Lin *et al.*, 2011). However, there is no extensive investigation on Se accumulation in Chinese milk vetch varieties to determine the best species for farming. The present study focused on Se accumulation on Chinese milk vetch grown in different cities from Anhui Province, which will provide new information for local species screening.

2 MATERIALS AND METHODS

A total of 24 Chinese milk vetch samples and their corresponding surface (0–3 cm) soil samples were collected from 8 cities in southern Anhui province (Fig. 1). The roots, stems, leaves and flowers were separated and their total Se concentrations were determined by Atomic Fluorescence Spectrometer (AFS-9230, Titan Co, Beijing) (Mazej *et al.*, 2008).

Figure 1. Sampling sites from 8 cities in southern Anhui, China.

3 RESULTS AND DISCUSSION

The determined concentrations of Se in roots, stems, leaves and flowers of Chinese milk vetch (*Astragalus sinicus* L.) and their underlying soils are listed in Table 1. Overall, the total Se of soils ranged from 192 ± 23 to 3708 ± 95 μg/kg. According to Tan (1989), almost all cities in southern Anhui are Se-adequate areas because Se contents in soils are in a range of 175–3000 μg/kg. However, Ningguo city is a Se-excessive area with 3708 ± 95 μg Se/kg in soils. The highest Se concentrations (in μg/kg) in Chinese milk vetch were observed in Ningguo City (Se-excessive area) with 775 ± 89 in roots, 378 ± 41 in stems, 447 ± 50 in leaves and 1039 ± 55 in flowers. However, in the Se-adequate areas, the contents of plant tissues were much lower than those in Se-excessive area, *i.e.*, Ningguo City, with 150–300 μg Se/kg in roots, 30–100 μg Se/kg in stems, 50–200 μg Se/kg in leaves, and 100–200 μg Se/kg in flowers. Overall, concentrations of Se in tissues of Chinese milk vetch from Se-adequate areas ranged as follows: roots > flowers > leaves > stems.

The highest bio-concentration factors of tissues were observed in Nanling City (Se-adequate areas) with 0.96 ± 0.10 of roots, 0.33 ± 0.04 of stems, 0.57 ± 0.06 of flowers. Meanwhile, the highest bio-concentration factors of leaf reached 0.55 in Qianshan city and Tongling city. Overall,

Table 1. Concentrations of Se in different tissues of Chinese milk vetch (*Astragalus sinicus* L.) and soil Se concentrations in different sampling areas in Southern Anhui, China. Values shown are mean ± standard deviation (n = 3).

		Concentrations of Se (μg/kg)				Bioconcentration factors (C_{tissue}/C_{scil})			
Location	Soil	Root	Stem	Leaf	Flower	Root	Stem	Leaf	Flower
Chizhou	192 ± 23[†]	141 ± 12	29.9 ± 4.1	90.1 ± 6.3	110 ± 10	0.73 ± 0.06	0.16 ± 0.02	0.47 ± 0.05	0.57 ± 0.06
Qianshan	208 ± 15	186 ± 23	46.8 ± 5.1	115 ± 12	103 ± 8	0.90 ± 0.09	0.23 ± 0.03	0.55 ± 0.06	0.50 ± 0.04
Qimen	425 ± 38	153 ± 13	45.7 ± 6.3	55.1 ± 7.2	99.0 ± 9.2	0.36 ± 0.04	0.11 ± 0.01	0.13 ± 0.01	0.23 ± 0.02
Tongling	437 ± 54	252 ± 24	70.1 ± 9.4	163 ± 8.3	129 ± 6	0.59 ± 0.05	0.16 ± 0.01	0.55 ± 0.05	0.30 ± 0.03
Chaohu	505 ± 32	294 ± 43	60.0 ± 9.6	78.3 ± 6.5	110 ± 8	0.58 ± 0.07	0.12 ± 0.01	0.13 ± 0.01	0.22 ± 0.02
Ningguo	3708 ± 95	775 ± 89	378 ± 41	447 ± 50	1039 ± 55	0.21 ± 0.02	0.10 ± 0.01	0.38 ± 0.03	0.28 ± 0.03
Fanchang	379 ± 27	209 ± 26	96.1 ± 7.7	95.9 ± 8.0	97.9 ± 9.5	0.55 ± 0.04	0.25 ± 0.02	0.16 ± 0.02	0.26 ± 0.03
Nanling	301 ± 21	290 ± 31	99.2 ± 9.2	116 ± 21	172 ± 20	0.96 ± 0.10	0.33 ± 0.04	0.12 ± 0.01	0.57 ± 0.06

the bio-concentration factors of roots were higher than those of other tissues.

4 CONCLUSIONS

Ningguo City was recognized as Se-excessive area with more than 3000 μg Se/kg in soils. The Chinese milk vetch (*Astragalus sinicus* L.) growing at this location had Se concentrations (μg/kg) as follows in the different plant parts: 775 ± 89 in roots, 378 ± 41 in stems, 447 ± 50 in leaves and 1039 ± 55 in flowers. These values were 3–10 times that of other Chinese milk vetch growing in the other tested cities of Southern Anhui. Our results indicate that it may be plausible to use Chinese milk vetch in Ningguo City as a bio-fertilizer (biofortification) for potentially producing Se-enriched crops.

ACKNOWLEDGEMENTS

This study was financially supported by Natural Science Youth Foundation of Jiangsu Province for Youth, China (BK 2012202, BK 2012195), and Jiangsu Research-Industry Joint Innovation Project (BY 2011186).

REFERENCES

Davis, M. 1972. Selenium accumulation in Astragalus species. *Agronomy Journal* 64(6): 751–754.

Lin, X.Y., Cao, W.D., Wu, Y.Q., Zhang, H., Qiu, X.X., Zhang, W.G. *et al.* 2011. Advance in *Astragalus sinicus* research. *Prataculturał Science* 28(1): 135–140. (in Chinese)

Mazej, D., Falnoga, I. & Michael, V. 2008. Determination of selenium species in plant leaves by HPLC-UC-HG-AFS. *Talanta* 68(3): 558–568.

Tan, J.A. (ed.) 1989. *The Atlas of Endemic Diseases and their Environments in the People's Republic of China.* Beijing: Science Press (in Chinese).

Selenium in the Environment and Human Health – Bañuelos, Lin & Yin (Eds)
© 2014 Taylor & Francis Group, London, ISBN 978-1-138-00017-9

Use of selenium-contaminated plants from phytoremediation for production of selenium-enriched edible mushrooms

Z.-Q. Lin, S. Haddad, J. Hong & J. Morrissy
Environmental Sciences Program, Southern Illinois University, Edwardsville, Illinois, USA

G.S. Bañuelos
USDA-ARS, Parlier, California, USA

L.Y. Zhang
Institute of Medicinal Fungi Research, Nanjing Xiaozhuang University, Nanjing, China

1 INTRODUCTION

A mushroom is a general term for the fruiting body of a fungus. Edible mushrooms are very popular food crops in the world because of their special texture, flavor, and nutritional values. Typically consumed as a vegetable, mushrooms provide a good source of vitamins, minerals, and digestible proteins (Mattila *et al.*, 2001; Kalač & Svoboda, 2000). In the U.S., per capita mushroom consumption has quadrupled since the 1960s, and in 2001 U.S. consumption of all mushrooms was 21% greater than in 1991 (Lucier *et al.*, 2003). According to the National Agricultural Statistics Service (NASS, 2010), the sale of mushrooms in the U.S. was 357 million kg in 2010. Edible mushrooms are most popular in the Western and Midwestern U.S., with two times more consumption among upper income consumers than the lower income individuals.

There are about 35 mushroom species cultivated commercially worldwide, and approximately 20 of those species on an industrial production scale (Sadler, 2003). The U.S. mushroom industry is currently the second largest producer of mushrooms in the world. The white button mushroom (*Agaricus bisporus*) accounts for 87% of all domestic mushroom sales in the U.S., while baby bella and portabella mushrooms (*A. bisporus*) account for 11% of the domestic sales (Lucier *et al.*, 2003). With mushrooms seeing such a rise in popularity and with the current scientific interest in functional foods with high nutritional values, edible mushrooms offer an ideal choice for selenium (Se) biofortification study.

2 CAN MUSHROOMS BE AN IMPORTANT SOURCE OF DIETARY SELENIUM?

To determine concentrations of Se in mushrooms that are most commonly produced in the U.S., 10 different kinds of mushrooms were purchased from grocery stores in the St. Louis and San Francisco areas, including White Button (*Agaricus bisporus*), Portabella (*A. bisporus*), Baby Bella (*A. bisporus*), Shiitake (*Lentinus edodes*), Oyster mushroom (*Pleurotus ostreatus*), White Beech (*Hypsizygus tessulatus*), Brown Beech (*H. tessulatus*), Enoki (*Flanmmulina velutipes*), Maitake (*Grifola frondosa*), and King Trumpet (*Pleurotus erungli*). Our study also shows that concentrations of Se in commercially-produced mushrooms varied significantly not only among different species, but also among different growers who have used different sources of growth substrate materials in mushroom production. Most of the commonly-available mushroom species/varieties contained relatively low Se concentrations of <0.25 mg/kg (dry weight), except for the three varieties of *Agaricus bisporus* which contained more Se than the others. The highest Se concentration, 2.68 mg/kg, was observed in Baby Bella mushrooms. These findings generally concur with those from previous studies (Falandysz, 2008). Thus, there is a strong need to develop and produce Se-biofortified edible mushrooms for the U.S. market.

3 LIMITING FACTORS FOR SELENIUM ACCUMULATION IN MUSHROOMS

3.1 *Levels of Se in substrate*

Concentrations of Se in growth substrates significantly affected Se accumulation in mushroom fruit bodies. When the Se treatments in the substrates were 2, 4, and 8 mg/kg, concentrations of Se in Oyster mushrooms were 20.6 ± 1.9 mg/kg, 45.1 ± 7.3 mg/kg, and 65.1 ± 6.9 mg/kg, respectively. The accumulation of Se in mushrooms increased significantly with increasing substrate Se concentration. In a different experiment with *A. bisporus*, the accumulation of Se in White Button increased with increasing Se treatment level in the substrate, while Baby Bella showed a similar steady increase below the Se treatment of 50 mg/kg. Above this high level of Se treatment, increasing the substrate Se concentration did not result in a significant increase of Se concentration in Baby Bella (Haddad & Lin, 2011).

3.2 *Chemical forms of Se in substrate*

When the growth substrate was treated with 3 mg/kg Se of different chemical forms, including selenate (SeO_4), selenite (SeO_3) and selenomethionine (SeMet), the highest Se concentration of 31.2 ± 2.6 mg/kg was observed in Oyster mushroom with the growth substrate treated with selenate, followed by the SeMet treatment (24.8 ± 2.9 mg/kg) and the selenite treatment (21.2 ± 2.5 mg/kg). However, when the growth substrate was treated with 5 mg/kg Se of selenate, selenite or SeMet, the Se accumulation in White Button showed a different descending order of SeMet (26.9 ± 1.6 mg/kg) > selenite (20.0 ± 0.8 mg/kg) > selenate (6.0 ± 0.4 mg/kg).

4 FEASIBILITY OF GROWING SELENIUM ENRICHED MUSHROOMS USING PLANT WASTES FROM PHYTOREMEDIATON

4.1 *Growing Se-biofortified White Button and Baby Bella*

Because White Button and Baby Bella mushrooms are secondary decomposers, these mushrooms need to be grown in the substrates that have been partially decomposed previously by other microorganisms. Therefore, Se-laden Elephant grass (*Pennistum purpureum*) and Rye grass (*Leymus triticoides*) leaves collected from Se phytoremediation field sites were composted first and then mixed with horse manure and cottonseed meal. The Se concentration in this growth substrate was 0.8 mg/kg, and in which concentrations of Se in biofortified White Button and Baby Bella mushrooms were 3.2 mg/kg and 3.3 mg/kg, respectively, which is about 2-fold higher than the mushroom Se concentrations using the industrial growth substrates.

Prince Plum (*Stanleya pinnata*) has been successfully used for Se phytoremediation in central California. From the field site, *S. pinnata* leaf contained approximately 3000 mg/kg Se. When the industrial growth substrate was mixed with *S. pinnata* leaf powder to reach a Se treatment level of 5 mg/kg, the Se concentration in White Button increase from approximate 1.5 mg/kg in the control to 30 mg/kg in the growth substrates treated with the Se-laden plant powders.

4.2 *Growing Se-biofortified Oyster mushroom*

Selenium-laden Eucalyptus wood and rye grass tissues were used directly as the growth substrate materials to produce Se-enriched Oyster mushroom, as a primary decomposer. When the Se concentration in the growth substrate, *i.e.*, the mixed grass and wood tissues, was 1.98 ± 0.16 mg/kg, the Se-biofortified Oyster mushrooms accumulated 2.9 mg/kg Se in fruit body from the Se-contaminated plant materials containing 1.98 mg/kg Se, compared with 0.2 mg/kg in commercially produced Oyster mushrooms in the U.S. The Se biofortification treatment resulted in a 13-times increase of Se accumulation in Oyster mushrooms, during a growth period of four weeks.

5 CONCLUSIONS

Phytoremediation of selenium-contaminated agricultural soil and water can be environmentally sustainable and cost-competitive. However, the management of Se-contaminated plant materials is challenging. We proposed that Se-phytoremediation wastes could be used as a valuable organic source of Se for the production of Se-biofortified edible mushrooms. Concentrations of Se in commercially produced mushrooms vary significantly among different growers. While the common button mushroom (*A. bisporus*) has a reasonably high Se content (approximately 2.5 mg/kg), most other mushrooms that are commonly produced in the U.S. contain relatively low levels of Se (<0.25 mg/kg). Thus, to increase Se dietary intake from mushroom crops, there is a strong need to produce Se-biofortified edible mushrooms.

ACKNOWLEDGEMENTS

The authors are grateful for the funding support from US EPA (SU836036, to Lin) and for industrial mushroom production materials provided by Monterey Mushroom, Inc. (Watsonville, California).

REFERENCES

Falandysz, J. 2008. Selenium in edible mushrooms. *J. Environ. Sci. and Health Part C* 26: 256–299.

Haddad, S. & Lin, Z.-Q. 2011. Tolerance and accumulation of selenium by White Button and Baby Bella (*Agaricus bisporus*) mushrooms. In G.S. Banuelos, Z.-Q. Lin, X.B. Yin & N. Duan (eds.), *Selenium: Global Perspectives of Impacts on Humans, Animals and the Environment*. 99–100. Hefei, China: USTC Press.

Kalač, P. & Svoboda, L. 2000. A review of trace element concentrations in edible mushrooms. *Food Chem.* 69: 273–281.
Lucier, G., Allshouse, J. & Lin, B. 2003. *Factors Affecting U.S. Mushroom Consumption*. Washington, D.C.: Economic Research Service, USDA.
Mattila, P., Könkö, K., Eurola, M., Pihlava, J.-M., Astola, J., Vahteristo, L. *et al.* 2001. Contents of vitamins, mineral elements, and some phenolic compounds in cultivated mushrooms. *J. Agric. Food Chem.* 49: 2343–2348.
NASS (National Agricultural Statistics Service), 2010. *Mushroom Crop Report*. Washington, D.C.: USDA.
Sadler, M. 2003. Nutritional properties of edible fungi. *Nutrition Bulletin* 28: 305–308.

Selenium in the Environment and Human Health – Bañuelos, Lin & Yin (Eds)
© 2014 Taylor & Francis Group, London, ISBN 978-1-138-00017-9

Selenium biofortification differences on thin and fat types of edible fungi

Lei Luo
SOA Key Laboratory for Polar Science, Polar Research Institute of China, Shanghai, China
Jiangsu Bio-Engineering Research Centre on Selenium, Suzhou, Jiangsu, China

Linxi Yuan
Jiangsu Bio-Engineering Research Centre on Selenium, Suzhou, Jiangsu, China

Xuebin Yin
SOA Key Laboratory for Polar Science, Polar Research Institute of China, Shanghai, China
Advanced Lab for Selenium and Human Health, Suzhou Institute for Advanced Study, University of Science and Technology of China, Suzhou, Jiangsu, China

1 INTRODUCTION

Selenium (Se) is an essential trace element for human body, and plays important roles on anti-oxidant, anti-aging, enhancing immunity, preventing diabetes and liver disease (Rayman, 2000). However, it is estimated that about 1 billion people in the world are suffering Se-deficiency, called "Se-hidden hunger" (Thavarajah, 2008). Edible fungi not only has high contents of proteins, polysaccharides and other physiologically active ingredients, but they also can accumulate micronutrients (*e.g.*, Se) (Manzi, 1999). Especially, previous studies displayed that edible fungi could transform inorganic Se into organic Se which was more superior in safety, physiological activity and absorption rate (Bhatia, 2013). However, few edible fungi have been selected to fortify as Se supplementation for Se-deficiency populations.

2 MATERIALS AND METHODS

Na_2SeO_3 was used as Se source to uniformly mix with cultivation materials, and the edible fungi species and designed Se groups were summarized in Table 1. According to the normal management, the Se-fortified fruiting bodies of edible fungi were collected after about 3 months. The total Se contents and the Se speciation were determined on first fruiting bodies by AFS and LC-HG-AFS, respectively. (Ogra *et al.*, 2004). In the present study, edible fungi were divided into thin type (*Flammulina velutiper*, *Agrocybe cylindracea*) and fat type (*Pleurotus eryngii*, *Copyinds comatus*) based on their appearance. The regularity of Se accumulation and transformation will be studied to provide references for large-scale Se biofortification on edible fungi.

Table 1. Edible fungi species and Se concentrations in cultivation materials.

Species	Se concentration (mg/kg)			
Flammulina velutiper	0.5	5	10	50
Agrocybe cylindracea	1.5	5	10	50
Copyinds comatus	1.5	5	10	20
Pleurotus eryngii	1.5	2.5	5	10

3 RESULTS AND DISCUSSION

The total Se contents in first fruiting bodies were shown in Figure 1. For thin type, the Se contents gradually increased from 1.87 mg/kg with addition of 0.5 mg Se/kg to 30–50 mg/kg with addition of 50 mg Se/kg. For fat type, the Se contents also gradually increased from 10–20 mg/kg with addition of 1.5 mg Se/kg to 50 mg/kg with addition of 20 mg Se/kg, which were much higher than those from thin type. To characterize the Se accumulation ability, the bioconcentration factor was calculated as the ratio of Se concentration of fruiting body and the Se concentration of cultivation material (Data was not shown). The average bioconcentration factor in fat type (*Copyinds comatus,* 1.41 and *Pleurotus eryngii,* 3.10) were greater than those in thin type (*Flammulina velutiper,* 0.57 and *Agrocybe cylindracea,* 0.84). Furthermore, the dominant Se was selenomethionine (SeMet), and the proportion of organic Se was more than 70% of total Se in both types of fungi (Data was not shown).

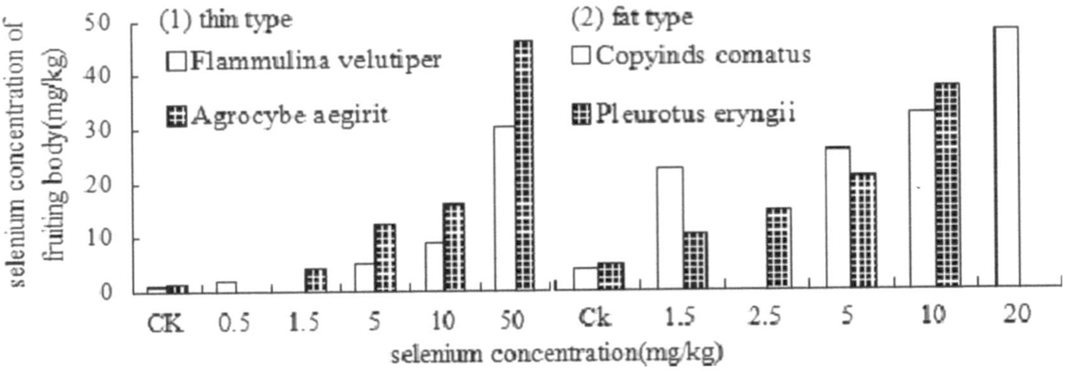

Figure 1. The total Se concentrations of first fruiting bodies in thin type (1) and fat type (2) of edible fungi after Se-biofortification.

Hu *et al.* (2012) reported that *Pleurotus eryngii* (fat type) could grow in a liquid medium containing more than 50 mg Se/kg, while the tolerance level of *Copyinds comatus* (fat type) is 50 mg Se/kg. However, *Flammulina velutiper* (thin type) have been severely suppressed under growing conditions with 40 μg/g Se concentration (Shang, 1999). It has been reported that Se could significantly increase the protein contents of edible fungi (Ling, 2008), and preferentially combine with proteins (Zhao, 2004). Since fat types (*Pleurotus eryngii, Copyinds comatus*) had much higher contents of proteins (over 20%) than those of thin types (Khan, 2009; Sadler, 2003; Khalil, 1997), the fat fungus likely accumulates more Se in their fruiting bodies.

ACKNOWLEDGEMENTS

This study was supported by Open Fund of SOA Key Laboratory for Polar Science, the Polar Research Institute of China (KP201106), the Jiangsu Research-Industry Joint Innovation Project (BY2011186) and the Suzhou Leading Talent Project (ZXL2010045).

REFERENCES

Bhatia, P., Aureli, F., D'Amato, M., Prakash, R., Cameotra, S.S., Nagaraja, T.P. *et al.* 2013. Selenium bioaccessibility and speciation in biofortified Pleurotus mushrooms grown on selenium-rich agricultural residues. *Food Chemistry* 140(2013): 225–230.

Hu, H.T., Yuan, L.X., Zhen, P. & Yin, X.B. 2012. Comparison on selenium-accumulation ability and selenium speciation in four kinds of edible fungi. *Edible Fungi of China* 31(3): 38–41. (in Chinese)

Khalil, I.A., Shah, H. & Jabeen, S. 1997. Nutritional composition and protein quality of Pleurotus mushroom. *Sarhad Journal of Agriculture* 13(6): 621–626.

Khan, M.A., Khan, L.A., Hossain M.S., Tania, M. & Uddin, M.N. 2009. Investigation on the nutritional composition of the common edible and medicinal mushrooms cultivated in Bangladesh. Bangladesh *J. Mushroom* 3(1): 21–28.

Ling, H.T., Song, B., Lin, Q.Y. & Li, T.H. 2008. Advances in Selenium-Enriched Edible Fungi Study. *Journal of Mircobiology* 28(4): 78–84. (in Chinese)

Manzi, P., Gambelli, L., Marconi, S., Vivanti, V. & Pizzoferrato, L. 1999. Nutrients in edible mushrooms: an inter-species comparative study. *Food Chemistry* 65(4): 477–482.

Ogra, Y., Ishiwata, K., Ruiz Encinar, J., Łobiński, R. & Suzuki, K.T. 2004. Speciation of selenium in selenium-enriched shiitake mushroom, Lentinula edodes. *Analytical and Bioanalytical Chemistry* 379: 861–866.

Rayman, M.P. 2000. The importance of selenium to human health. *The lancet* 356(9225): 233–241.

Sadler, M. 2003. Nutritional properties of edible fungi. *Nutrition Bulletin* 28(3): 305–308.

Shang, D.J. & Wang, G.L. 1996. Comparison of selenium accumulation ability in the mycelia of four edible fungi. *Acta Edulis Fungi* 6(3): 17–20. (in Chinese)

Thavarajah, D., Ruszkwoski, J. & Vandenberg, A. 2008. High potential of selenium biofortification of lentils (Lens culinaris L.). *Journal of Agricultural and Food Chemistry* 56: 10747–10753.

Zhao, L., Zhao, G., Zhao, Z., Chen, P., Tong, J. & Hu, X. 2004. Selenium distribution in a Se-enriched mushroom species of the genus Ganoderma. *Journal of agricultural and food chemistry* 52(12): 3954–3959.

Selenium in the Environment and Human Health – Bañuelos, Lin & Yin (Eds)
© 2014 Taylor & Francis Group, London, ISBN 978-1-138-00017-9

Plants used in the phytomanagement of selenium may be useful for producing selenium-biofortified plant products

G.S. Bañuelos
ARS-USDA, Parlier, California, USA

1 INTRODUCTION

The use of green plants and associated microbes for environmental remediation is called *phytoremediation* (Chaney *et al.*, 2000). This green technology has been developed for the cleanup/management of metalloid-contaminated soils and waters via the processes of phytoextraction, phytovolatilization, and phytostabilization (Zayed *et al.*, 2001). Based upon these processes a phytomanagement strategy for Se remediation has been developed and is in operation in the Westside of central California. Importantly, to develop a widely acceptable and sustainable selenium (Se) phytoremediation strategy requires a plan for the utilization of the harvested plants grown under Se-laden conditions. In this regard, Bañuelos and colleagues will discuss the production of Se-biofortified products. These products containing different species of organic Se may be useful in Se-deficient areas.

2 MATERIALS AND METHODS

Field Se phytoremediation studies were conducted between 2002–2013 on 50 to 100 ha field sites located at Red Rock Ranch (RRR), Five Points, CA and at the USDA-ARS research facility on Parlier, CA. The soil in the experimental area at RRR was classified as an Oxalis silty clay loam (fine montmorillonitic, thermic Pachic haploxeral with a well-developed salinity profile). Soil salinity (EC = Electrical Conductivity) was <10 dS/m and water-extractable soil Se and B concentrations ranged from 0.50 to 1.5 and 8 to 13 mg/L, respectively, to a soil depth of 90 cm. Drainage water produced from this soil had a range of soluble Se from 0.100 to 0.150 mg/L, soluble B from 4 to 7 mg/L, and a sodium sulfate-dominated salinity from 5 to 8 dS/m. Selected plants were exposed to Se growing conditions as follows: (1) cabbage (*B. oleracea Var. capitata*) varieties and *Stanleya pinnata* grown in Se-laden soils (approximately total Se of 2.5 mg/kg; (2) pomegranate (*Punica granatum*) irrigated with saline drainage water containing Se (0.25 mg/L) and either 3 or 6 dS/m and 4 mg B/L; (3) broccoli (*B. oleracea*) and carrots (*Daucus carota*) grown in soil amended with Se-enriched *S. pinnata* (approximately 400 mg Se/kg). After harvest, all samples were prepared for Se analysis as described in detail by Bañuelos *et al.* (2011). Concentrations of total Se were measured by Inductively-Coupled Plasma Mass Spectrometry (ICP-MS), while soluble Se compounds in aqueous extractions were determined using Liquid Chromatography Mass Spectrometry (LC-MS) in conjunction with ICP-MS (Bañuelos *et al.*, 2011). The Se standard compounds included selenite, selenite, selenocysteine (SeCys), selenomethionine (SeMet), methylselenocysteine (MeSeCys), and selenodiglutathione (GS-Se-SG).

3 RESULTS

We have shown that some common and alternative crops are able to remove Se from different sources of Se, *e.g.*, soil, drainage water, or soils amended with organic Se, despite the high sulfate-salinity conditions. Figure 1 shows the accumulation and removal of Se from natural-occurring soil Se by different vegetables, while Figure 2 shows the percentages of soluble Se in aqueous extracts. Figure 3 shows total Se in fruit juice of pomegranates irrigated with Se-containing drainage water at different salinity and boron levels, while Figure 4 shows the percentages of soluble Se in aqueous extracts from two treatments. Figure 5

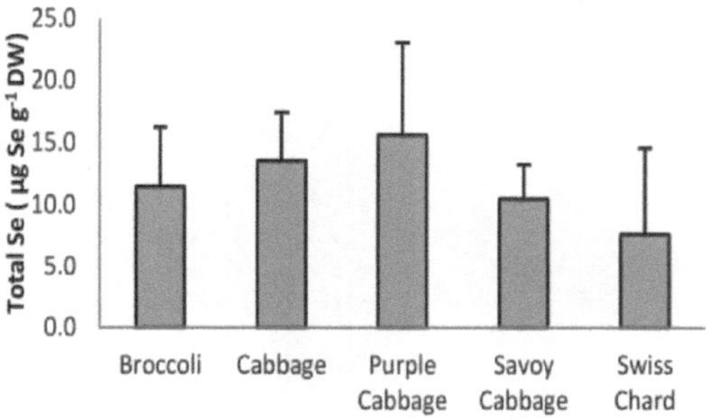

Figure 1. Total Se concentration in vegetables grown in Se-laden soils.

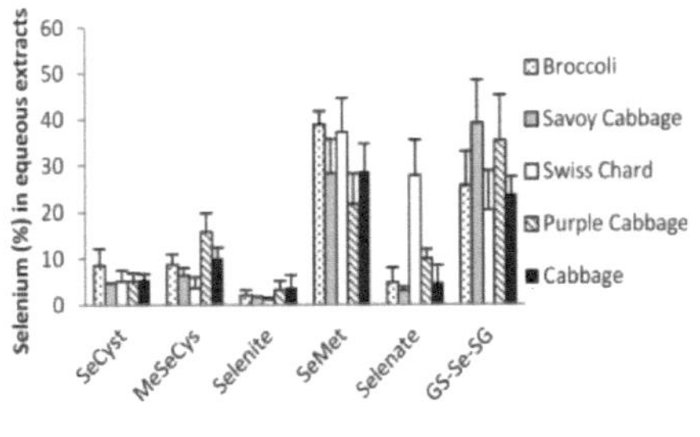

Figure 2. Percentages of soluble Se in aqueous extracts from vegetables grown in Se-laden soils.

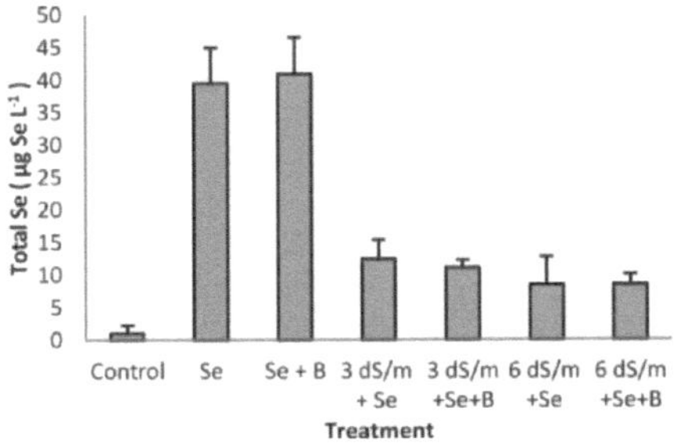

Figure 3. Total Se concentrations in fruit juice from pomegranates irrigated with Se-laden salt and boron water.

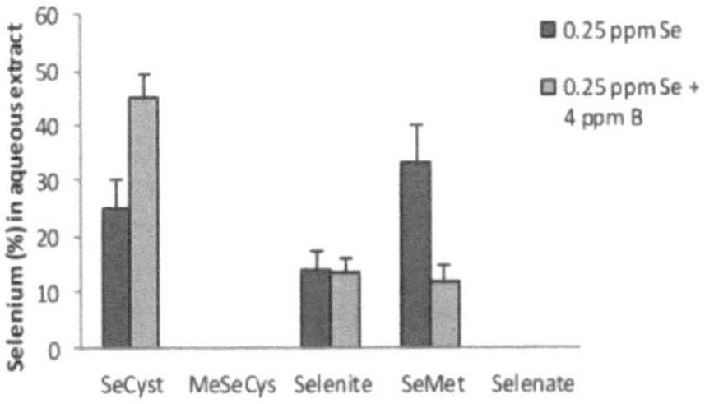

Figure 4. Percentage of soluble Se in aqueous extracts from pomegranates irrigated with Se-laden salt and boron water.

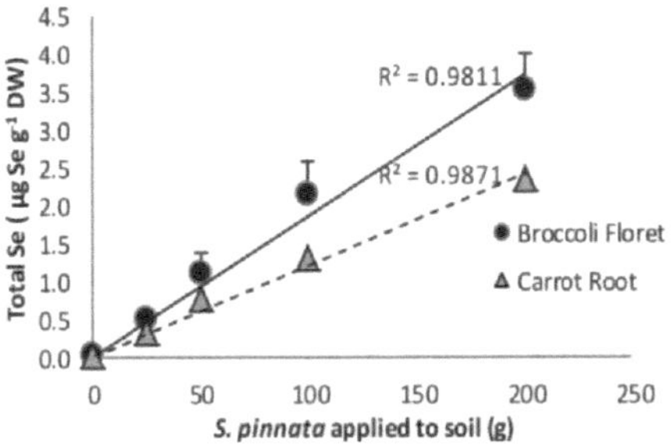

Figure 5. Total Se concentrations in broccoli and carrots grown in soils amended with different rates of Se-enriched *S. pinnata*.

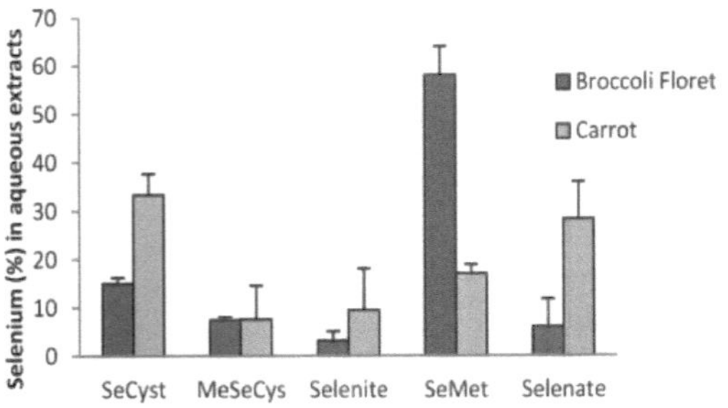

Figure 6. Percentages of soluble Se in aqueous extracts from broccoli and carrots grown in soils amended with *S. pinnata*.

shows the total accumulated Se in broccoli and carrots removed from a soil amended with different rates of Se-enriched *S. pinnata*, while Figure 6 shows the general percentages of their soluble Se in aqueous extracts from all treatments (Bañuelos and Freeman, unpublished).

4 CONCLUSIONS

Sustainable operation of an agronomic-based phytomanagement system is dependent upon developing a phytoremediation system for managing Se compounds, while producing bio-based products that may have value in Se biofortification strategies. Successful phytomanagement of Se-contaminated water and soil may also provide growers with new and innovative Se-enriched commodities.

REFERENCES

Bañuelos, G.S., Fakra, S.C., Walse, S.S., Marcus, M.A., Yang, S.I., Pilon-Smits, E.A.H. et al. 2011. Selenium accumulation, distribution and speciation in spineless prickly pear cactus: a salt, boron, and drought tolerant, selenium-enriched nutrecutical fruit crop. *Plant Physiology*, 155: 315–327.

Chaney, R.L., Li, Y.M. & Brown, S.L. 2000. Improving metal hyperaccumulator wild plants to develop commercial phytoextraction systems: Approaches and Progress. In N. Terry & G.S. Bañuelos (eds.), *Phytoremediation of Contaminated Soil and Water*, 129–159. Boca Raton, Florida: CRC Press.

Zayed, A., Pilon-Smits, E., de Souza, M., Lin, Z.Q. & Terry, N. 2000. Remediation of selenium polluted soils and waters by phytovolatilization. In N. Terry & G.S. Bañuelos (eds.), *Phytoremediation of Contaminated Soil and Water*, 61–83. Boca Raton, Florida: CRC Press.

Selenium in the Environment and Human Health – Bañuelos, Lin & Yin (Eds)
© 2014 Taylor & Francis Group, London, ISBN 978-1-138-00017-9

Effects of selenium fertilization on selenium accumulation, nutrient contents and yield in baby pak choi (*Brassica rapa* subsp. chinensis)

G.F. Xi & C.-X. Yao
Institute for Agro-Product Quality Standards and Testing Technologies, Shanghai Academy of Agricultural Sciences, Shanghai, China

S.B. Zhou & J.J. Kong
College of Environmental Science and Engineering, Anhui Normal University, Wuhu, Anhui, China

1 INSTRUCTION

Selenium (Se) is an essential nutrient for humans and animals. Previous studies showed that both deficient and excess concentrations of Se can lead to a variety of diseases in humans and animals (Wu *et al.*, 2000). The consumption of vegetables accounts for a significant portion of human dietary Se intake and therefore, the production of Se-enriched vegetables or produces is an effective source of Se intake. The objectives of this present study were to evaluate the effects of Se fertilization on Se and nutrient contents in baby pak choi and to determine the effects of the Se fertilization on the vegetable growth or yield. Our research findings will provide a better scientific understanding for the production of Se-biofortified green vegetables.

2 MATERIALS AND METHODS

Baby pak choi or shanghai green (*Brassica rapa* subsp. *chinensis*) was selected for the field trials containing soils treated with different amounts of Se fertilizer (1000 mg Se/kg). The treatments included: control (CK, 0 g Se/hm^2 or 10000 m^2), T1 (37.125 g Se/hm^2), T2 (74.925 g Se/hm^2), T3 (112.725 g Se/hm^2), T4 (149.85 g Se/hm^2). All Se fertilizers and basic fertilizers were applied when plowing. The physio-chemical properties of the experimental soil were: pH 6.54, organic carbon 2.69 g/kg, available N 209 mg/kg, available P 167 mg/kg, and available K 178 mg/kg.

After harvest, concentrations of Se in plant leaf tissues were measured by using Atomic-Fluorescence Spectroscopy (AFS), nitrate concentrations by UV spectrophotometry, Vitamin C by the 2,6-dichloro indophenol titration method (Li, 2000), soluble sugar content by the Orman method (Orman, 1980), protein assay by the Coomassie brilliant blue G-250-binding method (Bradford, 1976), and nitrate reductase by the Sabulal method (Sabulal, 2007).

3 RESULTS AND DISCUSSIONS

3.1 *Effects of Se fertilizers on Se content*

Selenium concentrations in the leaves of baby pak choi were shown in Table 1. The Se content increased from 8.2 ± 1.6 μg/kg in the control group to 1417.7 ± 8.5 μg/kg in the 112.725 g/hm^2 Se-fertilizer treated group. There was, however, no difference on Se contents in upper plant parts between T3 and T4, which indicates the appropriate Se-fertilizer dosage might be 110–150 g/hm^2.

3.2 *Effects of Se fertilizers on yield*

The yields of the vegetable gradually decreased from 10,923 kg/hm^2 in control to 10,430 kg/hm^2 in the soil treatment of 149.85 g Se/hm^2. The high Se fertilizer treatment could affect the growth of the plant and likely reduce its yields, but the difference was not significant ($P > 0.05$).

Table 1. Concentrations of Se in leaves of baby pak choi treated with different levels of soil Se fertilizer application.

Treatment	Se concentration (μg/kg)
Control	8.2 ± 1.6[a†]
T1	96.7 ± 2.4[b]
T2	917.9 ± 78.4[c]
T3	1417.7 ± 8.5[c]
T4	1373.6 ± 268.2[c]

[†]Means with different letters show statistically significant difference ($P < 0.05$).

Table 2. Effect of Se fertilizer on Vitamin C, protein, and soluble sugar content in baby pak choi.

Treatment	Vitamin C (mg/kg)	Protein (mg/kg)	Soluble sugar (mg/kg)
Control	37.0 ± 0.67	10.77 ± 0.59	3.38 ± 0.36[a†]
T1	37.5 ± 5.00	10.66 ± 0.73	2.68 ± 0.39[ab]
T2	37.5 ± 0.00	9.77 ± 1.08	2.95 ± 0.72[ab]
T3	42.5 ± 6.67	10.64 ± 0.69	3.05 ± 0.31[ab]
T4	40.0 ± 3.33	10.86 ± 0.93	2.37 ± 0.4[b]

†Means with different letters show statistically significant difference ($P < 0.05$).

3.3 *Effects of Se fertilizers on other nutrient content*

The contents of Vitamin C, proteins and soluble sugars are presented in Table 2. Overall, there were no significant differences in Vitamin C and protein content among the Se treatments, indicating that that the Se treatments in this study did not significantly affect the nutritious quality of baby pak choi. However, the Se treatment significantly affected the soluble sugar content in the plant tissue. Previous studies have reported that Se fertilizer had a significant role on enhancing crop quality (Lu *et al.*, 2009; Li *et al.*, 2007). With the application of Se fertilizers, the nitrate content had decreased by 37%, Vitamin C and soluble sugar content increased by 54% and 16.3%, respectively.

3.4 *Effects of Se fertilizers on mineral element accumulation in baby pak choi*

Concentrations of Ca, Mg, Fe, K, Zn and Mn in leaves are shown for the different Se fertilizer treatments (Table 3). Generally, the contents of Ca, Mg, Fe, K, Zn and Mn first increase and then decrease, with increasing Se dosage. Selenium displayed significant synergistic effects on the accumulation of Fe, Zn, Ca, K and Mn, especially with treatment of Se of < 112.725 g/hm². In contrast, slight increases of Mg content were observed at Se treatments less than 112.725 g/hm², and significant antagonistic effects at the two highest Se treatments.

3.5 *Effects of Se fertilizer on nitrate reductase activity and nitrate contents*

Results in Table 4 show a decrease in nitrate content and an increase in nitrate reductase activity with increasing the Se treatment. The nitrate accumulation and the nitrate reductase activity had a significantly negative correlation ($y = -0.0055x + 57.733$, $R^2 = 0.777$), indicating that Se likely played an important role on reducing nitrate accumulation in the edible leaves.

Table 3. Effects of Se fertilizer on concentrations of mineral elements in baby pak choi.

	Element concentrations (mg/kg)		
Treatment	Cd	Mg	Fe
Control	16.81 ± 0.60[a†]	2.15 ± 0.16[ac]	7.48 ± 0.32[a]
T1	29.54 ± 2.25[b]	2.89 ± 0.50[ab]	38.10 ± 9.47[bc]
T2	30.67 ± 2.26[b]	2.54 ± 0.08[b]	35.77 ± 1.09[b]
T3	30.31 ± 2.43[b]	1.92 ± 0.12[c]	49.65 ± 1.97[c]
T4	22.80 ± 0.46[c]	1.46 ± 0.46[c]	34.18 ± 0.81[b]
	K	Zn	Mn
Control	10.13 ± 1.00[a†]	5.84 ± 0.13[a]	8.75 ± 0.31[a]
T1	16.24 ± 4.43[ab]	13.36 ± 1.64[b]	13.22 ± 0.69[b]
T2	16.30 ± 0.68[bc]	24.40 ± 1.32[c]	13.48 ± 1.44[b]
T3	11.83 ± 2.99[ac]	19.12 ± 0.96[d]	16.54 ± 1.18[c]
T4	12.27 ± 0.31[a]	14.66 ± 0.66[c]	10.27 ± 0.60[d]

†Means with different letters show statistically significant difference ($P < 0.05$).

Table 4. The effect of Se fertilizer treatments on accumulation of nitrate and nitrate reductase activity in baby pak choi leaves.

Treatment	$NaNO_3$ (mg/kg)	Enzyme activity (μgNO_2/g/h)
Control	7226.88 ± 96.42	19.60 ± 0.55
T1	7124.61 ± 130.55	18.93 ± 1.61
T2	6272.39 ± 248.95	19.71 ± 1.81
T3	5829.23 ± 171.58	27.08 ± 1.39
T4	5590.61 ± 136.36	28.42 ± 1.76

4 CONCLUSIONS

The application of soil Se fertilizer significantly increased the accumulation of Se and other mineral nutrients in above-ground vegetable tissues, as well as enhanced nitrate reductase activity, and reduced leaf nitrate contents. The application of soil Se fertilizer did not, however, significantly affect the content of plant Vitamin C and proteins. The content of soluble sugars was reduced at a high level of Se fertilizer (T4). Lastly, increased levels of Se fertilizer will reduce yields in baby pak choi.

ACKNOWLEDGEMENTS

This research was supported by the Natural Science Foundation of Shanghai, China (No.12ZR1420200) and Special Research of State Administration of Quality and Public service sectors (No. 201310269).

REFERENCES

Bradford, M.M. 1976. A rapid and sensitive method for the quantitation of microgram quantities of protein utilizing the principle of protein-dye binding. *Analytical Biochemistry* 72: 248–254.

Li, H.S. 2000. *The Experimental Principles and Techniques of Plant Physiology and Biochemistry*, 246–247. Beijing: Higher Education Press.

Li, J., Xia, J.G., Li, T.X. & Gong, F.Y. 2007. The effects of spraying sodium selenite on selenium content and chemical quality of tea. *Jiangsu Agricultural Sciences* (3): 186–187.

Lu, M.M. & Hong, J.P. 2009. The effect of different concentrations of selenium on the quality of hydroponic Chinese chive. *Journal of Shanxi Agricultural University* (Natural Science edition) 29(4): 340–342.

Orman, R.G. 1980. Peroxide levels and the activities of catalase, peroxidase, and indoleacetic acid oxidase during and after chilling cumber seedlings. *Plant physiol.* 65: 407–408.

Sabulal, B., Dan M. & John, A. 2007. Phenybutaniod-rich rhizome oil of Zingiber neesanum from Western Ghats, southern India. *Flavour and Fragrance Journal* 22: 521–524.

Wu, Q.L., Yang, Y.A. & Xie, Z.M. 2000. *Trace Elements and Biological Health,* 1–20. Guiyang: Guizhou Science and Technology Publishing House.

Selenium in the Environment and Human Health – Bañuelos, Lin & Yin (Eds)
 ISBN 978-1-138-00017-9

Selenium biofortified maize in typical selenium deficient areas of China

Yuanyuan Zhu & Xuebin Yin
SOA Key Laboratory for Polar Science, Polar Research Institute of China, Shanghai, China

Shoufeng Liu & Linxi Yuan
Jiangsu Bio-Engineering Research Centre on Selenium, Suzhou, Jiangsu, China

1 INTRODUCTION

Selenium (Se) is an essential trace element for humans and animals (Rayman, 2000). In China, there is a Se deficiency belt that distributes from northeast to southwest crossing 22 provinces (Tan, 1989). Many studies have demonstrated that there were many diseases related to Se deficiency (Coppinger *et al.*, 2001; Gore *et al.*, 2010). Today, Se biofortified food is regarding as a simple and safe way to supplement people's requirement for Se in deficient areas. In the present study, three typical Se deficient areas, including Heze city in Shangdong province, Mengcheng city in Anhui province and Taixing city in Jiangsu province, were selected to carry out Se biofortification on the main crop maize (*Zea mays*).

2 MATERIALS AND METHODS

Field plots of 13.4, 1.34 and 1.34 ha were created with maize in Heze city in Shangdong province, Mengcheng city in Anhui province, and Taixing city in Jiangsu province, respectively (Fig. 1). Selenium fertilizer (Se content: 1000 mg/kg) purchased from Suzhou Setek company was applied near the maize roots (or cultivation layer) at the jointing growth stage.

The amounts of Se fertilizer applied were 90 to 225 kg/ha, and the maize varieties at each of the three experimental sites were Zhengdan 958 in Heze and Mengcheng, Suyu 26 in Taixing, and Eyu18 in Enshi. Maize samples were collected after harvest, and total Se and Se speciation were measured by AFS and LC-UV-AFS (Titan, Beijing), respectively.

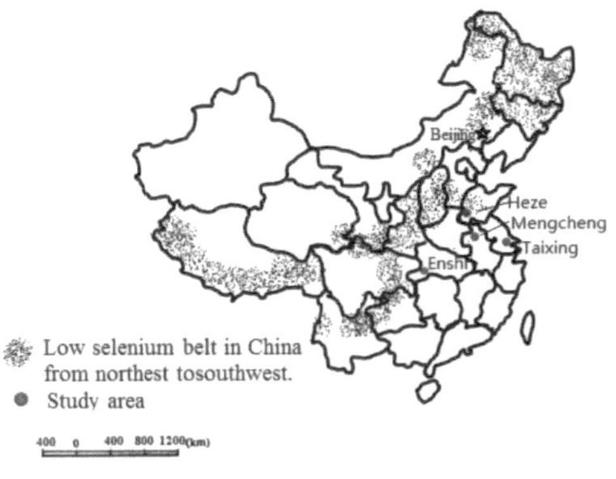

Figure 1. Selenium biofortified maize in Se deficient areas, of Heze, Mengcheng and Taixing, compared with Enshi located in a seleniferous area in China.

3 RESULTS AND DISCUSSION

The total Se contents of biofortified maize in Heze city, Mengcheng city and Taixing city ranged from 1.491 to 12.81, 0.327 to 16.10 and 0.235 to 15.97 mg/kg respectively (Table 1). The Se contents of biofortified maize were 10–1000 times higher than that of non-biofortified maize (Se content: 0.021 ± 0.018 mg/kg). The main Se speciation in maize was selenomethionine (SeMet) that accounted for over 65% of the total Se. This observation indicates that most Se applied in the inorganic form could be eventually transformed into organic Se.

To determine the best Se content for biofortification, the distribution patterns were illustrated in Figure 2. It is obvious that 30–60% Se contents stayed in a range of 2–5 mg/kg. However, about 60% of Se contents were lower than 2 mg/kg in Mengcheng city.

Table 1. Concentrations of Se and dominant SeMet contents in Se-biofortified maize in different agricultural zones in China.

Site	Total Se (mg/kg)	SeMet (%)
Heze	1.491–12.81 ($n = 31$)	73 ± 19 ($n = 22$)
Mengcheng	0.327–16.10 ($n = 24$)	68 ± 16 ($n = 14$)
Taixing	0.235–15.97 ($n = 30$)	65 ± 11 ($n = 13$)
Enshi	0.239–23.40 ($n = 42$)	72 ± 17 ($n = 16$)

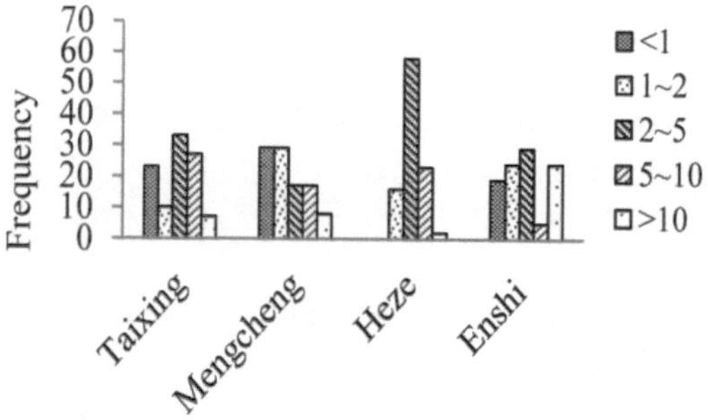

Figure 2. The frequency of different total Se concentrations observed in Se-biofortified maize grain at four different study locations in China.

Because the Se content of the soil from the city of Enshi is already at a high level (Zhu *et al.*, 2008), the frequency of high selenium concentration in maize samples of this area (more than 10 mg/kg) is much higher than the Se biofortified areas. Among the three typical Se deficient areas, it is much more effective via biofortification in Heze city, since about 60% of Se contents could reach up to 2–5 mg /kg in Se-biofortified maize.

4 CONCLUSIONS

The Se contents of biofortified maize were 0.235–16.10 mg/kg, which were 10–1000 times higher than that in non-Se-biofortified one (0.021 ± 0.018 mg/kg). Furthermore, more than 65% of total Se was transformed into organic Se. There was no significant difference in Se accumulation between the Se-biofortified maize and the maize grown in the naturally occurring seleniferous soil in Enshi, suggesting that Se-biofortification would be an effective way to increase Se contents in crops grown in Se deficient soils or areas.

ACKNOWLEDGEMENTS

This study was supported by Open Fund of SOA Key Laboratory for Polar Science, the Polar Research Institute of China (KP201106), Jiangsu Research-Industry Joint Innovation Project (BY2011186), and Suzhou Leading Talent Project (ZXL2010045).

REFERENCES

Coppinger, R.J. & Diamond, A.M. 2001. Selenium deficiency and human disease. In D.L. Hatfield, M.J. Berry, V.M. Gladyshev (eds.), *Selenium: Its Molecular Biology and Role in Human Health*, 219–331. Dordrecht, The Netherlands: Kluwer Academic Publisher.

Gore, F., Fawell, J. & Bartram, J. 2010. Too much or too little? A review of the conundrum of selenium. *Journal of water and health* 8(3): 405–416.

Rayman, M.P. 2000. The importance of selenium to human health. *The Lancet*. 356: 233–241.

Tan, J.A. 1989. The Atlas of Endemic Disease and Their Environment. Beijing, China: Science Press.

Zhu, J.M., Wang, N. & Li, S.H. 2008. Distribution and transport of selenium in Yutangba, China: Impact of human activities. *Sci. Total Environ.* 392(2–3): 252–261.

Selenium in the Environment and Human Health – Bañuelos, Lin & Yin (Eds)
© 2014 Taylor & Francis Group, London, ISBN 978-1-138-00017-9

The selenium speciation in the seeds of the common wheat genotypes tending to accumulate high concentrations of selenium

Tao Li, An Wang, Pan Gong & Shiliang Gu
Jiangsu Provincial Key Laboratory of Crop Genetics and Physiology; Key Laboratory of Plant Functional Genomics of Ministry of Education, Yangzhou University, Yangzhou, China

Linxi Yuan, Fei Li, Xuebin Yin & Wenwen Guan
Suzhou SeTek Ltd., Suzhou, China
Advanced Lab for Selenium and Human Health, Suzhou Institute of University of Science and Technology of China, Suzhou, China

1 INTRODUCTION

Selenium (Se), as a constituent of selenoproteins and an essential element in human and animal nutrition, is deficient in many countries. Consequently Se supplementation to the diet is nessesary, which can be achieved by the biofortification of wheat, a staple crop that accounts for 20% of the calories consumed by humans (Brenchley *et al.*, 2012). The biological effects of selenocompounds are highly dependent on its chemotype (Johansson *et al.*, 2005). Analysis of the Se species, however, has been performed only in very few wheat genotypes (Cubadda *et al.*, 2010). In this paper, the Se speciation in the seeds was determined in the wheat cultivars accumulating high amounts of total Se.

2 MATERIALS AND METHODS

2.1 *Materials and experimental design*

One hundred and twenty five wheat cultivars (or inbred lines) were planted in plastic pots filled with 10 kg of soil. Three levels (treatments) of Se nanoparticle fertilizer (supplied by Suzhou SeTek Ltd., Suzhou, China) and two replicates (pots) per treatment were performed. CK: the control (without applying Se fertilizers); Treatment 1 (Trt1): 1 g Se fertilizer per pot, and treatment 2 (Trt2): 1.5 g Se fertilizer per pot. The fertilizer was top-dressed at the elongation period of wheat in the spring season of 2012.

2.2 *Sample preparation, total Se determination and speciation analysis*

A 0.5 g dry wheat seeds was digested in a solution mixture of 4 mL HNO_3, 3 mL ddH_2O and 2.5 mL H_2O_2. The digestion process was performed in CEM-MARS 5 microwave oven (USA) following the recommended protocol. Total Se concentration was analyzed by ICP (Solar S4+GraphiteFurnace System 97, Thermo Elemental, USA). Ten top ranking cultivars in Trt2 samples in total Se content were then subjected to selenium speciation determination by LC-UV-AFS in Suzhou SeTek, Ltd., following the protocol recommended by Liang *et al.* (2006).

3 RESULTS AND DISCUSSION

3.1 *Total selenium determination*

Selenium application significantly increased the total Se concentration. The more fertilizer applied, the higher the Se concentration in the seeds had. The increase of Se varied among genotypes. The average of total Se concentration were 795 μg/kg in CK seeds, 16586 μg/kg in Trt1 and 24956 μg/kg in Trt2, respectively, The ratios of Trt1/CK, Trt2/CK and Trt2/Trt1 were 4.18–13.54, 8.29–17.69 and 1.05–2.73, respectively. In Trt 2, two accessions, X359 with a Se of concentration: 30699.43 μg/kg and Shan89150 with a Se of concentration: 31210.10 ug kg^{-1}, were the two top accessions in total Se concentrations.

3.2 *Selenium speciation in the high-Se seeds*

Five species of Se were detected in Trt2 samples. They were SeMet, SeCys, SeMeCys, $Se4^+$ and $Se6^+$, however, the percentages of each strongly depended on the genotypes (Fig. 1). SeMet accounted for the largest proportion of total Se, ranging from 47.76% (cv. Zhen9023) to 83.83% (cv. Ximai1376), with an average of 70.44%. The content of SeCys and its methylated form—SeMeCys—varied from 4.22% (Shan89150) to 23.85% (X359), with an average of 15.36%. Both $Se4^+$ (<0.37%) and $Se6^+$ (<3.42%) were found in trace amounts in the seeds. SeCys

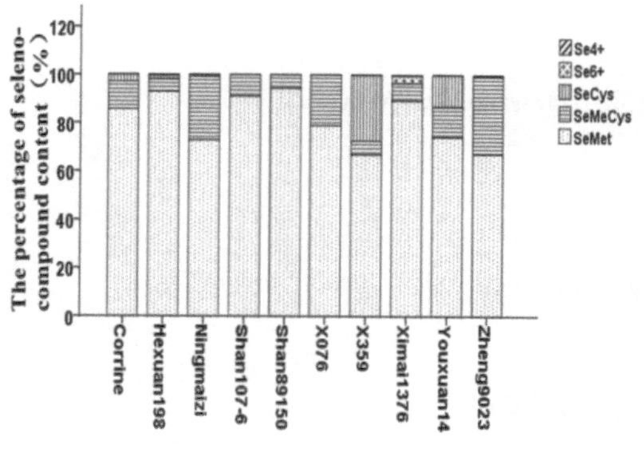

Figure 1. Selenium species in the ten wheat cultivars.

and SeMeCys are considered to be more favorable forms of Se to human and animal health, since they are not incorporated into proteins. In this regard, five cultivars had higher percentages of SeCys and SeMeCys than the average of 15.36%. They were X359 (23.85%), Zhen9023 (22.90%), Hexuan198 (22.67%), Ningmaizi18 (20.37%) and X076 (17.86%). Even though cv. Shan89150 had high total Se concentration, it had only contained 4.22% of both SeCys and SeMeCys, and 65.01% of SeMet. These percentages were significantly lower than the averages of the ten cultivars.

4 CONCLUSIONS

Both total Se concentration and Se species varied among wheat genotype. Only those genotypes (*i.e.*, cv. X359, an elite inbred wheat line developed in Yangzhou University, China) accumulated both high total Se and favorable selenocompounds. This cultivar would be an ideal choice for Se biofortification program and could be valuable in breeding programs, baking, and milling industry.

ACKNOWLEDGEMENT

This work was supported by grants from the Priority Academic Program Development of Jiangsu Higher Education Institutions, and by the Innovative Research Team of Universities in Jiangsu Province.

REFERENCES

Brenchley, R., Spannagl, M., Pfeifer, M., Barker, G.L., D'Amore, R., Allen, A.M. *et al.* 2012. Analysis of the bread wheat genome using whole-genome shotgun sequencing. *Nature*, 491: 705–710.

Cubadda, F., Aureli, F., Ciardullo, S., D'Amato, M., Raggi, A., Acharya, R. *et al.* 2010. Changes in selenium speciation associated with increasing tissue concentrations of selenium in wheat grain. *J. Agric. Food Chem.* 58: 2295–2301.

Johansson, L., Gafvelin, G. & Arnér, E.S. 2005. Selenocysteine in proteins-properties and biotechnological use. *Biochimica et Biophysica Acta (BBA)-General Subjects* 1726: 1–13.

Liang, L., Mo, S., Zhang, P., Cai, Y., Mou, S. *et al.* 2006. Selenium speciation by high-performance anion-exchange chromatography-post-column UV irradiation coupled with atomic fluorescence spectrometry. *J. Chromatography* 1118: 139–143.

Selenium in the Environment and Human Health – Bañuelos, Lin & Yin (Eds)
© 2014 Taylor & Francis Group, London, ISBN 978-1-138-00017-9

Microbial-enhanced selenium biofortification of wheat (*Triticum aestivum* L.)

M. Yasin & M. Faisal
Department of Microbiology and Molecular Genetics, University of the Punjab, Quaid-e-Azam Campus, Lahore, Pakistan

E.A.H. Pilon-Smits
Colorado State University, Biology Department, Fort Collins, CO, USA

1 INTRODUCTION

Selenium (Se), which is ubiquitous in the environment, is an essential micronutrient that can potentially be toxic, depending on its chemical form and dose. Unlike other trace elements that are also essential, it can be toxic for human health such as manganese and chromium. Selenium has garnered strong consideration from the scientific community and the public as a potential anti-carcinogen and antioxidant. Several laboratory studies have shown that Se may interact with other elements such as arsenic and mercury and that this may reduce the toxicity of both the Se and the other elements.

Foods (*i.e.*, vegetables, grains, cereals and other consumable plant products) can contain various chemical forms and concentrations of Se. For human beings an estimated and adequate Se intake is 50–60 μg/day, while toxic levels for Se intake are 350–700 μg/day. In different organisms inorganic selenite can be converted into organic forms, which are considered safer and more effective dietary sources of Se. Selenium may also be complexed via binding with various polysaccharides and proteins. For human beings, Se consumption or supplementation with low doses has substantial health benefits, among which the prevention of cardiovascular disease and cancer is of special importance (Sanmartín *et al.*, 2012).

Soil Se is very uneven in distribution as well as chemical availability. So, the Se content of the soil, from which crops and other consumable plant foods are derived, have major influence on dietary intake of Se; plants are major sources of Se. Around the world Se concentration in most of the soils is 0.1–2 μg Se/kg, with a mean of ~0.4 mg/kg; however, in seleniferous areas, higher concentrations (>10 mg/kg) can also occur. Selenium deficient soils in Australia, central Siberia, New Zealand, northeast to south central China, Denmark, parts of Bangladesh and India, produce crops with very low Se contents (Combs, 2001).

A better understanding of Se uptake and metabolism in plants is highly necessary to help find a solution for curtailing nutritional Se deficiencies. This goal may be achieved by elevating soil Se levels or by enhancing uptake efficiency. Bacteria are known to affect elemental bioavailabilty and Se uptake by plants. In this study, we tested a Se-tolerant *Bacillus pichinotyi*-YAM2 strain for its capacity to enhance growth, as well as Se uptake in wheat (*Triticum aestivum* L.).

2 MATERIALS AND METHODS

These studies were carried out using *T. aestivum* (variety Seher 2006) and bacterial strain *B. pichinotyi* -YAM2, which was isolated from industrial drainage sediment and can tolerate high concentrations of sodium selenite (>20 mg/ml of LB-agar). The plants were grown to maturity in 8 kg pots (n = 3 pots with 15 plants each) and treated with/without Se and bacterial inoculum. Selenium treatment consisted of two additions of 3 mg/kg sodium selenate. Inoculation was performed at the seedling level with 3 mL culture (OD600 = 1) per pot. Elemental analysis included acid digestion followed by inductively-coupled plasma atomic emission spectrometry (ICP-AES).

3 RESULTS AND DISCUSSION

3.1 *Isolation of Strain YAM2*

Strain YAM2 was isolated from the sediments of a Roohi Nala drain, sub branch of Hudaira drain that receives domestic as well as industrial effluents of Lahore city. Strain YAM2 was identified using 16rDNA sequencing; it shows 99% homology with *Bacillus pichinotyi*. YAM2 can tolerate Na_2SeO_3 to levels >20 mg/ml of LB-agar.

3.2 *Effect of inoculation on plant growth*

In the absence of Se, YAM2 stimulated wheat plant growth: Inoculated plants showed a significant increase in dry wt/ pot (41%) and shoot length (14%) compared to un-inoculated control (Table 1).

Hayat (2010) have also reported that beneficial soil bacteria can promote plant growth.

3.3 *Effect of Se treatment on plant growth*

Selenium treatment at 3–6 mg Se/kg soil showed a positive effect on *T. aestivum* growth parameters. Plants treated with Se showed significant increases in germination % (9.4%), dry wt (38%), shoot length (5%) and spike length (8%) compared to untreated control plants (Table 2). Kong *et al.* (2005) have also reported that at low concentration Se promote plant growth. But we have found that even at high concentration Se can promote wheat plant growth.

In the presence of Se, YAM2-inoculated plants showed a significant increase in dry wt. (35%), shoot length (16%) and spike length (11%) compared to un-inoculated Se-treated control plants (Table 3). From literature it is proved that microorganisms can enhance plant growth and plant biomass production under heavy metal stress conditions (Sheng & Xia, 2006).

Interestingly, Se-treated inoculated plants showed a significant increase in tissue Se concentration in both stem (252%) and kernels (167%) compared to Se-treated control plants (Table 3). de Souza *et al.* (1999) have also reported that rhizosphere bacteria can enhance Se accumulation in Indian mustard. Thus, YAM2 positively affected both plant growth and Se concentration, two important traits for biofortification.

Table 1. Effect of YAM2 strain inoculation on selected physical growth parameters of *Triticum aestivum*. The data shown are mean ± one standard error in parenthesis (n = 3). Mean values followed by different letters in a column for the same parameter are significantly different.

Treatment	Germination (%)	Dry Wt (g/pot)	Shoot length (cm)	Spike length (cm)
Control plants	88 (1.7)[a]	27.7 (2.5)[a]	57.7 (0.6)[a]	7.6 (0.2)[a]
Inoculated plants	95 (2.9)[a]	39.0 (2.1)[b]	65.6 (0.5)[b]	8.3 (0.3)[a]
P-value	0.134	0.025	0.000003	0.136

Table 2. Effect of Selenium treatment over growth parameters % germination (n = 3), dry weight/pot (n = 3), shoot length (n = 5) and spike length (n = 5) of *Triticum aestivum*. The data shown are mean ± one standard error in parenthesis. Mean values followed by different letters in a column for the same parameter are significantly different.

Treatments	Germination (%)	Dry Wt (g/pot)	Shoot length (cm)	Spike length (cm)
Control plants	88 (1.7)[a]	27.67 (2.49)[a]	57.7 (0.6)[a]	7.6 (0.2)[a]
Se-treated plants	96 (1.7)[b]	38.25 (1.63)[b]	60.3 (0.4)[b]	8.2 (0.2)[b]
P-value	0.024	0.037	0.007	0.04

Table 3. Effect of YAM2 strain inoculation on the physical growth parameters of plant treated with Se: dry weight per pot (n = 3), shoot length (n = 5) and spike length (n = 5) as well as Se content (n = 3) in *Triticum aestivum* treated with Se. The data shown are mean ± one standard error in parenthesis. Mean values followed by different letters in a column for the same parameter are significantly different.

Treatment	Dry Wt (g/pot)	Length (cm) Shoot	Spike	Se concentration (mg/kg) Stem	Kernel
Control	38.3 (1.6)[a]	60.3 (0.4)[a]	8.15 (0.2)[a]	10.8 (0.7)[a]	34.77 (1.7)[a]
Inoculated plants	51.4 (3.2)[b]	69.8 (0.7)[b]	9.07 (0.4)[b]	37.9 (1.8)[b]	92.79 (9.0)[b]
P-value	0.03	<0.001	0.05	0.001	0.02

4 CONCLUSIONS

YAM2 significantly stimulated plant Se accumulation, both via enhanced plant growth and higher tissue Se levels. Se levels were higher in the kernel as well as the stem. The enhanced kernel Se level is important for human consumption and the stem for animal consumption. Thus, YAM2 is showing very promising results with respect to Se biofortification of wheat. It may show similar results for other important crop species.

REFERENCES

de Souza, M.P., Chu, D., Zhao, M., Zayed, A.M., Ruzin, S.E., Schichnes, D. & Terry, N. 1999. Rhizosphere bacteria enhance selenium accumulation and volatilization by Indian mustard. *Plant Physiol.* 119: 565–574.

Hayat, R., Ali, S., Amara, U., Khalid, R. & Ahmed, I. 2010. Soil beneficial bacteria and their role in plant growth promotion: A review. *Annals Microbiol.* 60: 579–598.

Kong, L., Wang, M., & Bi, D. 2005. Selenium modulates the activities of antioxidant enzymes, osmotic homeostasis and promotes the growth of sorrel seedlings under salt stress. *Plant Growth Regul.* 45: 155–163.

Sanmartin, C., Plano, D., Sharma, A.K. & Palop, J.A. 2012. Selenium compounds, apoptosis and other types of cell death: an overview for cancer therapy. *Int. J. Mol. Sci.* 13: 9649–9672.

Sheng, X.F., & Xia, J.J. 2006. Improvement of rape (*Brassica napus*) plant growth and cadmium uptake by cadmium-resistant bacteria. *Chemosphere*, 64: 1036–1042.

Selenium in the Environment and Human Health – Bañuelos, Lin & Yin (Eds)
© 2014 Taylor & Francis Group, London, ISBN 978-1-138-00017-9

Selenium biofortification and its effects on flour paste viscosity properties in wheat

Pan Gong, Tao Li, An Wang, Fayu Sun & Shiliang Gu
Jiangsu Provincial Key Laboratory of Crop Genetics and Physiology, Key Laboratory of Plant Functional Genomics of Ministry of Education, Yangzhou University, Yangzhou, China

Xuebin Yin
Advanced Lab for Selenium and Human Health, Suzhou Institute of USTC, Suzhou, China

Wenwen Guan
Innovation Center of Rice Cultivation Technology in the Yangtze Valley, Ministry of Agriculture, Yangzhou University, Yangzhou, China
Suzhou SeTek, Ltd., Suzhou, China

1 INTRODUCTION

Wheat is an important dietary source of selenium (Se), which is an integral component for normal cell metabolism in humans and animals, and may have antioxidant, anti-carcinogenic, and antiviral effects (Rayman, 2012). The pasting viscosity properties of wheat flour are closely related to the eating quality and processing of food or cooking quality (Panozzo & McCormick, 1993; Zhang *et al.*, 2005). The responses of flour viscosity to Se application, however, have been seldomly addressed. In this study, the variation of Se concentrations in 59 wheat varieties was investigated by the application of Se nanoparticles through basal-dressing and application of selenate fertilizer through foliar spray. The relationship between Se fertilization and flour paste viscosity was also investigated.

2 MATERIALS AND METHODS

2.1 *Materials and experimental design*

Fifty-nine wheat varieties, including wheat landraces, commercial varieties and introduced varieties, were biofortified with Se fertilizers. In the experiment with nanoparticle fertilizer (E1), all cultivars were planted in plastic pots filled with 10 kg of soil. Three levels (treatments) of Se nanoparticle fertilizer (supplied by Suzhou SeTek, Ltd, Suzhou China) and two replicates (pots) per treatment were performed. CK: without applying Se fertilizers; treatment 1 (Trt1): 1 g Se fertilizer per pot, and treatment 2 (Trt2): 1.5 g Se fertilizer per pot. The fertilizer was basal-dressed at the elongation period of wheat in the spring season of 2012. In the experiment with selenate fertilizer (E2), the same set of cultivars were planted in the wheat nursery in Yangzhou University, with twenty plants per cultivar (row) and two replicates per cultivar. Four levels (treatments) of selenate fertilizer were performed at the wheat heading stage. CK: no fertilizer applied; Trt1, Trt2 and Trt3 were applied with 10 mg, 20 mg and 30 mg Na_2SeO_4 in a volume of 100 mL ddH_2O per row, respectively.

2.2 *Total Se determination*

A 0.5 g dry wheat seeds were digested in a solution mixture of 4 mL HNO_3, 3 mL double deionized water (ddH_2O) and 2.5 mL H_2O_2. The digestion was performed using a microwave oven (CEM-MARS 5, USA) following the recommended protocol. Total Se concentration was analyzed on ICP (Solar S4+Graphite Furnace System 97, Thermo Elemental, USA).

2.3 *Evaluation of flour pasting viscosity*

A 3.5 g wheat flour in a volume of 25 mL ddH_2O was used for paste viscosity measurement on Rapid Visco Analyser (RVA, Super 3, Australia). Three major parameters (*i.e.*, peak viscosity, trough viscosity and final viscosity) were recorded.

3 RESULTS AND DISCUSSION

3.1 *Variations of Se content in the wheat grain under Se fertilization*

Selenium concentration in the grains increased with the elevated amount of fertilizers applied. In E1, the average Se content in the grain of CK, Trt1 and Trt2 was 1.90, 13.17, and 17.83 mg/kg, respectively. In E2, the average Se contents in the grains of CK, Trt1, Trt2 and Trt3 were 1.57, 5.99, 10.11, and 13.40 mg/kg, respectively. The Se density between any two treatments were significantly different over the two experiments ($P < 0.001$). The increase gradients between two adjacent treatments, however, decreased as the dosages of Se fertilizers increased. In E1, the increase gradients of Se concentration were 11.27 mg/kg (Trt1-CK) and 4.65 mg/kg (Trt2-Trt1). In E2, those of the gradients were 4.42 mg/kg (Trt1-CK), 4.12 mg/kg (Trt2-Trt1) and 3.29 mg/kg (Trt3-Trt2), respectively. These results imply that capacity for Se density in wheat was limited and unlikely increased linearly. The ratios of Trt2/Trt1 in E1 and Trt3/Trt2 in E2 had the equal value of 1.5 in fertilization dosage, and their corresponding ratios in Se-concentration were also very close, 1.35 and 1.33, respectively. High concentration of selenate (Trt3 in E2), however, caused a macroscopic necrosis on the leaves in one-third of cultivars, whereas no similar symptoms were observed in E1. This observation suggests that the application of nanoparticle fertilizer could be a safer strategy for Se biofortification than using selenate.

3.2 *The effect of Se application on wheat flour paste properties*

The response of these three major RVA parameters to Se biofortification varied among the different genotypes (Fig. 1). After Se application, ten cultivars (*i.e.*, Yangmai1, Youxuan14 and Qinmai11, *etc.*) showed a consistent increase and eight cultivars (*i.e.*, Yangfu9798, Zhen9023, Luomai8 and 313, *etc.*) had a simultaneous decrease in these three parameters over the two experiments. The other cultivars demonstrated inconsistent changes, depending on either fertilization methods and/or genotypes. Peak viscosity was reported to be positively correlated to noodle elasticity, toughness and eating quality. Trough viscosity was positively correlated to noodle appearance, viscoelasticity and taste, while final viscosity was positively correlated to noodle smoothness (Nagamine *et al.*, 2003; Panozzo & McCormick, 1993; Zhang, 2005). These results indicate that Se fertilization may modify physiochemical properties of grain starch and the end-use quality of wheat.

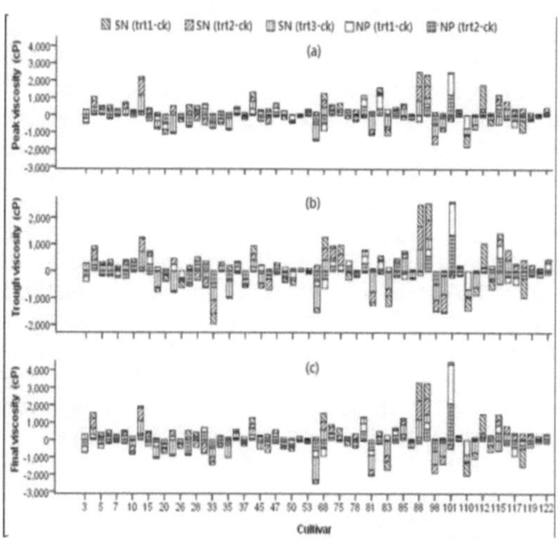

Figure 1. The response of the three RVA parameters to Se application. SN: foliar spray of selenate; NP: basal dressing of nanoparticle fertilizer.

4 CONCLUSIONS

This study showed Se application significantly but differentially improved the total Se density in wheat grains, irrespective of fertilizer-application method, although application of nanoparticle fertilizer might be a safer strategy than using selenate for Se biofortification. In addition, Se-biofortification may also alter the flour paste viscosity properties and end-use quality and have implications in milling industry. Genotypic differences of wheat should be taken into account in Se-biofortification strategies used for improving Se nutrition, and the possible effects of Se-biofortification on end-use food quality should not be ignored as well.

ACKNOWLEDGEMENT

This work was supported by grants from the Priority Academic Program Development of Jiangsu Higher Education Institutions, and by the Innovative Research Team of Universities in Jiangsu Province.

REFERENCES

Nagamine, T., Ikeda T., Yanagisawa T., Yanaka M. & Ishikawa N. 2003. The Effects of Hardness Allele Pinb-D1b on the Flour Quality of Wheat for Japanese White Salty Noodles. *Journal of Cereal Science*, 37, 337–342.

Panozzo, J. & McCormick, K. 1993. The rapid viscoanalyser as a method of testing for noodle quality in a wheat breeding programme. *Journal of Cereal Science* 17: 25–32.

Rayman, M.P. 2012. Selenium and human health. *Lancet* 379: 1256–1268.

Zhang, Y., Nagamine, T., He, Z., Ge, X., Yoshida, H. & Pena, R.J. 2005. Variation in quality traits in common wheat as related to Chinese fresh white noodle quality. *Euphytica* 141: 113–120.

Selenium in the Environment and Human Health – Bañuelos, Lin & Yin (Eds)
 ISBN 978-1-138-00017-9

Screening wheat genotypes for selenium biofortification in Brazil

M.F. Moraes
Federal University of Mato Grosso, Barra do Garças-MT, Brazil
Federal University of Parana, Curitiba-PR, Brazil

P.L. Scheeren
Embrapa Wheat, Passo Fundo-RS, Brazil

J.A.L. Pascoalino, C.R.S. Domingos, G. Orso & H.L. Villetti
Federal University of Parana, Curitiba-PR, Brazil

F.A. Franco & A. Evangelista
Center for Agricultural Research, Cascavel-PR, Brazil

A.R. Reis
Federal University of Lavras, Lavras-MG, Brazil

1 INTRODUCTION

Among the mineral elements, iron (Fe), iodine (I), selenium (Se) and zinc (Zn) are of the greatest concern for human health, especially in developing countries. Deficiencies of these micronutrients have been an increasing focus for policy discussions on food security. One way to minimize this problem is through biofortification, which can be an important process in programs designed to reduce malnutrition. There is evidence of Se-deficiency in some regions of Brazil, according to studies assessing the levels of Se in soils, agricultural products and diets (Moraes *et al.*, 2009). In regards to the world Se status, Combs Jr. (2001) estimated there are between 0.5 and 1.0 billion people in the world with a probable deficiency of this element in their diets.

For Brazil, there is a lack of information on the Se status and the possibility of its deficiency in various regions. For this reason, it is important to investigate the potential use of improved wheat genotypes and their Se and nutritional content in foods made from wheat. The aim of this study was to assess wheat genotypes regarding their Se concentration in the grains and the interaction of Se with yield and levels of Fe and Zn.

2 MATERIALS AND METHODS

The experiment was conducted under field conditions on the experimental farm of Cooperative Center for Agricultural Research (COODETEC), located in the municipality of Palotina, Parana State-Brazil, in the winter of 2012. Soil in the experimental area was classified as Eutroferric Red Latosol (Oxisol), with 72% of clay. We evaluated 186 wheat genotypes from the germplasm banks of the Embrapa Wheat and COODETEC, based off a previous study, and selected for agronomic performance, technological qualities, their potential for Se-biofortification and high levels of tissue Fe and Zn.

A total of 186 genotypes were arranged in two blocks. Plots for each cultivar were chosen at random, for a total of 372 plots. Each plot consisted of 6 rows with 4 m in length. The rows were spaced 17 cm apart, with density of 60 to 80 seeds per linear meter. Lime was applied before sowing to raise the base saturation to 70%, and fertilization was carried out as recommended by the Brazilian Commission on Wheat and Triticale Research.

Yields were measured and quality of the grains was tested for levels of Fe, Zn, and Se. Grain samples were dried, ground, and digested in a nitric-perchloric acid solution. Concentrations of Fe and Zn in the grains were determined by ICP-AES, while the level of Se was determined according to the UAEPA Method 3051A. To validate the method employed in the nutrient level measurements, a certified reference sample was used (White Clover BCR-420). Results validated with the Pearson correlation test (SAS) to assess the correlation between yield and levels of Fe, Zn and Se in the grains.

3 RESULTS AND DISCUSSION

Among the 186 genotypes evaluated, only 46 had levels above the limits of detection (<8.13 μg/kg) for Se. Of these 46 genotypes, we selected the 15 most representative from the spectrum of variation.

According to the results, there was great variation in the Se concentration in the grains, ranging from <8.36 to 55.44 μg/kg. The same pattern was observed for Fe and Zn, with ranges from 33 to 76 mg/kg and 50 to 80 mg/kg, respectively (Table 1). These variations can be explained by the findings of White & Broadley (2005), who reported the existence of genotypes that are naturally more efficient in accumulating minerals and vitamins in the edible part. Additionally, the traits for high levels of minerals in grains might be genetically regulated and associated with a large number of genes.

The evaluation of the interaction of Se with yield showed a negative but not statistically significant correlation. There was an inversely proportional tendency between the two parameters (Se content versus yield). Similar behavior was noted for the other elements (Fe and Zn), but this negative correlation (-0.56) was only significant in the case of Fe (Table 1). Similar results were observed by Murphy et al. (2008), who reported an inverse relation between productivity and micronutrient concentrations in wheat grains. These results can be explained by the effect of dilution of micronutrients in cultivars improved to boost yield (Garvin *et al.*, 2006).

Table 1. Yield and concentrations of Fe, Zn and Se in wheat genotypes and their respective interactions.

Cultivar	Yield (kg/ha)	Fe (mg/kg)	Zn	Se (μg/kg)
Ocepar 10-Garça	1,549.02	69	72	34.54
BR 17	1,681.37	49	59	48.27
RS 8-Westphalen	1,730.39	65	78	22.75
BRS 264	1,882.35	46	69	28.27
Iapar 28	1,887.25	52	68	16.96
Ocepar 12-Maitaca	1,965.69	51	62	15.29
BR 8	1,980.39	42	74	11.41
CEP 24-Industrial	2,205.88	58	63	31.01
Mentana	2,254.90	76	80	55.44
Embrapa 24	2,553.92	50	69	52.07
IPR 118	2,598.04	47	72	23.42
CD 115	2,725.49	46	60	24.08
BRS 194	2,936.27	48	69	26.01
BRS 210	2,980.39	42	50	8.36
IPR 111	3,274.51	33	69	26.09
	Pearson´s Correlation			
Yield	1	-	-	-
Fe concentration	$-0,56^{***}$	1	-	-
Zn concentration	$-0,30^{NS}$	$0,51^{***}$	1	-
Se concentration	$-0,16^{NS}$	$0,48^{***}$	$0,34^{*}$	1

***, **, *: Level of significance at 0.001, 0.05 and 0.10, respectively; and NS: not statistically significant at $P > 0.10$.

There was a positive and significant interaction for all possible combinations of Se, Fe and Zn (Table 1). This finding suggests the existence of a directly proportional pairwise relation between the three nutrients, indicating the possibility of synergism.

4 CONCLUSIONS

Concentrations of Se in the grains varied among the genotypes were assessed. There was an inversely proportional relationship between the concentration of Se and plant yield. This relationship was not observed Fe and Zn. These results are important to guide future studies of biofortification with Se in Brazil, and may help provide information to improve the nutritional quality of foods.

REFERENCES

Combs, G.F. Jr. 2001. Selenium in global food systems. *British Journal of Nutrition* 85(5): 517–547.

Garvin, D.F., Welch, R.M. & Finley, J.W. 2006. Historical shifts in the seed mineral micronutrient concentrations of US hard red winter wheat germplasm. *Journal of the Science of Food and Agriculture* 86(13): 2213–2220.

Moraes, M.F., Welch, R.M., Nutti, M.R., Carvalho, J.L.V. & Watanabe, E. 2009. Evidences of selenium deficiency in Brazil: from soil to human nutrition. In G.R. Banuelos, Z.-Q. Lin & X.B. Yin (eds), *Selenium: Deficiency, Toxicity and Biofortification for Human Health*: 73–74. Hefei: University of Science and Technology of China Press.

Murphy, K.M., Reeves, P.G. & Jones, S.S. 2008. Relationship between yield and mineral nutrient concentrations in historical and modern spring wheat cultivars. *Euphytica* 163(3): 381–390.

White, P.J. & Broadley, M.R. 2005. Biofortifying crops with essential mineral elements. *Trends in Plant Science* 10(12): 586–593.

Selenium in the Environment and Human Health – Bañuelos, Lin & Yin (Eds)
© 2014 Taylor & Francis Group, London, ISBN 978-1-138-00017-9

Effects of hog manure application on selenium accumulation in agricultural soil and grains

S. Singh & Z.-Q. Lin
Environmental Sciences Program, Southern Illinois University, Edwardsville, IL, USA

T.Q. Zhang
Agricultural and Agri-Food Canada, Greenhouse and Processing Crops Research Centre, Harrow, Ontario, Canada

1 INTRODUCTION

Concentrations of selenium (Se) in soil worldwide are generally low, ranging from 0.01 to 2.0 mg Se/kg (Broadley *et al.*, 2006). Low dietary Se intake or Se deficiency has been linked with several important health problems, including Kaschin Beck and Keshan diseases (Seppanen *et al.*, 2010; Sun *et al.*, 2009). Therefore, the strategy of producing Se-enriched agricultural products via agronomic approaches (*i.e.*, Se biofortification) has become one of important research topics in recent years (Zhao & Shewry, 2011; Bañuelos & Lin, 2009). The biofortification technology sets the goal of increasing Se accumulation in crops and other agricultural products to the desired levels by manipulating the soil-plant system (Zhao & Shewry, 2011). Previous studies conducted mainly focused on the application of Se fertilizers to agricultural soils to increase Se content in crops. For example, Se concentrations in crops and animal products were substantially increased in Finland by the use of Se-fertilizers in agricultural soils, and resulted in a 70% increase of human serum Se concentration (Aro *et al.*, 1995).

Selenium is an essential nutrient for human and animals. Sager (2006) reported that livestock are oftentimes supplemented with Se in their feedstuff normally close to the upper permissible dosage of Se. Concentrations of Se in solid hog manure were about 1–7 mg/kg. The use of hog manure as organic fertilizer in agricultural soils might result in Se input of 0.2–2 g/ha (Sager, 2006; Reilly, 2010). Thus, the specific objective of this project was to (1) determine the effect of organic hog manure application on Se accumulation in soils and corn and soybean grains, compared with inorganic NPK fertilizer, and (2) compare the effects of different forms of hog manure on crop Se accumulation.

2 METHODS AND MATERIALS

2.1 *Field experiment*

The field experiment was conducted at the Research Farm of Agriculture and Agri-Food Canada in Harrow, Ontario, with a randomized complete block design. Corn planted in the first year (starting in 2004) followed by the soybean in the next year (*i.e.*, 2005), forming corn-soybean rotation in 2004 to 2007. The soil treatments included: inorganic N-P-K fertilizer (IP), Liquid hog Manure (LM), fresh Solid hog Manure with wheat straw (SM), and hog Manure Composted with wheat straw (MC). The plot without fertilization was used as control (CK). Concentrations of Se in different types of hog manure varied from 0.174 mg/L in liquid manure to 0.465–1 mg/kg in composted manure and to 0.574–1.318 mg/kg in fresh manure.

For direct comparison, the application rates of the inorganic fertilizer and the three forms of hog manure were determined based on the same P rate of 100 kg/ha. Fertilizers were only applied to the soil in the years with corn (*i.e.*, 2004 and 2006), but not for soybean. Manures were applied on soil surface and incorporated by disk, triple-K and pack. Each treatment had 3 replicates, with a total of 15 plots and each in size of 9 × 25 m. The composite soil samples of 12 cores from 0–15 cm depth per plot were collected using probe equipped with an acetate sleeve both at the time of pre-planting and post-harvest. Soil samples were air-dried and passed through a 2-mm sieve. A composite grain sample (corn or soybean) was randomly collected from each plot, and grain samples were oven-dried at 45°C to the constant weight.

2.2 *Sample preparation and chemical analyses*

The soil and grain samples were grounded and then acid-digested by following EPA method 3050B using HNO_3, H_2O_2 and HCl. Concentrations of Se in soil and plant acid-digestion solutions were measured using Inductively-Coupled Plasma Mass Spectroscopy (ICP-MS), while soil test (or available) S was measured using Mehlich III method (Mehlick, 1984). The S concentrations in extraction solution were determined by Inductively-Coupled Plasma Atomic Emission Spectroscopy (ICP-AES). NIST Standard Reference Material (SRM-8436,

Durum wheat flour), along with blanks and sample duplicates, were used as quality control samples.

3 RESULTS AND DISCUSSION

Concentrations of Se in control soils varied from 0.272 ± 0.021 to 0.311 ± 0.023 mg/kg during the 4-year time period. The application of different fertilizers in soil during the course of the field study did not show any consistent significant changes of soil Se content. The corn-soybean rotation also did not result in significant effects on soil Se concentration under field conditions (Fig. 1).

Compared with the control, the fertilizer treatments did not show significant increase in Se accumulation in corn and soybean grain. Selenium accumulation in soybean was generally greater than in corn (Fig. 2), even though no fertilization occurred during the soybean rotation years.

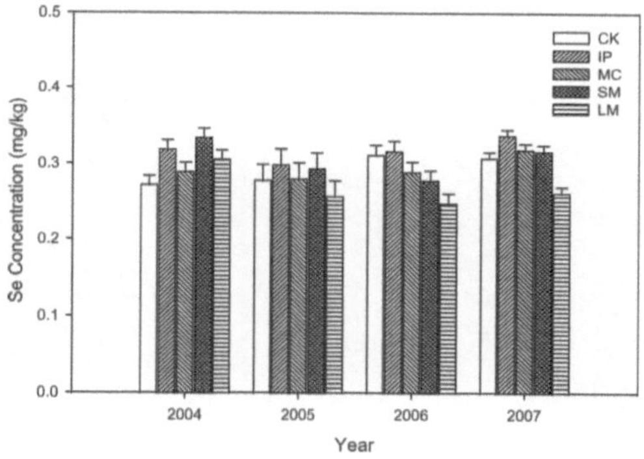

Figure 1. Soil Se concentration changes in a corn-soybean rotation cropping system with or without fertilization treatments during a 4-year field study. Data shown are means ± one standard error (n = 3).

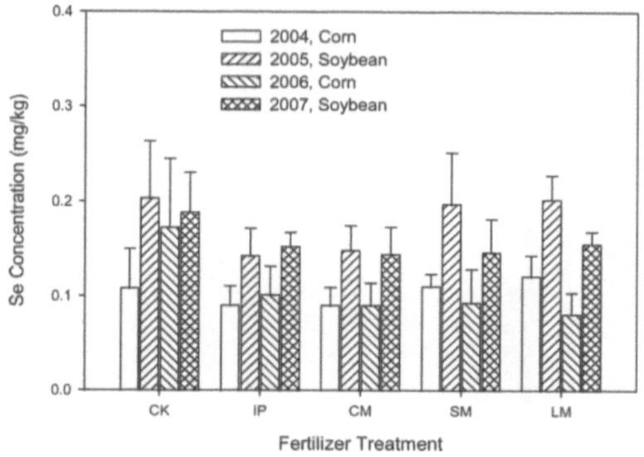

Figure 2. Concentrations of Se in corn and soybean grains in a corn-soybean rotation cropping system during a 4-year field study. Different fertilizers were only applied to the corn soil. Data shown are means ± one standard error (n = 3).

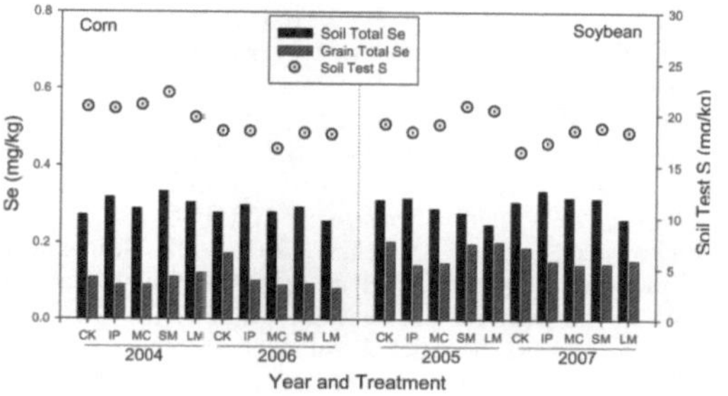

Figure 3. Relationship between the grain total Se and the soil total Se or the soil test S concentrations in a 4-year corn-soybean rotation system. Different fertilizers were only applied to the corn field in 2004 and 2006.

Different soil physiochemical and biological factors can affect Se accumulation in crop tissues. This study shows that corn and soybean in the control soil had similar or slightly higher Se concentrations than in the fertilizer treated soils. Previous laboratory studies demonstrated significant correlation between soil S and plant Se accumulation, due to their chemical analogs of S and Se (Mackowiak & Amacher, 2008). However, under the field conditions, no significant correlation between the soil S concentration and the corn or soybean Se concentration was observed (Fig. 3). Indeed, soil Se bioavailability can be further complicated by many other soil properties and various environmental factors, such as soil organic matter, pH and precipitation in this long-term field study.

4 CONCLUSIONS

Use of animal manure for agricultural production is environmentally sound and sustainable. Livestock wastes generally contain relatively high levels of Se, which could be important sources of Se in agricultural crops. This 4-year field experiment does not show a significant and consistent pattern with the effect of inorganic fertilizer and different forms of hog manure on Se accumulation in corn and soybean grains. Soybean accumulates more Se in grain than corn, regardless of the form of fertilization. Effects of soil S on crop Se accumulation might be confounded by other environmental factors under field conditions.

REFERENCES

Aro, A., Alfthan, G. & Varo, P. 1995. Effects of supplementation of fertilizers on human selenium status in Finland. *Analyst*. 120(3): 841–843.

Bañuelos, G.S. & Lin, Z.-Q. (eds.). 2009. *Use and Development of Biofortified Agricultural Products*. Boca Raton, Florida: CRC Press.

Broadley, M.R., White, P.J., Bryson, R.J., Meacham, M.C., Bowen, H.C., Johnson, S.E., *et al.* 2006. Biofortification of UK food crops with selenium. *Proceedings of the Nutrition Society* 65: 169–181.

Mackowiak, C.L. & Amacher, M.C. 2008. Soil sulfur amendments suppress selenium uptake by alfalfa and Western wheatgrass. *J. Environ. Qual.* 37(3): 772–779.

Mehlich, A. 1984. Mehlich 3 soil test extractant: A modification of Mehlich 2 extractant. *Commun. Soil Sci. Plant Anal.* 15: 1409–1416.

Reilly, C. 2010. *Selenium in Food and Health*, 2nd ed., New York, NY: Springer.

Sager, M. 2006. Selenium in agriculture, food and nutrition. *Pure and Applied Chemistry* 78: 111–133.

Seppanen, M., Kontturi, J. & Hartikainen, H. 2010. Biofortifcation of food chain with selenium. *NJF Report* 6(2): 6–9.

Sun, W., Huang, B., Zhao, Y., Shi, X., Darilek, J.L., Deng, X. *et al.* 2009. Spatial variability of soil selenium as affected by geologic and pedogenic processes and its effect on ecosystem and human health. *Geochemical Journal* 43: 217–225.

Zhao. F.-J. & Shewry, P.R. 2011. Recent development in modifying crops and agronomic practice to improve human health. *Food Policy* 36(1): S94–S101.

Selenium in the Environment and Human Health – Bañuelos, Lin & Yin (Eds)
© 2014 Taylor & Francis Group, London, ISBN 978-1-138-00017-9

Selenium bioavailability in typical selenium-biofortified foods

Jian Wang
Nano Science and Technology Institute and Advanced Lab for Selenium and Human Health, Suzhou Institute, for Advanced Study, University of Science and Technology of China, Suzhou, Jiangsu, China

Xiaoqi Lu
Advanced Lab for Selenium and Human Health, Suzhou Institute for Advanced Study, University of Science and Technology of China, Suzhou, Jiangsu, China
Jiangsu Bio-Engineering Research Centre on Selenium, Suzhou, Jiangsu, China

Linxi Yuan
Jiangsu Bio-Engineering Research Centre on Selenium, Suzhou, Jiangsu, China

Xuebin Yin
Advanced Lab for Selenium and Human Health, Suzhou Institute for Advanced Study, University of Science and Technology of China, Suzhou, Jiangsu, China

1 INSTRUCTION

Selenium (Se) as a nutritionally essential element is well established (Rotruck *et al.*, 1973; Arthur *et al.*, 1990). Selenium biofortified foods are often used to improve Se dietary uptake in Se-deficiency areas (Rayman, 2000). However, Se bioavailability in those Se-enriched foods is critical for recommended usage criterion. Gastrointestinal ecosystem *in vitro* is usually employed to evaluate nutrients bioavailability (Laird *et al.*, 2007). In the present study, six typical Se-biofortified foods were selected to study Se bioavailability via an *in vitro* gastrointestinal ecosystem, which many provide scientific guidance for Se supplementation.

2 MATERIALS AND METHODS

Se-biofortified foods including; grains (*i.e.*, soybean, wheat and corn), vegetable (*i.e.*, broccoli) and edible fungi (*Coprinus comatus, Pleurotus ostreatus*) were provided by Setek Co. and they each have a total Se concentration (mg/kg) as follows: 4.4, 25, 4.5, 68.2, 27, and 60.5, respectively. Those samples were powdered through a 70 mesh and 6 g powder were added to 250 ml cups (Dissolution Tester RC-8, Tian Jin Optical Instrument Factory, China) for Se bioavailability evaluation. The detailed processes were shown in Figure 1 according to Laird *et al.* (2007). Total Se contents were determined by atomic fluorescence spectrometer (AFS-9230, Tian Co, Beijing).

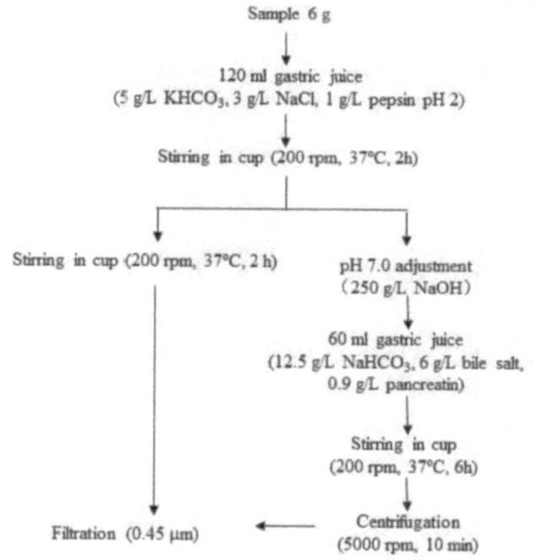

Figure 1. The detailed processes of an *in vitro* Gastrointestinal ecosystem.

3 RESULTS AND DISCUSSION

3.1 *Dynamic dissolution rate of Se-biofortified corn, wheat and soybean*

The dissolution processes of Se-biofortified corn, wheat and soybean an *in vitro* gastrointestinal ecosystem are displayed in Figure 2. Overall, during stomach phase, the Se dissolution rate of corn reached up to about 65%, higher than those of wheat (55%) and soybean (40%). However, the Se dissolution rate became gradually stable in next 2 h and was as high as 78–90% after 6 hours digestion in intestine.

A similar *in vitro* gastrointestinal enzymolysis procedure was used to extract Se from cooked fresh cod (Crews *et al.*, 1996). The extractionn ratio of Se in gastric (4 h) and gastrointestinal (4 h) stages were 50.5% and 60.1%, respectively, suggesting that the dissolution of Se mainly occurred in the first 4 hours.

3.2 *Dissolution rate of typical Se-biofortified foods*

The dissolution rates determined by an *in vitro* gastrointestinal ecosystem were ranked in Figure 3, which displayed that 78–90% Se in those Se-biofortified foods could be dissolved and released in the simulated digest system, except *Pleurotus ostreatus,* it had a bit lower dissolution rate (60%). It is reported that 90% of the total Se contents of Se(VI)-enriched green onions and chives became bioaccessible (Kapolna & Fodor, 2007). For selenized yeast, 89% of total Se was extractable after gastrointestinal digestion (Hinojosa *et al.*, 2006). Overall, it is concluded that Se in Se-biofortified products is easy to release in the simulated *in vitro* gastrointestinal system, suggesting that they are a good source of Se supplementation for Se-deficiency populations.

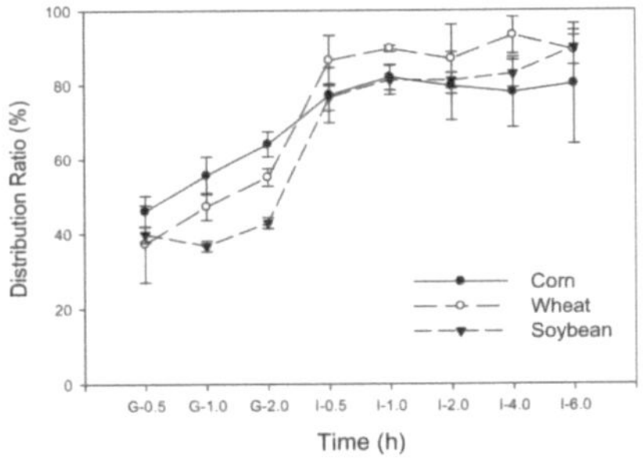

Figure 2. Dissolution rate of Se-biofortified grain (corn, wheat and soybean) an in *vitro* gastrointestinal ecosystem. G: stomach phase and I: intestine phase.

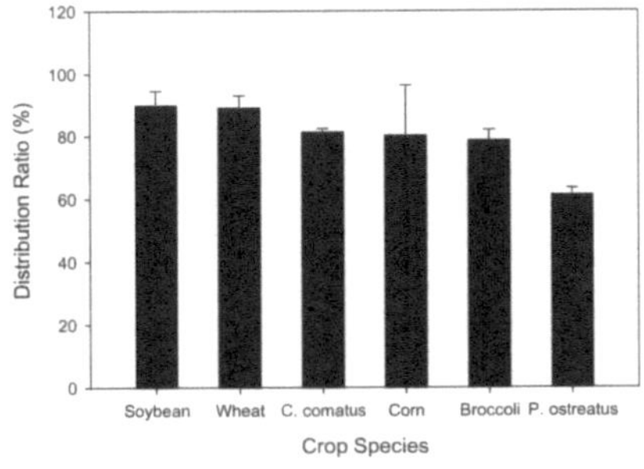

Figure 3. Dissolution rate of typical Se-biofortified foods in *in vitro* Gastrointestinal ecosystem.

4 CONCLUSIONS

Selenium biofortified foods are often used to improve Se dietary uptake in Se-deficiency areas. To provide scientific guidance for Se supplementation, we evaluated selenium bioavailability of six typical Se-biofortified foods via an *in vitro* gastrointestinal ecosystem. The system shows that 78–90% Se in those Se-biofortified foods could be dissolved and released in the simulated digest system, except *Pleurotus ostreatus* (60%), suggesting that these foods could provide a good source of Se supplementation in our everyday diet.

ACKNOWLEDGEMENT

This study was supported by the Jiangsu Research-Industry Joint Innovation Project (BY2011186), Natural Science Youth Foundation of Jiangsu Province for Youth, China (BK2012202), and Open Fund of State Key Laboratory of Soil and Sustainable Agriculture, Institute of Soil Science, Chinese Academy of Sciences (Y052010020).

REFERENCES

Arthur, J.R. & Nicol, F. 1990. Hepatic iodothyronine 5'-deiodinase. *Biochem. J.* 272: 537–540.

Crews, H.M. & Clarke, P.A. 1996. Investigation of selenium speciation in in vitro gastrointestinal exracts of cooked cod by high-performance liquid chromatography-inductively coupled plasma mass spectrometry and electrospray mass spectrometry. *J. Anal. Atom. Spectrom.* 11: 1177–1182.

Hinojosa, R.L. & Ruiz, E.J. 2006. Selenium bioaccessibility assessment in selenized yeast after 'in vitro' gastrointestinal digestion using two dimensional chromatography and mass spectrometry. *J. Chromatogr. A* 1110: 108–116.

Kapolna, E. & Fodor, P. 2007. Bioavailability of selenium from selenium-enriched green onions (*Allium fistulosum*) and chives (*Allium schoenoprasum*) after 'in vitro' gastrointestinal digestion. *International Journal of Food Sciences and Nutrition* 58(4): 282–296.

Laird, B.D. & Corriveau, M.C. 2007. Gastrointestinal microbes increase arsenic bioaccessibility of ingested mine tailings using the simulator of the human intestinal microbial ecosystem. *Environ. Sci. Technol.* 41(15): 5542–5547.

Rayman, M.P. 2000. The importance of selenium to human health. *Lancet* 356: 233–241.

Rotruck, J.T. & Ganther, H.E. 1973. Selenium: biochemical role as a component of glutathione peroxidase. *Science* 179: 588–590.

Selenium in the Environment and Human Health – Bañuelos, Lin & Yin (Eds)
© 2014 Taylor & Francis Group, London, ISBN 978-1-138-00017-9

Total selenium contents and selenium speciation in typical selenium-biofortified materials

Dan Li & Xuebin Yin
Nano Science and Technology Institute and Advanced Lab for Selenium and Human Health, University of Science and Technology of China, Suzhou, Jiangsu, China

Yuanyuan Zhu & Linxi Yuan
Jiangsu Bio-Engineering Research Centre on Selenium, Suzhou, Jiangsu, China

1 INTRODUCTION

Selenium (Se) is an important trace element for human health and Se deficiency plays a risk factor for some diseases such as Keshan Disease, Kashin-Beck Disease, and prostate cancer (Rayman, 2000; Fenech *et al.*, 2007). Diet intake of some Se-enriched products would help to decrease such risk of diseases (Rayman, 2005; Coppinger *et al.*, 2001). Currently, Se-biofortication technology is adopted to produce various Se-enriched products. In the present study, some typical Se-biofortified materials including; fungi, cereals and vegetables developed by Setek Co, were selected to determine total Se content and Se speciation in them. This information will provide a data base for further implementation of the Se-biofortification strategy.

2 MATERIALS AND METHODS

Yeast (fungi), mushroom (*Pleurotus ostreatus*) (fungi), Chinese milk vetch (*Astragalus sinicus* L.) (vegetable), broccoli (*Brassica oleracea* L. *var. botrytis*) (vegetable), corn (*Zea mays*) (cereal), soybean (*Glycine max*) (cereal) were Se-biofortified and provided by Setek Co. (Suzhou, China). All samples were acid-digested (HNO_3:$HCLO_4$, 4:1, v/v), and the total Se concentration was determined by Atomic Fluorescence Spectrometer (AFS-9230, Titan Co, Beijing) (Mazej *et al.*, 2008). For the preparation of Se speciation analysis, enzymolysis, centrifugation, and filtration were necessary. Finally, Se speciation was determined by HPLC-UV—HG-AFS (AFS-9230, Titan Co, Beijing).

3 RESULTS AND DISCUSSION

Concentrations of the total Se accumulated in the selected Se-biofortified materials are shown in Table 1. Overall, the Se concentrations ranged from 104.2 to 1297 μg/g in yeast and mushroom (fungi), 54.3 to 68.2 μg/g in Chinese milk vetch and broccoli, and 28.1 to 29.2 μg/g in corn and soybean. This study indicated that yeast and fungi are the best organisms for Se-biofortification.

Table 1. Total Se contents in typical Se-biofortified fungi, vegetables, and cereals.

Sample	Total Se (μg/g)
Yeast	1297 ± 48.2
Mushroom	104.2 ± 2.1
Chinese milk vetch	68.2 ± 3.6
Broccoli	54.3 ± 4.3
Corn	29.2 ± 1.1
Soybean	28.1 ± 2.8

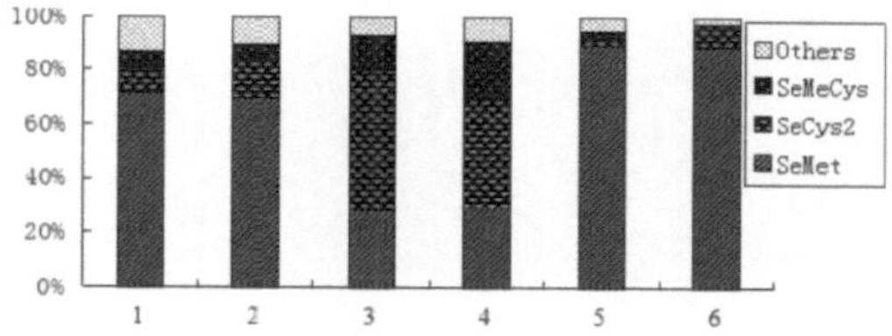

Figure 1. Selenium speciation in different Se-biofortified materials: (1) yeast; (2) mushroom; (3) Chinese milk vetch; (4) broccoli; (5) corn; and (6) soybean.

In regards to Se speciation the proportion of SeMet compound was 90–95% in corn and soybean, 70–75% in mushrooms, and 30–35% in broccoli (Fig. 1).

However, $SeCys_2$ and SeMeCys accounted for more than 60% of the total Se accumulated in vegetables, which is the highest proportion among the tested materials. This result might be related to the metabolic mechanisms of different species. More research is needed to determine the reason why $SeCys_2$ and SeMeCys are present in vegetables.

4 CONCLUSIONS

Generally speaking, fungi, especially yeast, will be an ideal material for Se-biofortification with Se as high as 104.2–1297 μg/g. However, Se-enriched vegetables, such as Chinese milk vetch and broccoli, had much higher proportions of $SeCys_2$ and SeMeCys (60–70%). All the studied Se-biofortified products can accumulate most organic Se and the biological effects of Se are highly dependent on organic Se (Darja *et al.*, 2008). Therefore, these types of Se-enriched materials would be a good source of Se supplement.

ACKNOWLEDGEMENTS

This study was supported by the Jiangsu Research-Industry Joint Innovation Project (BY2011186), Natural Science Youth Foundation of Jiangsu Province for Youth, China (BK2012202), and Open Fund of State Key Laboratory of Soil and Sustainable Agriculture, Institute of Soil Science, Chinese Academy of Sciences (Y052010020).

REFERENCES

Coppinger, R.J. & Diamond, A.M. 2001. Selenium deficiency and human disease. In D.L. Hatfield, M.J. Berry, V.N. Gladyshev (eds.), *Selenium: Its Molecular Biology and Role in Human Health,* 219–331. Dordrecht, The Netherlands: Kluwer Academic Publisher.

Darja, M., Joze, O. & Vekoslava, S. 2008. Selenium species in leaves of chicory, dandelion, lamb's lettuce and parsley. *Food Chemistry* 107: 75–83.

Fenech, M. 2007. Nutrition and genome health, *Forum of Nutrition* 60: 49–65.

Mazej, D., Falnoga, I. & Michael, V. 2008. Determination of selenium species in plant leaves by HPLC-UC-HG-AFS. *Talanta* 68(3): 558–568.

Rayman, M.P. 2000. The importance of selenium to human health. *The Lancet* 356: 233–241.

Rayman, M.P. 2005. Selenium in cancer prevention: A review of the evidence and mechanism of action. *Proceedings of the Nutrition Society* 64: 527–542.

Selenium in the Environment and Human Health – Bañuelos, Lin & Yin (Eds)
 ISBN 978-1-138-00017-9

Ethanol extraction of selenoproteins in selenium-enriched yeast in winemaking

Qiankun Zhou & Shoubiao Zhou
College of Life Science, Anhui Normal University, Wuhu, Anhui, China

Linxi Yuan & Xuebin Yin
Advanced Lab for Selenium and Human Health, Suzhou Institute for Advanced Study, University of Science and Technology of China, Suzhou, Jiangsu, China
Jiangsu Bio-Engineering Research Centre on Selenium, Suzhou, Jiangsu, China

1 INTRODUCTION

Selenium (Se), as an essential micronutrient for human and animals, has been shown to exert its biological function mainly as selenoproteins *in vivo* (Lu & Holmgren, 2009). However, Se intake, especially in China, is below the level suggested by WHO for adults (55 μg/day). Hence supplementation with Se-enriched foods could be adopted in such areas to reduce Se deficiency (Peñasa *et al.*, 2012). Since proteins cannot be dissolved in water or alcohol, Se-enriched soft drinks or wines have not been produced as a novel Se supplementation. It is known that proteins can be divided into a water-soluble fraction, an alcohol-soluble fraction, and an alkali-soluble fraction. Thus, the present study focused on the extraction of alcohol-soluble Se-containing proteins in Se-enriched yeast that maybe used in developing Se-enriched wine.

2 MATERIALS AND METHODS

Selenium-enriched yeast with 1500 μg Se/kg dry weight was purchased from Angle Co. Ltd. Fifty g of yeast was extracted in 3% NaCl solution with pH 5.5 for 20 h at 55 °C. The mixed solution was then thrice homogenized at 10,000 psi for 3 min (Shan *et al.*, 2001). A series of C_2H_5OH solutions (30, 40, 50, 60, and 70%) were used to extract alcohol-soluble Se-containing proteins from homogenized Se-enriched yeast solution for 18 h. In addition, 0.2% papain (by weight) was added into the above solutions, adjusted pH to 5.5, and was extracted another 18 h at 45 °C. The extracted Se in alcohol solution and papain solution were determined by AFS (AFS-9230, Titan Co, Beijing).

The above extracted Se-containing selenoprotein solutions were concentrated to 80 μg/g by vacuum batch distillation and then was distributed into wine to make Se-enriched wine with 200, 400, 500, 1000, 1500, and 2000 μg Se/L. One month later, visible deposits were observed and filtered through a 0.22 μm filter. The total Se content in the filtered wine was determined by AFS.

3 RESULTS AND DISCUSSION

The extraction rates of alcohol-soluble Se-containing proteins from Se-enriched yeast are shown in Table 1. In C_2H_5OH-based solutions, the highest extraction rate was 12% observed in 30%-alcohol solution, and the lowest extraction rate was 6% observed in 70%-alcohol solution. These results show a decreasing Se extraction rate with increasing alcohol content. However, after treating with papain, the Se extractions rates sharply increased to 31–35%. Thus, it will be better to choose 30–40% alcohol and 0.2% papain to extract alcohol-soluble Se-containing proteins from Se-enriched yeast.

After one month, some visible deposits could be observed on produced Se-enriched wines, and the Se contents were determined after filtering (Table 2). The results revealed that the highest loss percentage of Se (approximately 50%) was observed

Table 1. Extraction rates of Se in Se-enriched yeast by different C_2H_5OH solutions.

C_2H_5OH (%) (No. of samples)	Extraction rate of total Se (%)	
	– Papain	+ Papain
30 (n = 3)	12 ± 0.9	34 ± 1.2
40 (n = 3)	11 ± 0.4	35 ± 1.4
50 (n = 3)	8 ± 0.3	32 ± 1.7
60 (n = 3)	7 ± 0.5	31 ± 1.1
70 (n = 3)	6 ± 0.4	35 ± 1.8

Table 2. Selenium concentrations and loss percentages of Se in the filtered Se-enriched wine after one-month storage. The data shown are means ± standard deviation (n = 3).

Se treatment (μg/L)	Filtered wine	
	Se (μg/L)	Se loss (%)
200	106 ± 9.2	47 ± 0.05
500	310 ± 20	38 ± 0.04
1000	588 ± 40	41 ± 0.04
1500	1099 ± 80	27 ± 0.05
2000	1599 ± 99	20 ± 0.05

at 200 μg Se/L. However, the highest Se content of Se-enriched wine (2000 μg/L) only lost 20% Se (the lowest losses). The stability of Se-enriched wine depended on the interaction of water, alcohol and selenoprotein (Yoshikawa *et al.*, 2012).

4 CONCLUSIONS

Se-enriched wine could be made via adding alcohol-soluble Se-containing amino-acids, and 30–40% alcohol and 0.2% papain could extract 31–35% alcohol-soluble Se-containing amino-acids from Se-enriched yeast. To minimize losses of Se, the Se content in wine should be at least 2000 μg/L. After filtering through a 0.22 μm filter, the Se-enriched wine is stable without visible deposits after half a year.

ACKNOWLEDGEMENTS

This study was supported by the Jiangsu Research-Industry Joint Innovation Project (BY 2011186), Natural Science Youth Foundation of Jiangsu Province for Youth, China (BK 2012202), and Open Fund of SOA Key Laboratory for Polar Science, the Polar Research Institute of China (KP 201106).

REFERENCES

Lu, J. & Holmgren, A. 2009. Selenoproteins. *The Journal of Biological Chemistry* 284(2): 723–727.

Peñasa, E., Martinez-Villaluengaa, C., Friasa, J., Sánchez-Martínez, M.J., Pérez-Corona, M.T., Madrid, Y. *et al.* 2012. Se improves indole glucosinolate hydrolysis products content, Se-methylselenocysteine content, antioxidant capacity and potential anti-inflammatory properties of sauerkraut. *Food Chemistry* 132(2): 907–914.

Shan, Z.X., Jiang, L. & Wang, Y.L. 2001. A comparative study of two methods for the shive of the cell wall of selenium—rich brewer's yeast. *Journal of Southwest Agricultural University* 23(4): 365–373.

Yoshikawa, H., Hirano, A., Arakawa, T. & Shiraki, K. 2012. Mechanistic insights into protein precipitation by alcohol. *International Journal of Biological Macromolecules.* 50: 865–871.

Selenium in the Environment and Human Health – Bañuelos, Lin & Yin (Eds)
© 2014 Taylor & Francis Group, London, ISBN 978-1-138-00017-9

The antioxidant activity in vitro of selenoprotein in selenium-enriched plants

Hongling Lei & Dazhai Zhou
Key Laboratory of Biological Resources Protection and Utilization of Hubei Province, China
Biological Scientific and Technical College, Hubei University for Nationalities, Enshi, Hubei, China

1 INSTRUCTIONS

Selenium (Se) is an essential trace element in humans and animals. Although it has a variety of physiological effects, its most important biological role is involved with free radical scavenging and antioxidant activity (Chan & Berry, 2003). Others have shown that organic Se is more effective than inorganic Se, in regards to free radical scavenging capacity. Selenoprotein has greater ability of removing radicals than Se-polysaccharide (Tan *et al.*, 2010).

Cardamine violifolia is a Se accumulator and it was fund grown in high-Se soils in Enshi. The Se content of its mature leaves is 628 mg/kg (DW), with 86 mg/kg (DW) as soluble protein-bound Se. In order to explore the biological activity of purified selenoprotein and provide a reference for the use of plant selenoproteins, we examined the free radical scavenging ability of purified selenoprotein, and its in vitro antioxidant activity using Kunming mice as the experimental animal model.

2 MATERIALS AND METHODS

Wild *Cardamine violifolia* was collected from high-Se areas in Yutangba, Shuanghe, Enshi. Selenoprotein was extracted from mature leaves.

2.1 *$O_2^{-}\cdot$ and $\cdot OH$ scavenging capacity measurement*

The scavenging capacity measurement for $O_2^{-}\cdot$ and $\cdot OH$ was based on hypoxanthine/xanthine oxidase system and Fenton reaction system, respectively, using $O_2^{-}\cdot$ and $\cdot OH$ kit produced by produced Nanjing Jiancheng Bioengineering Institute (following the manufacturer's instructions).

2.2 *Anti-ipid peroxidation measurement*

Anti-lipid peroxidation was measured by preparing 1:25 egg yolk suspension with PBS buffer and using TBA assay.

2.3 *The inhibition of selenoprotein on peroxidation*

2.3.1 *Measurement on malondialdehyde (MDA) of mitochondria and liver in mouse*

The selenoprotein was dissolved into five different Se concentrations of protein solution. 0.2 mL of each was used to investigate the inhibitory impact of selenoprotein on MDA generation in mitochondria and liver tissue homogenate in mouse, which was supplemented with $FeSO_4$ (5 mmol/L) and H_2O_2 (10 mmol/L) induction. The normal saline was used as control. The analysis used TBA assay following the kit's instructions.

2.3.2 *The analysis on TBHP-induced erythrocyte lipid peroxidation damage*

2.0 mL of protein solution buffer or 2.0 mL of different concentrations of protein solution were added into 3.0 mL 1% erythrocyte suspension. After incubation at 37°C for 15 min, 1.0 mL 2.5 mmol/L tertbutyl hydroperoxide (TBHP) was added. Protein lysate was added to the control group. The mixture was incubated at 37°C for 30 min, immediately transferred to ice bath, and centrifuged at 4500 r/min, 4°C for 10 min. The OD values of supernatant at 540 nm and 630 nm were measured, and the hemolysis rate and methemoglobin formation rate were calculated, respectively. Another 0.2 mL supernatant was used to measure MDA levels. MDA content was calculated referring to the kit's instructions.

3 RESULTS AND DISCUSSION

3.1 *Free radical scavenging of selenoprotein*

The $\cdot OH$ scavenging ability of different Se concentrations of protein solution and its stability were shown that the purified protein solution has good ability to remove $\cdot OH$. All treatments showed significant difference compared to the control treatment, with Se concentrations ranging from 0.38 to 5.96 μg/mL. With Se concentration at 0.95 μg/mL, the $\cdot OH$ scavenging rate was over 50%.

The inhibitory effect of selenoprotein on $O_2^{-}\cdot$ was represented by the decrease in absorbance. The data show that selenoprotein solution has obvious inhibitory effect on $O_2^{-}\cdot$, and the inhibition of 0.95 μg/mL selenoprotein solution can reach 50.26%. There were significant differences among treatments. One of the most important physiological roles of Se for humans and animals is to enhance the body's antioxidant capacity. Nan *et al.* (1999) used the EPR technique to investigate the antioxidant effect of selenoprotein in Se-enriched soybean. When the concentration of selenoprotein was 800 mg/L, the ·OH removal effect is over 50%. Du *et al.* (2007) isolated a selenoprotein with molecular weight of 36,600 from Se-enriched *Ganoderma lucidum*, and it also has a high radical scavenging ability.

3.2 *Anti-lipid peroxidation activities*

MDA is generally used to measure the level of lipid peroxidation level. The results show that selenoprotein had a certain lipid peroxidation scavenging ability, and protein solutions with different Se concentrations exhibited significant differences comparing to control, especially at 5.96 μg/mL exerting 62.7% inhibition.

3.3 *In vitro antioxidant activity of selenoprotein*

3.3.1 *The impact of selenoprotein on MDA of mitochondria and liver*

Liver and mitochondria are the most active organs in metabolism and are most susceptible to peroxidation damage. All treatment groups exhibited significant differences compared to the control. The inhibitory effects enhanced with the increase of Se concentration. When the Se concentration in protein solution was 0.15, 0.38, 0.95, 2.38 and 5.96 μg/mL, the inhibition rate was 26.78%, 40.85%, 51.37%, 63.13% and 75.71%, respectively.

Fe^{2+} and H_2O_2 both have inductive roles on the MDA generation in liver homogenate, while Fe^{2+} caused a significant increase of MDA in liver homogenate. Selenoprotein showed inhibitory effects on the MDA induction by both materials, but only high Se concentration exhibited obvious inhibition on Fe^{2+} induction. For the saline group and H_2O_2-induced group, all Se-treated groups had significant difference compared to the control. Thus, selenoprotein plays a protective role in liver damage.

3.3.2 *The elimination of selenoprotein on TBHP—induced peroxidation*

As shown in experiment results, selenoprotein solution has a role in eliminating TBHP induced hemolysis, MDA generation and methemoglobin production. 0.95 μg/mL selenoprotein solution showed significant inhibition on MDA formation in red blood cells, with an inhibition rate of 51.85%; the inhibition rate at 5.96 μg/mL reached 98.2%. When the Se concentration of selenoprotein solution was > 0.95 μg/mL, the inhibition rates on red blood cell hemolysis and methemoglobin production could be over 60%.

4 CONCLUSIONS

The purified selenoprotein in Se-enriched plants with Se concentrations ranging from 0.95 to 5.96 μg/mL in protein solution can inhibit the formation of free radicals, lipid peroxidation, and reduce the damage to red blood cells to maintain their normal physiological function.

ACKNOWLEDGEMENTS

The authors are grateful for grant support from NNSF of Hubei Province, China (2012FFC02501).

REFERENCES

Chen, J. & Berry, M.J. 2003. Selenium and selenoproteins in the brain and brain diseases. *Journal of Neurochemistry* 86: 1–12.

Du, M. & Zhao, L., Li, C.R., Zhao, G.H. & Hu, X.S. 2007. Purification and properties of a novel Se-containing protein from Se-enriched *Ganoderma lucidumand* its antioxidant activity. *Chemical Journal of Chinese Universities* 28(1): 75–78.

Li, G.P., Liu, Z.C. & Wang, M.L. 2001. Effects of organic selenium and inorganic selenium on biochemical criterions in mice. *Journal of Shanxi College of Traditional Chinese Medicine* 2(1): 19–21.

Nan, P., Gong, F.J., Chen, C.Y., Wu, X.J. & Yuan, P. 1999. Investigation with esr on the hydroxyl radical scavenging effect of selenonic protein. *Journal of Wuhan Botanical Research* 17(1): 87–90.

Tian, J.M., Zhang, D., Fu, R.J., Zhang, R. & Wei, H. 2010. Comparison of the bioavailability of soybean selenoprotein and sodium selenite. *Soybean Science* 29(3): 534–536.

Selenium in the Environment and Human Health – Bañuelos, Lin & Yin (Eds)
© 2014 Taylor & Francis Group, London, ISBN 978-1-138-00017-9

Selenium biofortification in North America: Using naturally selenium-rich feeds for livestock

J.B. Taylor
United States Department of Agriculture, Agricultural Research Service, U.S. Sheep Experiment Station, Dubois, Idaho, USA

J.S. Caton
Center for Nutrition and Pregnancy, Animal Sciences Department, North Dakota State University, ND, USA

R. Larsen
Intermountain Farmers Association, Salt Lake City, Utah, USA

1 INTRODUCTION

For many years, researchers throughout North America have challenged the food industry to manufacture and market products that can be made from naturally Se-rich grains. Likewise, several researchers have taken this challenge to the livestock feed industry. However, these challenges seem to have fallen on deaf ears, with the common argument, at least in the United States, that regulatory issues and logistics were prohibitive. Furthermore, the presumed value of Se-rich raw grains/forages from the growers' perspective has been much greater than the manufactures and(or) end-users were willing to pay. Recently, through persistent action of researchers and extension educators, some North America feed manufactures are making a concerted effort to test the marketing potential of natural specialized products for the labeled purpose of biofortifying livestock with Se. Presented is this abstract are the preliminary test results from a cooperative effort to manufacture and test the efficacy of a specialized Se-rich feed product on biofortifying livestock in the production system with Se.

2 MATERIALS AND METHODS

In 2010 and 2011, we requested an audience with several North America-based international and regional feed manufactures to demonstrate the utility of using naturally Se-rich feeds as tools to biofortify livestock with Se. We presented data (*e.g.*, Lawler *et al.*, 2004; Meyer *et al.*, 2010; Taylor, 2013) on the efficacy of Se biofortification to avert Se deficiency and create Se-rich food products. We discussed availability of naturally Se-rich products for product manufacturing, and considered potential regulatory issues related to product labeling. We promoted the use of Se-rich nonfood coproducts and forages to avoid direct competition for raw grains with the food and energy industries. A regional feed manufacture (IFA; Intermountain Farmers Association, Salt Lake City, Utah, USA) agreed to launch a three-year test on the feasibility that a specialized Se-rich product line could be marketable to various livestock industry segments.

We selected the Intermountain West-Pacific Coast (North America) sheep producers as the test producer group. Much of the forages and feeds grown in the Intermountain West and Pacific Coast regions are notoriously deficient in Se, which is manifested as moderate to severe Se-deficiency in livestock (*e.g.*, Muth and Allwaway, 1963). Early-lactating ewes were selected as the test animal; based on recent data, lactation is the most opportune time to biofortify simultaneously the dam and nursing offspring with Se.

A specialize Se-rich product (SePR) was manufactured, which would meet nutrient needs of early-lactating ewes. SePR was made with Se-rich alfalfa hay (approximately 16 mg/kg, dry matter basis), high-energy feedstuffs, and other micronutrients (no additional Se supplements were added). SePR was pelleted (5 mm) and contained 160 g crude protein/kg, 2.7 Mcal ME/kg, and 6 mg Se/kg (dry matter basis).

The Oregon Sheep Commission (Salem, Oregon, USA) solicited for volunteers to test SePR; five producers (Locations A, B, C, D, and E) volunteered. Locations A and B were shed-lambing production systems (flock > 3,000 ewes), while Location C was a pasture-lambing production system (flock = 500 ewes). Only data from locations A, B, and C will be discussed in this abstract; data from all locations will be presented in the oral presentation.

Participating producers were instructed to feed SePR daily for 20 to 40 days during early lactation and to collect whole blood and new-growth wool (near follicle) samples from ewes and lambs before feeding SePR and within 7 days after SePR feeding was stopped. The recommended daily allowance was 0.7 to 0.9 kg (as-fed basis) per ewe. At Locations A and B, SePR was included as a part of the total mixed ration. At Location C, SePR was fed daily as a pasture supplement.

Blood and wool samples were analyzed for Se. Data were analyzed (ANOVA) within location. An effect was declared when the probability of a greater F-test statistic was $P < 0.05$. Selenium concentrations of samples collected before SePR was fed were compared with Se concentrations of samples collected after SePR feeding stopped. For purposes of discussion, differences are expressed as fold greater (*e.g.*, increased 2 fold).

3 RESULTS AND DISCUSSION

SePR was fed during early lactation for 22, 20, and 40 days at locations A, B, and C, respectively. SePR enhanced the Se status of ewes and nursing lambs. Selenium content of whole blood in ewes increased ($P < 0.01$) 1.2, 1.2, and 1.5 fold for locations A, B, and C, respectively. Note that the increase was greatest for location C, which is an effect of feeding SePR twice as long as was fed at other locations. This response is consent with the response we observed in a laboratory setting (Taylor, 2005; Taylor, 2013).

Selenium content of whole blood in lambs nursing the test ewes increased ($P < 0.01$) 1.3 and 2.2 fold for locations B and C, respectively; blood samples were not collected from lambs at Location A. As with the ewes, duration of feeding SePR to lactating ewes had a notable effect on whole blood Se in lambs, with lambs from Location C having the greatest response. Lambs did not consume SePR, and thus, lambs received the benefit of SePR through the ewes' milk. This response is consistent with our previous results (Taylor, 2013), and further demonstrates that the best time to enhance Se status of lambs is to feed SePR to the ewes during early lactation when milk accounts for >90% of the daily nutrition of lambs.

Selenium content of new-growth wool in ewes and lambs increased ($P < 0.01$) 1.7 to 2.1 fold at all locations after SePR was fed to the ewes. Regardless of location and animal, the magnitude of response in wool was much greater than what was observed in whole blood. This indicates that much of the Se in SePR was selenomethionine, which readily enriches the body's general protein pool with Se. Based on previous data, the rate of Se incorporation in wool is similar to skeletal muscle when selenomethionine-rich feeds are fed (Taylor, 2005). Therefore, the Se response in wool is an indicator of Se biofortification of skeletal muscle, the optimal tissue for Se-enrichment in livestock production.

4 CONCLUSIONS

These preliminary results provide credence to the concept that specialized Se-rich feed products, formulated from Se-rich coproducts and forages, can be manufactured and subsequently used for the specific purpose on biofortifying livestock with Se. We will conclude a second year of product testing in fall of 2013. In 2013 and 2014, marketing surveys will be conducted to estimate the potential demand for naturally Se-rich products from various livestock industry segments in North American.

REFERENCES

Lawler, T.L., Taylor, J.B., Finley, J.W. & Caton, J.S. 2004. Effect of supranutritional and organically bound selenium on performance, carcass characteristics, and selenium distribution in finishing beef steers. *J. Anim. Sci.* 82: 1488–93.

Meyer, A.M., Reed, J.J., Neville, T.L., Taylor, J.B., Hammer, C.J., Reynolds, L.P. *et al.* 2010. Effects of plane of nutrition and selenium supply during gestation on ewe and neonatal offspring performance, body composition, and serum selenium. *J. Anim. Sci.* 88: 1786–1800.

Muth, O.H., & Allaway, W.H. 1969. The relationship of white muscle disease to the distribution of naturally occurring selenium. *J. Am. Vet. Med. Assoc.* 142: 1379–84.

Taylor, J.B. 2005. Time dependent influence of supranutri-tional organically bound selenium on selenium accumulation in growing wether lambs. *J. Anim. Sci.* 83: 1186–93.

Selenium in the Environment and Human Health – Bañuelos, Lin & Yin (Eds)
© 2014 Taylor & Francis Group, London, ISBN 978-1-138-00017-9

Effects of organic selenium on pork quality and carcass traits: A meta-analysis

Qing Zheng
State Key Laboratory of Soil and Sustainable Agriculture, Institute of Soil Science, Chinese Academy of Sciences, Nanjing, Jiangsu, China
Jiangsu Bio-Engineering Research Centre on Selenium, Suzhou, Jiangsu, China

Liqiang Qin
Department of Nutrition and Food Hygiene, School of Public Health, Soochow University, Suzhou, Jiangsu, China

Linxi Yuan & Xuebin Yin
Advanced Lab for Selenium and Human Health, Suzhou Institute for Advanced Study, University of Science and Technology of China, Suzhou, Jiangsu, China

1 INTRODUCTION

As an essential trace element, the biological function and nutritional benefits of inorganic and organic selenium (Se) have been discussed elsewhere (Payne & Southern, 2005). Although the utilization efficiency of organic Se was higher than that of inorganic Se in animals (Mahan *et al.*, 1996; Mateo *et al.*, 2006), it is still debated on which form of Se affects pork quality and carcass traits (Wolter *et al.*, 2004; Clyburn *et al.*, 2002). The purpose of the present study was to clarify the differences on pork quality and carcass traits between organic Se and inorganic Se using Meta-analysis of Randomized Controlled Trials (RCTs). This information will provide references for pork Se nutrition.

2 MATERIALS AND METHODS

After a systematic review of current studies, the qualified articles were searched according to the inclusion criteria in Web of Knowledge Databases. Detailed criteria were made as follows: (1) research object: feeding pigs, (2) treatment group: organic Se supplementation (0.3 mg/kg) in basal diets and inorganic Se used as the control, (3) a randomized controlled trial, (4) indexes of pork quality and carcass traits including drip loss, *p*H value, loin eye area, back fat and proportion of lean meat (%), and (5) calculations on the effect value, weight, variance, and the overall effect value using Stata 11.0 software.

The meta-analysis reported the following results: (1) the heterogeneity between studies with the Q-test (subject $\chi 2$ distribution using the random-effects model), (2) the publication bias with a funnel plot with Begg's rank correlation method and the Egger regression, and (3) the model sensitivity.

3 RESULTS AND DISCUSSION

Heterogeneity (Q-test) showed that all the *P* values of indexes were less than 0.05, indicating that the random-effects model can be used to count and conflate the Standard Mean Difference (SMD) values. Compared with the control groups, the SMD values were –1.78 for drip loss, –0.79 for *p*H, 0.23 for loin eye area, –0.60 for back fat, and –0.60 for lean meat percentage, respectively. The 95% CI ranges were [–2.78, –0.79] for drip loss, [0.15, 1.43] for *p*H, [–0.12, 0.57] for loin eye area, [–2.24, 1.04] for back fat, and [–2.24, 1.04] for lean meat percentage, respectively (Fig. 1). The Q-test results revealed that the drip loss was significantly decreased, while the *p*H value and lean meat percentages were significantly improved using organic Se, although the loin eye area and back fat displayed no differences. Begg's rank correlation test displayed that the *P* values of drip loss, *p*H value, loin eye area, back fat and lean meat percentage were 1.000, 1.000, 0.734, 1.000 and 0.734, respectively. The Egger regression test revealed that the *P* values of *p*H, loin eye area, backfat and lean meat percentage were 0.837, 0.679, 0.778, and 0.626, respectively. These tests showed that there was no significant publication migration on selected articles.

To observe the maximum weight of the studies on the impact of the merger SMD, we removed the largest weight of all indexes. The merger SMDs were as follows: –1.98 (95% CI: –2.97, –0.99); 0.96 (95% nCI: 0.36, 1.56); 0.19 (95% CI: –0.25, 0.63);

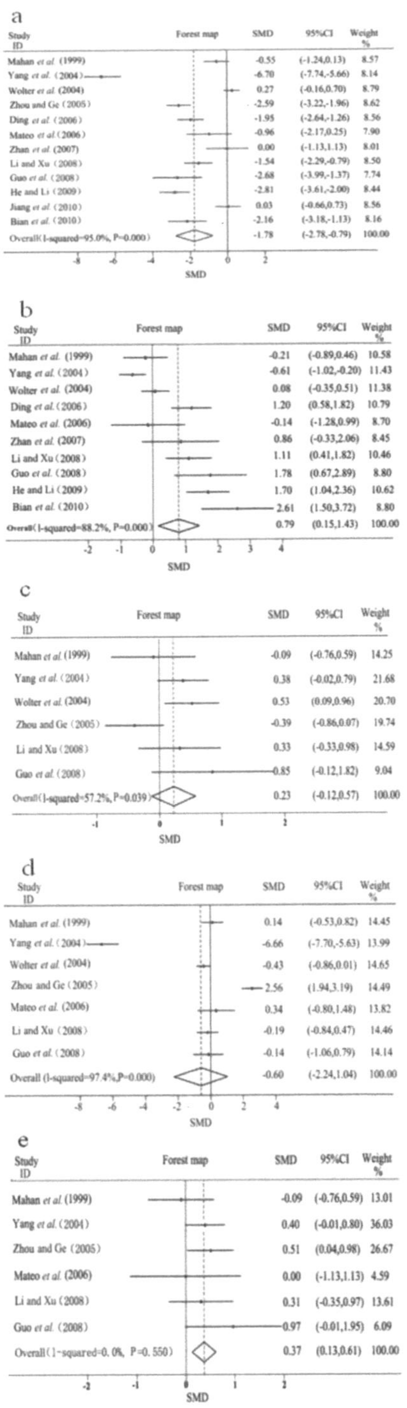

Figure 1. The forest maps of five indexes above of pork quality and carcass traits, including a: drip loss, b: *p*H. c: loin eye area, d: backfat, and e: lean meat percentage.

–0.64 (95% CI: –2.81, 1.53); 0.35 (95% CI: 0.05, 0.66) respectively, indicating the sensitivity of the selected articles were low.

4 CONCLUSIONS

Meta-analysis demonstrated that drip loss of pork was significantly decreased, and *p*H value and lean meat proportion (%) were significantly improved by organic Se. However, there were no significant differences on loin eye area and back fat between organic Se and inorganic Se treatments. Thus, this study suggests that it might be better for farmers to select organic Se compounds as the Se supplement source to improve pork quality and carcass traits.

ACKNOWLEDGEMENTS

This study was supported by Open Fund of State Key Laboratory of Soil and Sustainable Agriculture, Institute of Soil Science, Chinese Academy of Sciences (Y 052010020), the Jiangsu Research-Industry Joint Innovation Project (BY 2011186) and the Suzhou Leading Talent Project (ZXL 2010045).

REFERENCES

Clyburn, B.S. 2002. *Effects of sel-plex (organic selenium) and vitamin E on performance, immune response, and beef cut shelf life of feedlot steers*. Ph.D. thesis, Texas: Texas Tech University.

Mahan, D.C. & Parrett, N.A. 1996. Evaluating the efficacy of selenium enriched yeast and sodium selenite on tissue swine selenium retention and serum glutathione peroxidase activity in grower and finisher. *Journal of Animal Science* 74: 2967–2974.

Mateo, R.D., Spallholz, J.E., Elder, R., Yoon, I. & Kim, S.W. 2006. Efficacy of dietary selenium sources on growth and carcass characteristics of growing-finishing pigs fed diets containing high endogenous selenium. *Journal of Animal Science* 85: 1177–1183.

Payne, R.L. & Southern, L.L. 2005. Comparison of inorganic and organic selenium sources for broilers. *Poultry Science* 84(6): 898–902. 384–390.

Spallholz, J.E. 1994. On the nature of selenium toxicity and carcinostatic activity. *Free Radical Biol. Med.* 17: 45–64.

Wotter, B., Ellis, M., Mckeith, F.K. & McKeith, F. 2004. Influence of dietary selenium source on growth performance, carcass and meat quality characteristics in pigs. *Journal of Animal Science* 79: 119–121.

Selenium in the Environment and Human Health – Bañuelos, Lin & Yin (Eds)
© 2014 Taylor & Francis Group, London, ISBN 978-1-138-00017-9

Biofortification of maternal diets with selenium: Postnatal growth outcomes

J.S. Caton, T.L. Neville, L.P. Reynolds, C.J. Hammer & K.A. Vonnahme
Center for Nutrition and Pregnancy, Animal Sciences Department, North Dakota State University, ND, USA

A.M. Meye
Division of Animal Sciences, University of Missouri, Columbia, MO, USA

J.B. Taylor
USDA-ARS, US Sheep Experiment Station, Dubois, ID, USA

1 INTRODUCTION

Selenium is an essential trace element needed for normal growth and development. Dietary biofortification is the provision of additional or elevated levels of dietary nutrients with expectation of specific biological outcomes. Regarding Se, dietary biofortification occurs either naturally because of Se enriched ingredients or from direct dietary additions. Our laboratories have investigated Se biofortification of maternal diets during gestation with a focus on both maternal and offspring outcomes. Our overarching hypothesis has been that supranutritional Se in maternal diets would alter molecular, cellular, and tissue events resulting in mitigation of compromised offspring outcomes resulting from restricted or excess nutrient supply during gestation. Dietary supranutritional Se represent concentrations that are above known requirements but below toxicity. Gestation is a critical time for both livestock and humans with the delivery of healthy offspring and a quick recovery of the mother being highly desired outcomes. Additionally, recent data in the area of developmental programming indicates that maternal nutrition plays a major role in postnatal outcomes and lifelong consequences. Herein we report on the impacts of dietary biofortification during gestation with organically bound Se on postnatal growth and outcomes.

2 MATERIALS AND METHODS

Animal care and use was approved by the Institutional Animal Care and Use Committee at North Dakota State University (NDSU), Fargo, and the USDA, ARS, US Sheep Experiment Station (USSES) in Dubois, ID. A full description of the breeding and feeding programs has been previously published (Swanson *et al.*, 2008; Meyer *et al.*, 2010; Neville *et al.*, 2010; Hammer *et al.*, 2011) but will be described briefly in the sections below.

In both experiments, supranutritional Se treatments were initiated at breeding by collaborators at the USSES. At approximately 40 d of age 84 pregnant ewe lambs for each experiment were shipped to Fargo where animals were individually housed and fed until parturition. In both experiments the treatment design was a 2 × 3 factorial arrangement with two levels of Se and three planes of maternal nutrition. In Experiment 1, selenium was provided to High-Se (HSe) ewes in the form of Se-enriched yeast and treatments were adequate Se (ASe; 9.5 µg/kg of BW) vs. HSe (81.8 µg/kg of BW) and were delivered one time daily in pelleted form. Several research projects have been conducted in our laboratory using this amount of Se with no signs of toxicity (Neville *et al.*, 2008; Swanson *et al.*, 2008; Carlson *et al.*, 2009). In Experiment 2, Se treatments were ASe, 11.5 µg/kg of BW and HSe, 77.0 µg/kg of BW. In both experiments planes of maternal nutrition were implemented on approximately d 45 and ewes within Se treatments were assigned randomly to 1 of 3 nutritional planes: 60% (RES), 100% (control, CON), or 140% (HI) of requirements (NRC, 1985) excluding Se. This arrangement resulted in the following six treatments: ASe-RES; ASe-CON; ASe-HI; HSe-RES; HSe-CON; and HSe-HI. Daily allotment of diets were calculated based on BW, ME requirements, and dietary ME and Se concentrations and were previously published (Meyer *et al.*, 2010, 2011; Neville *et al.*, 2010; Hammer *et al.*, 2011).

Initial care of lambs in both Experiment 1 and Experiment 2 was similar. All births were observed, and lambs were immediately removed from their dams before suckling and cared for in a separate, clean room. Lambs were fed artificial colostrum and milk replacer. The amount of artificial colostrum was based on birth weight to achieve consumption of 10.6 g of IgG/kg of BW during the first 20 h postpartum. The dosage of IgG was calculated to provide 50 g of IgG/4.7 kg of lamb BW.

Lamb care procedures were similar in Experiment 1 and Experiment 2; except for experimental length. In Experiment 1, lambs were raised until weaning (57 d) on milk replacer and then on commercial diets until 180 d, whereas in Experiment 2, lambs were raised on milk replacer until 19 d of age.

For both experiments, data were analyzed as a completely randomized design with a 2 × 3 factorial arrangement of treatments using GLM procedures (SAS Inst. Inc., Cary, NC).

3 RESULTS AND DISCUSSION

3.1 *Experiment 1*

Male and female lambs from RES-fed ewes were lighter ($P \leq 0.03$) at birth than lambs from CON-fed ewes, with lambs from HI-fed ewes being intermediate. Although maternal nutritional intake influenced ($P < 0.06$) Body Weight (BW) gain before weaning on d 57, both maternal nutritional intake and sex of offspring influenced ($P \leq 0.09$) BW gain from d 57 to 180. Although maternal nutritional intake did not influence ($P \geq 0.35$) female lamb BW gain, male lambs from RES-fed ewes were lighter ($P \leq 0.09$) than those from CON-fed ewes until d 162. By d 180, male lambs from RES- and HI-fed ewes were lighter ($P \leq 0.09$) than those from CON-fed ewes. In a subset of lambs used in a feed efficiency study, namely, those born to ASe ewes, HI maternal nutritional intake decreased ($P \leq 0.09$) Average Daily Gain (ADG) and Gain:Feed (G:F) compared with lambs born to RES- and CON-fed ewes, which did not differ ($P \geq 0.60$). Conversely, when lambs were born to HSe ewes, HI maternal nutritional intake increased ($P \leq 0.01$) ADG and G:F compared with CON, with RES being intermediate. Moreover, lambs born to ASe-HI ewes had decreased ($P < 0.01$) ADG and G:F compared with lambs born to HSe-HI ewes. Furthermore, male lambs had a greater ($P < 0.01$) G:F than female lambs. Maternal diet did not affect ($P \geq 0.11$) N retention in male lambs. These data indicate that maternal nutrition during gestation and sex of the offspring alters postnatal growth and efficiency of growth in offspring despite similar postnatal management.

3.2 *Experiment 2*

From d 95 of gestation through parturition, ewe BW and body condition score (BCS; a measure of adiposity) were least ($P \leq 0.01$) in the RES treatment, intermediate in the CON treatment, and greatest in the HI treatment. Ewes fed HSe had a greater ($P \leq 0.05$) BCS increase than those fed ASe during mid—and late gestation. During gestation, ewes in the HI treatment had the greatest ($P < 0.001$) ADG and G:F, those in the CON treatment were intermediate, and those in the RES treatment were least, whereas ewes fed HSe had greater ($P < 0.001$) ADG and G:F than those fed ASe during midgestation. Ewe backfat and loin muscle area on d 135 of gestation were least ($P < 0.001$) in the RES treatment, intermediate in the CON treatment, and greatest in the HI treatment, with ewes fed HSe having greater ($P \leq 0.03$) backfat than those fed ASe.

At birth, lambs born to ewes in the RES group weighed less and had decreased curved crown rump lengths ($P = 0.08$) compared with those born to ewes in the CON and HI groups, and lambs from ewes in the ASe-RES treatment were lighter ($P < 0.08$) than those from ewes in the HSe-RES, ASe-CON, and ASe-HI treatments. Lambs from dams in the RES group had less ($P < 0.05$) BW from d 7 to 19 and decreased ($P < 0.07$) overall ADG compared with lambs from dams in the CON and HI groups. Additionally, lambs from dams in the RES group had less ($P \leq 0.08$) perirenal fat than their counterparts, and lambs from dams in the HI group had greater ($P = 0.01$) omental and mesenteric fat than lambs from dams in the RES group. Postpartum serum Se of ewes and lambs (birth and d 19) was increased ($P < 0.001$) by HSe feeding during gestation. Results indicate that BW differences in pregnant ewes attributable to nutritional plane are accompanied by changes in body composition and offspring BW, both of which may be affected by Se supply.

4 CONCLUSIONS

Taken together these data are interpreted to indicate that biofortification of maternal diets with Se inconsistently mitigates some of the negative outcomes associated with compromised plane of maternal nutrition. Additional research is needed to clarify these responses.

REFERENCES

Carlson, D.B., Reed, J.J., Borowicz, P.P., Taylor, J.B., Reynolds, L.P., Neville, T.L. *et al.* 2009. Effects of dietary selenium supply and timing of nutrient restriction during gestation on maternal growth and body composition of pregnant adolescent ewes. *J. Anim. Sci.* 87: 669–680.

Hammer, C.J., Thorson, J.F., Meyer, A.M., Redmer, D.A., Luther, J.S., Neville, T.L. *et al.* 2011. Effects of maternal selenium supply and plane of nutrition during gestation on passive transfer of immunity and health in neonatal lambs. *J. Anim. Sci.* 89: 3690–3698.

Meyer, A.M., Reed, J.J., Neville, T.L., Taylor, J.B., Hammer, C.J., Reynolds, L.P. *et al.* 2010. Effects of nutritional plane and selenium supply during gestation on ewe and neonatal offspring performance, body composition, and serum selenium. *J. Anim. Sci.* 88: 1786–1800.

Meyer, A.M., Reed, J.J., Neville, T.L., Thorson, J.F., Maddock-Carlin, K.R., Taylor, J.B. *et al.* 2011. Nutritional plane and selenium supply during gestation impact yield and nutrient composition of colostrum and milk in primiparous ewes. *J. Anim. Sci.* 89: 1627–1639.

Neville, T.L., Caton, J.S., Hammer, C.J., Reed, J.J., Luther, J.S., Taylor, J.B. *et al.* 2010. Ovine offspring growth and diet digestibility are influenced by maternal Se supplementation and nutritional intake level during pregnancy despite a common postnatal diet. *J. Anim. Sci.* 88: 3645–3656.

Neville, T.L., Ward, M.A., Reed, J.J., Soto-Navarro, S.A., Julius, S.L., Borowicz, P.P. *et al.* 2008. Effects of level and source of dietary selenium on maternal and fetal body weight, visceral organ mass, cellularity estimates, and jejunal vascularity in pregnant ewe lambs. *J. Anim. Sci.* 86: 890–901.

NRC. 1985. Nutrient Requirements of Sheep. 6th rev. ed. Washington, DC: Natl. Acad. Press.

Swanson, T.J., Hammer, C.J., Luther, J.S., Carlson, D.B., Taylor, J.B., Redmer, D.A. *et al.* 2008. Effects of gestational plane of nutrition and selenium supplementation on mammary development and colostrum quality in pregnant ewe lambs. *J. Anim. Sci.* 86: 2415–2423.

Selenium in the Environment and Human Health – Bañuelos, Lin & Yin (Eds)
© 2014 Taylor & Francis Group, London, ISBN 978-1-138-00017-9

The functional agriculture in China: 5-year practice

Qiguo Zhao
State Key Laboratory of Soil and Sustainable Agriculture, Institute of Soil Science, Chinese Academy of Sciences, Nanjing, China

Xuebin Yin
State Key Laboratory of Soil and Sustainable Agriculture, Institute of Soil Science, Chinese Academy of Sciences, Nanjing, China
Advanced Lab for Selenium and Human Health, Suzhou Institute, University of Science and Technology of China, Suzhou, China

1 INTRODUCTION

In "Agricultural Science and Technology in China: A Roadmap to 2050" (Zhao, 2011), the authors in 2009 presented a systematic frame for functional agriculture in China. In brief, half of agriculture products in China will contain one or more functional substances around 2050. This possibility is a large challenge for Chinese scientists and food industries. As an essential trace element for human and animals, selenium (Se) has received increasing attention in biofortified agriculture products. Adequate Se supplementation is proved to be significant in cancer prevention, anti-aging and in boosting a human's immune system for the people living in Se deficiency areas. In this report, the past 5-year practice of functional agriculture in China will be reviewed. To meet the theme of Se conference, progress on Se biofortification will be focused. Contents will be selected from the book "Ecological High Value Agriculture in China," edited by Zhao (2013).

2 DEMAND OF FUNCTIONAL AGRICULTURE IN CHINA

At present, there are two trends of agricultural science and technology in China. One is to improve the safety of food products and the other is to enhance the nutritional value using various advanced techniques. The Chinese National Nutrition Survey in 2002 reported that the residents were facing the dual challenges of nutritional deficiencies and structural imbalances, especially deficiencies with some micronutrients such as Se, calcium, iron and iodine (Vander *et al.*, 2001; Gao *et al.*, 2010; Yin & Yuan, 2012). Therefore, biofortified agriculture foods with trace essential elements are urgently needed in future (Banuelos & Lin, 2009). The annual meeting of Chinese National Nutrition Society in 2013 claimed to raise Recommended Nutrient Intake (RNI) of Se from 50 microgram per day to 60 microgram per day based on Se protein P (Chinese Nutrition Society, 2013). According to this new standard, the proportion of inadequate dietary Se intake for Chinese may reach 70–80%. Consequently, the Chinese urgently need to have an additional Se-enriched food supply.

3 BEGINNING OF FUNCTIONAL FOODS IN CHINA

The functional foods with distinctive nutritional characteristics were developed under guidance of nutritional science, which is different from that of conventional foods. The present research and application on the functional foods mainly focus on increasing the contents of trace elements (*e.g.*, Se, zinc, iron, calcium, *etc.*), essential amino acids (*e.g.*, lysine and tryptophan), vitamins (*e.g.*, Vitamin A and E), antioxidants (*e.g.*, polyphenols, flavonoids, carotenoids, and anthocyanidin), and unsaturated fatty acids. Foods enriched in certain nutrients (*e.g.*, Se) were widely produced by biofortification methods in some regions. These Se-enriched products were sold at higher prices for satisfying higher requirements of humans. The consumption of functional food in China is still relatively low, approximately less than 1%. This percentage may increase to 10% by 2020. Since 2008 Setek Co. began to develop the Se-biofortified foods in China by applying Se-fertilizer to soils. To guarantee Se-enriched food safety, systematic criterions and regional standards have been developed. COFCO Group and Beidahuang Group, two of the largest agricultural companies in China, have successfully cooperated with Setek Co. to produce Se-enriched rice and wheat.

4 CHALLENGES OF FUNCTIONAL FOODS IN CHINA

As a developing country, the Chinese people know little about mineral nutrition, especially Se. Health knowledge should be conducted propaganda among the masses with community science post, broadcast, meeting, internet and other different media. Because, China has a large area (9.6 million km^2), it is much more complicated to establish a Se supplementation strategy. Therefore, unlike Finland, the Chinese government cannot effectively implement the Se biofortification strategy throughout China in a short time period. It will take at least 10 to 20 years before the Se biofortification strategy will be successfully practiced and become popular in Se-deficient regions of China.

ACKNOWLEDGEMENTS

This study was supported by Open Fund of State Key Laboratory of Soil and Sustainable Agriculture, Institute of Soil Science, Chinese Academy of Sciences (Y 052010020), the Jiangsu Research-Industry Joint Innovation Project (BY 2011186) and the Suzhou Leading Talent Project (ZXL 2010045).

REFERENCES

Bañuelos, G.S. & Lin, Z.-Q. 2009. *Use and Development of Biofortified Agricultural Products*. Boca Raton, Florida: CRC Press.

Chinese Nutrition Society. 2013. *Proceedings of the 11th China Nutrition Science Congress & International DRIs Summit*. Hangzhou, China.

Gao, J., Liu, Y., Huang, Y., Lin, Z.-Q., Bañuelos, G.S., Lam, M.H.W. *et al*. 2011. Daily Se intake in a moderate Se deficiency area of Suzhou, China. *Food Chemistry* 126: 1088–1093.

Vander, A., Sherman, J. & Luciano, D. 2001. *Human Physiology: The Mechanisms of Body Function (8th edition)*. New York: McGraw-Hill Higher Education.

Yin, X.B. & Yuan, L.X. (eds). 2012. *Phytoremediation and Biofrotification: Two Sides of One Coin*. Berlin: Springer Press.

Zhao, Q.G. 2011. *Agricultural Science and Technology in China: A Roadmap to 2050*. 100–126. Beijing and Berlin: Science Press and Springer.

Zhao, Q.G. 2013. *Ecological High Value Agriculture in China*. Beijing: Science Press (in Chinese).

5 *Cellular and molecular functions of selenium*

Selenium in the Environment and Human Health – Bañuelos, Lin & Yin (Eds)
© 2014 Taylor & Francis Group, London, ISBN 978-1-138-00017-9

The selenoprotein thioredoxin reductase as a key regulator of cellular signaling pathways

Elias S.J. Arnér
Division of Biochemistry, Department of Medical Biochemistry and Biophysics, Karolinska Institutet, Stockholm, Sweden

1 INTRODUCTION

Isoforms of the mammalian selenoprotein thioredoxin reductase (TrxR) promote cellular functions of different thioredoxin (Trx) proteins by keeping their active site disulfides in the reduced dithiol state using NADPH. The major forms of Trx and TrxR that are predominantly found in the cytosol are Trx1 and TrxR1, while Trx2 and TrxR2 are predominantly mitochondrial enzymes. Key cellular functions carried out by the Trx system include promotion of cell viability and replication (*e.g.*, provides electrons for ribonucleotide reductase), antioxidant defense systems (*e.g.*, supports peroxiredoxins and methionine sulfoxide reductases) and redox modulation of signaling pathways (e.g., activates receptor-linked signaling or regulates transcription factor activities) (Arnér, 2009; Arnér & Holmgren, 2000; Arnér & Holmgren, 2006; Nordberg & Arnér, 2001; Rundlöf & Arnér, 2004). In this summary, special focus shall be given on the role and regulation of thioredoxin reductase 1 (TrxR1) in relation to cellular signaling events.

2 MATERIALS AND METHODS

Recently published work, as well as preliminary results from the laboratory and collaborations of the author is presented, summarized and discussed.

3 RESULTS AND DISCUSSION

3.1 *Regulation of expression and activity of TrxR1*

Human TrxR1 is encoded by the *TXNRD1* gene on chromosome 12, which displays a complex organization with a housekeeping-type promoter that leads to a basal expression of the enzyme in most cells (Rundlöf *et al.*, 2001). This in combination with a Nrf2-responsive element allows for strong induction upon electrophilic or oxidative stress (Furman *et al.*, 2004; Locy *et al.*, 2012; Sakurai *et al.*, 2005). Highly extensive alternative splicing furthermore results in at least 25 different transcripts that encode up to five splice variants of TrxR1 (Rundlöf *et al.*, 2004). In addition, the enzyme expression is also regulated by selenium availability, although it is among the most prioritized selenoproteins during expression under selenium deficiency, even if the molecular mechanisms for its higher hierarchy are not fully known (Anema *et al.*, 1999; Crosley *et al.*, 2007; Ebert *et al.*, 2006; Gasdaska *et al.*, 1999; Rundlöf & Arnér, 2004). Moreover, under severe selenium deficiencies a Sec-to-Cys substituted form of the enzyme is expressed in cells and tissues, which have significantly lower activity than the selenoprotein TrxR1 (Lu *et al.*, 2009; Peng *et al.*, 2012; Xu *et al.*, 2010). Finally, we recently found that TrxR1 uniquely responds to excessive oxidative stress through an evolutionarily conserved highly reactive amino acid residue located outside of the active site, which modulates the enzyme activity in relation to oxidation (Xu *et al.*, unpubl.).

The combination of these several intricate mechanisms for the modulation of TrxR1 expression and activity in human cells and tissues are likely to be of importance for the role and regulation of the enzyme in relation to cellular signaling pathways.

3.2 *Compartmentalization of TrxR1 isoforms*

The major form of TrxR1 (named TXNRD1_v1) is cytosolic, while the isoenzyme TrxR2 is its mitochondrial counterpart (for a review of different TrxR isoenzymes see Arnér, 2009). In addition, the less common splice variants of the TrxR1 protein (see above) seem to provide specific compartmentalized targeting of the enzyme. The TXNRD1_v2 form is expressed from the same core promoter as TXNRD1_v1 but has an extended N-terminus as a result of alternative splicing with a nuclear localization signal. This form modulates signaling through interactions with both major forms of the estrogen receptor (Damdimopoulos *et al.*, 2004). The TXNRD1_v3 splice variant carries an atypical glutaredoxin domain, and in contrast, it is targeted to cell membranes and membrane rafts as a result of myristoylation and palmitoylation at its N-terminus. Consequently, TXNRD1_v3 has the capacity to induce formation of filopodia

(Cebula *et al.*, 2013; Damdimopoulou *et al.*, 2009; Dammeyer *et al.*, 2007). These examples show that TrxR1 can exert a range of functions in different cell compartments as a result of specific subcellular targeting patterns for its different splice variants.

3.3 *Relation of TrxR1 regulation to cellular signaling pathways*

With the net effect of TrxR1 activities through reduction of either Trx1 or other substrates of the enzyme, as well as targeting to different cell compartments including cytosol, nucleus and membrane rafts, it is likely that TrxR1 and the Trx system as a whole have a large number of functions in relation to diverse cellular signaling pathways. It should also be pointed out that most of these pathways, if not all, will be dependent upon the selenium status, as a result of the fact that the activity of TrxR1 involves a catalytic selenocysteine residue.

Full understanding of the exact roles of TrxR1 in relation to cellular signaling pathways is, however, far from complete. The intricate redox networks of cells are exceedingly complex and may furthermore be redundant, with many pathways being overlapping between the Trx system and that dependent upon glutathione (GSH). For example, although the mouse knockout for TrxR1 is embryonically lethal (Bondareva *et al.*, 2007; Jakupoglu *et al.*, 2005), the conditional hepatocyte knockout is without apparent phenotype unless also the GSH system becomes compromised. Similarly, hepatocytes *in vivo* are apparently not dependent upon GSH, as long as a single allele for TrxR1 is maintained in a functional state (Prigge *et al.*, 2012; Rollins *et al.*, 2010). Importantly, even if no grave phenotypes are not noted upon conditional knockout of TrxR1 in the liver of mice, upon closer inspection they display an intriguing "metabolic switch" that promotes an extensive glycogen accumulation instead of lipogenesis (Iverson *et al.*, 2013). This finding shows that TrxR1 is indeed not only an antioxidant enzyme or regulator of cellular redox control, but also a master regulator of the metabolic state, at least in liver.

A highly simplified scheme is shown in Figure 1 that summarizes the many levels of regulation and cell signaling where different variants of TrxR1 seem to play diverse roles. Much of the molecular mechanisms for these pathways are still insufficiently understood.

4 CONCLUSIONS

It is clear that the selenoprotein TrxR1, encoded by the *TXNRD1* gene in human, is a central player

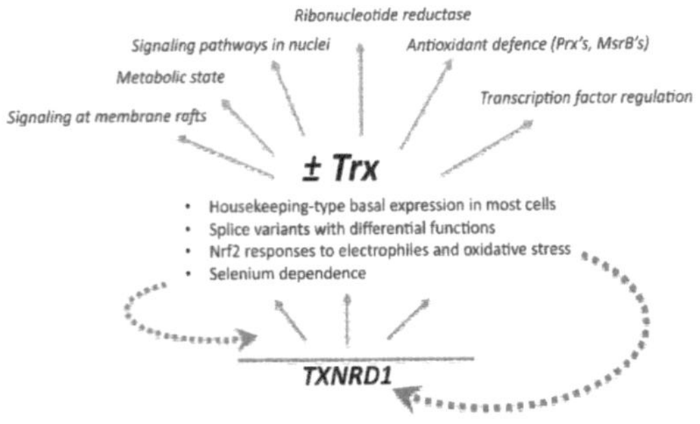

Figure 1. TrxR1 in signaling. Different splice variants of TrxR1 are encoded by the *TXNRD1* gene (lower arrows), in a housekeeping-type fashion but inducible upon stress (right dashed arrow) and enzyme activity being selenium dependent (left dashed arrow). The TrxR1 variants exert a number of effects in cells (top arrows) that are either dependent upon or independent from thioredoxin (±Trx).

for cellular redox control and serves as a master regulator of signaling pathways. The molecular mechanisms for the exact reactions by which these pathways are regulated must, however, be further studied if we are to understand how selenium status can control cellular function as well as health and disease, where TrxR1 is one important, yet only one of several, selenoproteins regulating these pathways.

REFERENCES

Anema, S.M., Walker, S.W., Howie, A.F., Arthur, J.R., Nicol, F. & Beckett, G.J. 1999. Thioredoxin reductase is the major selenoprotein expressed in human umbilical-vein endothelial cells and is regulated by protein kinase C. *Biochem. J.* 342: 111–117.

Arnér, E.S.J. 2009. Focus on mammalian thioredoxin reductases—important selenoproteins with versatile functions. *Biochim Biophys Acta* 1790(6): 495–526.

Arnér, E.S.J. & Holmgren, A. 2000. Physiological functions of thioredoxin and thioredoxin reductase. *Eur. J. Biochem.* 267: 6102–6109.

Arnér, E.S.J. & Holmgren, A. 2006. The thioredoxin system in cancer. *Semin. Cancer Biol.* 16(6): 420–426.

Bondareva, A.A., Capecchi, M.R., Iverson, S.V., Li, Y., Lopez, N.I., Lucas, O. *et al.* 2007. Effects of thioredoxin reductase-1 deletion on embryogenesis and transcriptome. *Free Radic. Biol. Med.* 43(6): 911–923.

Cebula, M., Moolla, N., Capovilla, A. & Arner, E.S. 2013. The rare TXNRD1_v3 ("v3") splice variant of human thioredoxin reductase 1 protein is targeted to membrane rafts by N-acylation and induces filopodia independently of its redox active site integrity. *J. Biol. Chem.* 288(14): 10002–10011.

Crosley, L.K., Meplan, C., Nicol, F., Rundlöf, A.K., Arnér, E.S., Hesketh, J.E. *et al.* 2007. Differential regulation of expression of cytosolic and mitochondrial thioredoxin reductase in rat liver and kidney. *Arch. Biochem. Biophys* 459(2): 178–188.

Damdimopoulos, A.E., Miranda-Vizuete, A., Treuter, E., Gustafsson, J.Å. & Spyrou, G. 2004. An alternative splicing variant of the selenoprotein thioredoxin reductase is a modulator of estrogen signaling. *J. Biol. Chem.* 279(37): 38721–38729.

Damdimopoulou, P.E., Miranda-Vizuete, A., Arner, E.S., Gustafsson, J.A. & Damdimopoulos, A.E. 2009. The human thioredoxin reductase-1 splice variant TXNRD1_v3 is an atypical inducer of cytoplasmic filaments and cell membrane filopodia. *Biochim Biophys Acta* 1793(10): 1588–1596.

Dammeyer, P., Damdimopoulos, A., Nordman, T., Jimenez, A., Miranda-Vizuete, A. & Arnér, E.S.J. 2007. Induction of cell membrane protrusions by the N-terminal glutaredoxin domain of a rare splice variant of human thioredoxin reductase 1. *J. Biol. Chem.* 283(5): 2814–2821.

Ebert, R., Ulmer, M., Zeck, S., Meissner-Weigl, J., Schneider, D., Stopper, H. *et al.* 2006. Selenium supplementation restores the antioxidative capacity and prevents cell damage in bone marrow stromal cells in vitro. *Stem Cells* 24(5): 1226–1235.

Furman, C., Rundlof, A.K., Larigauderie, G., Jaye, M., Bricca, G., Copin, C. *et al.* 2004. Thioredoxin reductase 1 is upregulated in atherosclerotic plaques: specific induction of the promoter in human macrophages by oxidized low-density lipoproteins. *Free Radic. Biol. Med.* 37(1): 71–85.

Gasdaska, J.R., Harney, J.W., Gasdaska, P.Y., Powis, G. & Berry, M.J. 1999. Regulation of human thioredoxin reductase expression and activity by 3'-untranslated region selenocysteine insertion sequence and mRNA instability elements. *J. Biol. Chem.* 274: 25379–25385.

Iverson, S.V., Eriksson, S., Xu, J., Prigge, J.R., Talago, E.A., Meade, T.A. *et al.* 2013. A Txnrd1-dependent metabolic switch alters hepatic lipogenesis, glycogen storage, and detoxification. *Free Radic. Biol. Med.* 63: 369–380.

Jakupoglu, C., Przemeck, G.K., Schneider, M., Moreno, S.G., Mayr, N., Hatzopoulos, A.K. *et al.* 2005. Cytoplasmic thioredoxin reductase is essential for embryogenesis but dispensable for cardiac development. *Mol. Cell Biol.* 25(5): 1980–1988.

Locy, M.L., Rogers, L.K., Prigge, J.R., Schmidt, E.E., Arner, E.S. & Tipple, T.E. 2012. Thioredoxin reductase inhibition elicits Nrf2-mediated responses in Clara cells: implications for oxidant-induced lung injury. *Antioxid. Redox Signal.* 17(10): 1407–1416.

Lu, J., Zhong, L., Lonn, M.E., Burk, R.F., Hill, K.E. & Holmgren, A. 2009. Penultimate selenocysteine residue replaced by cysteine in thioredoxin reductase from selenium-deficient rat liver. *Faseb. J.* 23(8): 2394–2402.

Nordberg, J. & Arnér, E.S.J. 2001. Reactive oxygen species, antioxidants, and the mammalian thioredoxin system. *Free Radic. Biol. Med.* 31: 1287–1312.

Peng, X., Xu, J. & Arner, E.S. 2012. Thiophosphate and selenite conversely modulate cell death induced by glutathione depletion or cisplatin: effects related to activity and Sec contents of thioredoxin reductase. *Biochem. J.* 447(1): 167–174.

Prigge, J.R., Eriksson, S., Iverson, S.V., Meade, T.A., Capecchi, M.R., Arnér, E.S. *et al.* 2012. Hepatocyte DNA replication in growing liver requires either glutathione or a single allele of txnrd1. *Free Radic. Biol. Med.* 52: 803–810.

Rollins, M.F., van der Heide, D.M., Weisend, C.M., Kundert, J.A., Comstock, K.M., Suvorova, E.S *et al.* 2010. Hepatocytes lacking thioredoxin reductase 1 have normal replicative potential during development and regeneration. *J. Cell Sci.* 123: 2402–2412.

Rundlöf, A.-K. & Arnér, E.S.J. 2004. Regulation of the mammalian selenoprotein thioredoxin reductase 1 in relation to cellular phenotype, growth and signaling events. *Antiox. Redox Signal.* 6: 41–52.

Rundlöf, A.-K., Carlsten, M. & Arnér, E.S.J. 2001. The core promoter of human thioredoxin reductase 1: Cloning, transcriptional activity and Oct-1, Sp1 and Sp3 binding reveal a housekeeping-type promoter for the ARE-regulated gene. *J. Biol. Chem.* 276: 30542–30551.

Rundlöf, A.-K., Janard, M., Miranda-Vizuete, A. & Arnér, E.S.J. 2004. Evidence for intriguingly complex transcription of human thioredoxin reductase 1. *Free Rad. Biol. Med.* 36: 641–656.

Sakurai, A., Nishimoto, M., Himeno, S., Imura, N., Tsujimoto, M., Kunimoto, M. *et al.* 2005. Transcriptional regulation of thioredoxin reductase 1 expression by cadmium in vascular endothelial cells: role of NF-E2-related factor-2. *J. Cell Physiol.* 203(3): 529–537.

Xu, X.M., Turanov, A.A., Carlson, B.A., Yoo, M.H., Everley, R.A , Nandakumar, R. *et al.* 2010. Targeted insertion of cysteine by decoding UGA codons with mammalian selenocysteine machinery. *Proc. Natl. Acad. Sci.* 107(50): 21430–21434.

Selenium in the Environment and Human Health – Bañuelos, Lin & Yin (Eds)
© 2014 Taylor & Francis Group, London, ISBN 978-1-138-00017-9

Molecular regulation of selenoprotein expression by local factors

Niels-Peter Becker, Janine Martitz & Lutz Schomburg
Institute for Experimental Endocrinology, Charité—Medical University Berlin, Germany

1 INTRODUCTION

Selenoprotein expression is tightly regulated by several local and systemic factors including selenium (Se) availability, signals affecting transcription of selenoprotein genes, and post-transcriptional mechanisms. Selenocysteine (Sec) biosynthesis is an energy consuming process involving a variety of Sec-specific enzymes and proteins specialized in Se metabolism and selenoprotein biosynthesis. Turnover of other trace elements like copper (Cu) or iron (Fe) is stringently regulated by tissue-specific conditions including oxygen availability or pro-inflammatory cytokine tone. Tissue hypoxia affects a major Fe-controlling hormone (*i.e.*, hepcidin), while IL-6 controls a serum transporter for Cu, *i.e.*, ceruloplasmin (Mukhopadhyay *et al.*, 2000; Conley *et al.*, 2005; Verga Falzacappa *et al.*, 2007; Pietrangelo, 2011). In humans, serum Se is transported by selenoprotein P (SePP) which is mainly produced in liver (Hill *et al.*, 2007; Renko *et al.*, 2008). Serum Se concentration declines under pathological conditions (Maehira *et al.*, 2002). Both hypoxia and high inflammatory IL-6 levels are observed in severe diseases like sepsis, trauma, or cancer. These two parameters may thus negatively affect selenoprotein and SePP biosynthesis causing the observed systemic Se deficit.

We thus hypothesized that both oxygen supply and IL-6 impair Se metabolism and selenoprotein expression at the transcriptional level or during translation. Our experiments with liver cells in culture support this notion and highlight these two parameters as important regulators for expression of both selenoproteins and biosynthetic factors.

2 MATERIALS AND METHODS

Human hepatocarcinoma HepG2 cells were cultivated under regular or low oxygen conditions (20% and 1% O2, respectively). Cells were harvested and expression levels of selenoproteins and biosynthesis factors were determined by qRT-PCR, Western blot analyses and suitable enzyme assays, respectively. Intracellular trace elements were quantified using total reflection X-ray fluorescence (TXRF) analysis. The same experimental setup and the same analytical procedures were performed after stimulation with IL-6 (100 ng/mL). Genomic DNA upstream of the transcriptional start site from a set of selenoprotein genes and genes encoding central factors of selenoprotein biosynthesis was amplified, subcloned and verified by sequencing. Reporter gene analyses with these promoter constructs were performed to test signaling cascades under normoxic or hypoxic conditions and after IL-6 stimulation, respectively.

3 RESULTS AND DISCUSSION

Secreted SePP measured in the medium of stimulated cells declined after hypoxic exposure and in response to IL-6 stimulation in a dose- and time-dependent manner. Assuming that similar molecular programs were responsible for the observed effects, we compared reporter gene activities of STAT and hypoxia responsive promoter constructs under our experimental conditions. Despite a strong induction of the hypoxia-responsive reporter, no parallel induction of the STAT reporter was observed; indicating the presence of different molecular executing pathways. We tested whether a decreased cellular Se uptake may be responsible for the inhibited SePP production using TXRF analysis. Under the experimental conditions no significant differences were observed. However, expression analysis of intracellular Se-binding proteins revealed strong induction of fatty-acid binding protein 1 (FABP1) after IL-6 stimulation. FABP1 is described as one of two central intracellular Se-binding proteins (Bansal *et al.*, 1990). On the protein level, hypoxia caused a decreased expression of selenoprotein S (SELS) and iodothyronine deiodinases type 1 (DIO1) and thereby potentially negatively affecting quality control of secreted proteins in the endoplasmic reticulum (Curran *et al.*, 2005) and intracellular thyroid hormone metabolism (Schomburg, 2011). Notably, some selenoproteins remained unaffected and one isoform of the glutathione peroxidase family even increased under these conditions. Our data thus indicate that SePP declines in response to both hypoxia and IL-6, albeit by different underlying mechanisms and in parallel to stimulus-specific intracellular changes in Se metabolism and selenoprotein expression. Collectively, Se transport and turnover as well as Sec-tRNA and selenoprotein biosynthesis thus constitute highly regulated processes influenced by a variety of external signals to dynamically adapt to environmental changes. This process secures survival of the cell, the organ and the individual by affecting global Se metabolism and the cell-specific expression of certain selenoproteins. The next major research aim will now be to elucidate how a given cell integrates these two signals, and at which checkpoints the pathway-specific rearrangement in selenoprotein expression takes place during disease.

4 CONCLUSIONS

Selenoprotein biosynthesis is controlled on several levels and in a gene-specific way. Different external stimuli are integrated and lead to cell-specific effects on the expression pattern of selenoproteins. The exact understanding of molecular mechanisms underlying the redirection of Se from systemic supply (away from SePP) towards intracellular selenoproteins implicated in central cell survival and protection pathways is a fundamental requisite for better understanding the beneficial effects of Se supplementation during severe or critical illnesses (Forceville *et al.*, 1998).

ACKNOWLEDGEMENTS

Research in the authors' laboratories is supported by the Deutsche Forschungsgemeinschaft (DFG), the Federal Ministry of Education and Research (BMBF) and Economics and Technology (BMWi).

REFERENCES

Bansal, M.P., Mukhopadhyay, T., Scott, J., Cook, R.G., Mukhopadhyay, R. & Medina, D. 1990. DNA sequencing of a mouse liver protein that binds selenium: implications for selenium's mechanism of action in cancer prevention. *Carcinogenesis* 11(11): 2071–2073.

Conley, L., Geurs, T.L. & Levin, L.A. 2005. Transcriptional regulation of ceruloplasmin by an IL-6 response element pathway. *Brain. Res. Mol. Brain Res.* 139(2): 235–241.

Curran, J.E., Jowett, J.B., Elliott, K.S., Gao, Y., Gluschenko, K., Wang, J. *et al.* 2005. Genetic variation in selenoprotein S influences inflammatory response." *Nat. Genet.* 37(11): 1234–1241.

Forceville, X., Vitoux, D., Gauzit, R., Combes, A., Lahilaire, P. & Chappuis, P. 1998. Selenium, systemic immune response syndrome, sepsis, and outcome in critically ill patients. *Crit Care Med* 26(9): 1536–1544.

Hill, K.E., Zhou, J., Austin, L.M., Motley, A.K., Ham, A.J., Olson, G.E. *et al.* 2007. The selenium-rich C-terminal domain of mouse selenoprotein P is necessary for the supply of selenium to brain and testis but not for the maintenance of whole body selenium. *J. Biol. Chem.* 282(15): 10972–10980.

Maehira, F., Luyo, G.A., Miyagi, I., Oshiro, M., Yamane, N., Kuba, M. *et al.* 2002. Alterations of serum selenium concentrations in the acute phase of pathological conditions. *Clin. Chim. Acta* 316(1–2): 137–146.

Mukhopadhyay, C.K., Mazumder, B. & Fox, P.L. 2000. Role of hypoxia-inducible factor-1 in transcriptional activation of ceruloplasmin by iron deficiency. *J. Biol. Chem.* 275(28): 21048–21054.

Pietrangelo, A. 2011. Hepcidin in human iron disorders: therapeutic implications. *J. Hepatol.* 54(1): 173–181.

Renko, K., Werner, M., Renner-Müller, I., Cooper, T.G., Yeung, C.H., Hollenbach, B. *et al.* 2008. Hepatic selenoprotein P (SePP) expression restores selenium transport and prevents infertility and motor-incoordination in Sepp-knockout mice. *Biochem. J.* 409(3): 741–749.

Schomburg, L. 2011. Selenium, selenoproteins and the thyroid gland: interactions in health and disease." *Nat. Rev. Endocrinol.* 8(3): 160–171.

Verga Falzacappa, M.V., Vujic Spasic, M., Kessler, R., Stolte, J., Hentze, M.W. & Muckenthaler, M.U. 2007. STAT3 mediates hepatic hepcidin expression and its inflammatory stimulation. *Blood* 109(1): 353–358.

Selenium in the Environment and Human Health – Bañuelos, Lin & Yin (Eds)
© 2014 Taylor & Francis Group, London, ISBN 978-1-138-00017-9

Human gene of 15-kDa selenoprotein is regulated by NF κB

C. Zou & J. Zheng
College of Life Science, Shenzhen Key Laboratory of Marine Bioresources and Ecology, Shenzhen, China

J. Li & J. Tian
College of Life Science, Shenzhen Key Laboratory of Microbial Genetic Engineering, Shenzhen University, Shenzhen, China

1 INTRODUCTION

15 kDa selenoprotein (Sep15) was first described in 1998 in human T cells (Andrew et al., 2006). The human gene of Sep15 is located at position p31 on human chromosome 1. This region is frequently associated with tumor progression, which suggests a potential of Sep15 to be associated with cancer pathogenesis. However, the regulation of Sep15 was rarely studied, especially the transcriptional level (Zoia et al., 2009). In this paper, the core regions of Sep15 promoter have been identified, and the results of Dual-Luciferase Reporter Assay, RT-PCR and Chromatin Immunoprecipitation assay suggested that NF-κB can improve transcription of Sep15 promoter and bind to its specific binding sites.

2 MATERIALS AND METHODS

2.1 *The Dual-Glo Luciferase assay*

HEK-293T cell lines were maintained in Dulbecco's Modified Eagle's Medium (DMEM, HyClone), supplemented with 10% FBS and 0.07 mg/ml penicillin-G potassium, and 0.14 mg/ml streptomycin sulfate. N2a cells were grown in a 1:1 mixture of DMEM: Opti media (HyClone), supplemented with 5% FBS and 0.07 mg/ml penicillin-G potassium, and 0.14 mg/ml streptomycin sulfate. All cell lines were maintained at 37°C in 5% CO_2. A series of Sep15 gene promoter deletion derivatives was generated by PCR methods and inserted into the pGL4.10 [Luc2] luciferase vector. A Renilla construct, pRL-TK vector (Promega), was used for normalizing transfection efficiency. Cells were transfected with plasmid DNA using Lipofectamine 2000. For transient transfections, cells were cultured in complete media until 90% confluent, and co-transfected with both 380 ng of human Sep15 reporter plasmid and 20 ng of pRL-TK expression vector per well. At least three independent transfections were carried out in triplicate. After 24 h of transfection, luciferase activity was assessed using the Dual-Glo Luciferase Assay System (Promega). TNF and BAY were added for the final 20 h before the luciferase assay. The luminescence of the Firefly and Renilla luciferase was measured by Luminoskan Ascent (Theromo Scientific).

2.2 *Real-time PCR*

Total RNA was isolated from HEK293T cells using TRI reagent (Sigma-Aldrich). Reverse transcription was carried out with M-MLV reverse transcriptase (Promega) and random primers. To compare the expression level of Sep15 between normal cells and overexpressing p65 cells PCR reactions were carried out using Real time PCR Master Mix (TOYOBO, Japan) in ABI PPRISM 7300 Real-Time PCR System (ABI). Each reaction was performed in triplicate with default conditions: 40 cycles of 95°C for 15 s, and 60°C for 1 min. The resulting ΔCt values were then used to calculate the relative change in mRNA expression as a ratio (R) according to the equation $R = 2^{-(\Delta Ct(treatment)-\Delta Ct(control))}$. Primers used in PCR were 5' CCTGGCACCCAGCACAAT 3'and 5'GCTGATCCACATCTGCTGGAA 3' (β-actin), 5'ATCAAGACATTGCTCCTC-CTGAGCGC 3' (β-actin detector), 5'AAGTC-CAAGCTTTTGTTAGGAGTGATA 3' and 5'AAGCTTTAATACAGGGTCTGAACCA 3' (Sep15), 5'ACCCAAACTGTTCAGAGGACT-GCAAATCA 3' (Sep15 detector).

2.3 *ChIP assay*

HEK293T cells were treated with 1% formaldehyde at room temperature to crosslink the protein to the DNA. ChIP assay was then performed using EZ ChIP Chromatin Immunoprecipitation Kit (Upstate Biotech), as recommended by the manufacturer's instructions. To co-immunoprecipitate the DNA, p65 antibodies were used. For negative controls, an aliquot of cross-linked chromatin was immunoprecipitated with normal rabit IgG instead of anti-p65 antibody (referred to as 'no

Ab' fraction). To confirm the same amounts of chromatins used in immunoprecipitation between groups, input chromatin was also used. The PCR product was visualized with ethidium bromide after electrophoretic separation on a 4% agarose gel. The eluted DNA was amplified by PCR.

3 RESULTS AND DISCUSSIONS

3.1 *The core regions of Sep15 promoter identified*

To facilitate analysis of the regulation of the Sep15 promoter, a series of Sep15 gene promoter deletion derivatives was generated by PCR methods and inserted into the pGL4.10 [Luc2] luciferase vector. The activities of different length promoter constructs were detected. Comparing the activities of different length Sep15 gene promoter constructs, the lack of the sequence from -818 to -248 led to near background reporter activity. This data suggests the particular importance in the region between -818 to -248, which is reported to contain a GC-island (Zoia et al., 2009) for sustaining Sep15 gene transcription in both HEK293T and N2a cells.

Furthermore, these results demonstrate that the Sep15 promoter contains multiple negative regulatory elements. The sequence between -2055 to -1430 contains a strong negative regulatory element and an anti-repressive element with opposing activities controlling Sep15 gene expression. The region of -248 to -166 and -1430 to -818 contain a weak negative regulatory element.

3.2 *Sep15 promoter activity was regulated by NF-κB*

P65, a subunit of NF-κB, has been used to study the functions of NF-κB (Siebenlist 1994). To assess the functional significance of NF-κB for the activity of human Sep15 promoter, p65 expression vectors were cotransfected with the Sep15 promoter construct pGL-818/-248, which was predicted to contain the binding site of NF-κB, into HEK293T and N2a cells. Comparing with empty vector control, the overexpression of p65 significantly enhanced the promoter activity.

To further confirm the role of p65 transcription factors on Sep15 expression, we examined the inhibition of Sep15 promoter activity by BAY11-7082, which is known to inhibit NF-κB transcription factor specifically (Mariana et al., 2005), and the activation by TNF-α (Karin and Ben-Neriah, 2000). As shown in Fig. 3, addition of BAY11-7082 to transfected cells led to a significant decrease in promoter activity. The addition of TNF-α led to a significant increase in promoter activity. Altogether, these results indicated that Sep15 promoter activity was regulated by NFκB. Furthermore, endogenous Sep15 gene mRNA level was detected in HEK293T cells. Overexpressing p65, Sep15 gene mRNA was increased three times compared with the negative control. The results suggest that p65, the subunit of NF-kB, increased Sep15 expression, and the ChIP assay also confirmed the physical binding of NF-kB to Sep15 promoter (data did not shown).

4 CONCLUSIONS

NFκB can bind to Human Sep15 promoter and enhance the Sep15 gene promoter activity and transcription.

ACKNOWLEDGEMENTS

This work was financially supported by the National Natural Science Foundation of China (No. 30901182, 31070731 and 21271131).

REFERENCES

Andrew, D.F. 2006. NMR structures of the selenoproteins Sep15 and SelM reveal redox activity of a new Thioredoxin-like family. *J. Biol. Chem.* 281(6): 3536–3543.

Karin, M. & Ben-Neriah, Y. 2000. Phosphorylation meets ubiquitination: the control of NF-kappaB activity. *Annu. Rev. Immunol.*18: 621–663.

Mariana, G., Garc'ıa, L.A., Lopes, E.C., Blanco, G., Hajos, S.E. & Alvarez, E. 2005. Inhibition of NF-κB activity by BAY 11–7082 increases apoptosis in multidrug resistant leukemic T-cell lines. *Leukemia Research.* 29: 1425–1434.

Siebenlist, U., Feunisa, G. & Brown, K. 1994. Structure, regulation and function of NF-κB. *Cell Biol*, 10: 405–455.

Zoia, R., Stoytcheva, Z.R., Marla, J. & Berry, M.J. 2009. Transcriptional regulation of mammalian selenoprotein expression. *Biochim. Biophys. Acta*, 1: 33–48.

Selenium in the Environment and Human Health – Bañuelos, Lin & Yin (Eds)
© 2014 Taylor & Francis Group, London, ISBN 978-1-138-00017-9

15-kDa selenoprotein (Sep15) interaction protein screened by a yeast two-hybrid system

J. Liu & J. Zheng
College of Life Science, Shenzhen Key Laboratory of Marine Bioresources and Ecology, Shenzhen, China

J. Li & J. Tian
College of Life Science, Shenzhen Key Laboratory of Microbial Genetic Engineering, Shenzhen University, Shenzhen, China

Z. You
College of Computer Science and Software Engineering, Shenzhen University, Shenzhen, China

1 INTRODUCTION

15-kDa selenoprotein (Sep15) is localized to the Endoplasmic Reticulum (ER) and other perinuclear compartments. Others have identified that Sep15 interacted with UDP-glucose: glycoprotein glucosylteransferase (UGGT) and forms a tight 1:1 complex. Sep15 contains an N-terminal cysteine-rich domain, which contains six conserved cysteine residues that form two CxxC motifs. Vyacheslav *et al.* (2005) proved that the cysteine-rich domain of Sep15 mediates protein-protein interactions with UGGT. UGGT is a large ER chaperone, and regulates the calnexin (CNX) cycle (Korotkov *et al.*, 2001). The CNX cycle is an ER-based quality control pathway that specifically assists in the folding of N-linked glycoproteins (Daniels *et al.*, 2003). The tight association of Sep15 and UGGT suggests that Sep15 probable is involved in controlling disulfide bonding or the thiol-disfulfide state of the UGGT substrates (Labunskyy *et al.*, 2005).

Sep15 was published high expessed in human brain (Labunskyy *et al.*, 2009), but there is no evidence about its function in the brain. To understand the role of Sep15 in central nervous system, a yeast two-hybrid system was employed to screen its interaction with proteins in human fetal brain cDNA library, and then the interaction rates between Sep15 and screened proteins were estimated by bioinformatics.

2 MATERIALS AND METHODS

The Sep15 gene was cloned from the human fetus brain cDNA library. Sep15' mutant was generated by site-directed mutation of the Sec residue in Sep15 to Cys (Tian *et al.*, 2010) and cloned into NpGBKT7 vector. Matchmaker TM Gold yeast two-hybrid system, yeast strains Y2HGold, plasmids pACT2 and NpGBKT7, and human fetal brain cDNA library (using pACT2 as the vector) were purchased from Clontech Laboratories (Mountainview, California). NpGBKT7-Sep15' was used to screen the human fetal brain cDNA library via the yeast two-hybrid system. Yeast transformation and library screening were performed following the procedures described in the user manuals (Yeastmaker TM Yeast Transformation System 2 and Matchmaker TM Gold Yeast Two-hybrid System, Clontech). The screened positive prey plasmids were cotransformed with the bait plasmid into yeast respectively for verification of the re-transformation.

The verified positive prey plasmids were then sequenced and the sequences were BLAST at human reference protein database (http://hprd.org/), and the interaction ratio was estimated by local descriptor scores (You *et al.*, 2013).

3 RESULTS AND DISCUSSION

The human Sep15 gene was mutated to Sep15' for yeast two hybridization by changing the Sec residue to a Cys (Tian *et al.*, 2010), because no selenoprotein was identified in yeasts and then cloned into vector NpGBKT. NpGBKT7-Sep15' containing yeast was then used to screen the fetal brain cDNA library. 28 colonies were grown on five selection plates of quadruple dropout (SD/-Ade/-His/ -Leu/ -Trp) containing X-a-Gal and aureobasidin A (Aba). All plasmids extracted from the yeasts of those colonies were separately transformed into the competent *E. coli* Top10 cells to obtain the cDNA coding for the putative interactive proteins (AD plasmid).

The BD plasmid and the screened AD plasmid were then re-transformed into yeast cells, Y2HGold. For the co-transformed Y2HGold cells that use X-a-Gal as the substrate of β-galactosidase

(MEL1expression product), large blue colonies were grown on SD/-Leu/-Trp/X-a-Gal/Aba selection plates. The AD plasmids in 16 positive colonies following re-transformation were sent out for DNA sequencing and bioinformatics analysis using human reference protein database. All the preys and their sub cellular localizations are shown in Table 1. The abbreviation of RDH11, SLIT1, FTH1, TUBB4, UBE2CBP, WTIP, AP2A2 respectively mean retinol dehydrogenase 11 (RDH11), slit homolog 1 (SLIT1), ferritin heavy chain 1 (FTH1), tubulin beta-4 (TUBB4), ubiquitin-conjugating enzyme E2C binding protein (UBE2CBP), Wilms Tumor 1 Interacting Protein (WTIP), adaptor related protein complex 2, and alpha2 subunit (AP2A2). Their interaction ratio with Sep15' was estimated and the results are shown in Table 2. The interaction ratio between Sep15' and SLIT1, RDH11, TUBB4, UBE2CBP and AP2A2 are greater than 80%, but only RDH11 localizes at ER. Sep15 residents at ER (Shchedrina *et al*,. 2010) and play some roles in redox hemeostatis (Kasaikina *et al.*, 2011). RDH11 is also a dehydrogenase/reductase. Considering the co-localization and functional similarity, RDH11 is the most possible interaction protein with Sep15.

Further work, such as Fluorescence Resonance Energy Transfer (FRET), co-immunoprecipitation (co-IP), or pull-down assays, should, however, be performed to verify the interaction between Sep15 and RDH11.

Table1. Positive preys and their sub cellular localization.

Preys	Sub cellular localization
SLIT1	Extracellular
RDH11	Endoplasmic Reticulum (ER), Nucleus
FTH1	Cytoplasm
TUBB4	Microtubule
UBE2CBP	Unknown
WTIP	ER, Cytosol, Nucleus, etc.
AP2A2	Plasma membrane

Table 2. The interaction ratio between positive preys and Sep15'.

Preys	Interaction ratio
SLIT1	89.2%
RDH11	84.6%
FTH1	78.9%
TUBB4	88.6%
UBE2CBP	82.8%
WTIP	56.7%
AP2A2	95.3%

4 CONCLUSIONS

Sixteen positive preys were isolated as Sep15" interaction proteins using yeast two-hybrid system and then the interaction was analyzed by bioinformatics. All results showed that RDH11 was the most possible interaction protein of Sep15.

ACKNOWLEDGEMENTS

This work was financially supported by the nationa natural science foundation of China (No. 30901182, 31070731, and 21271131).

REFERENCES

Daniels, R., Kurowski, B., Johnson, A.E. & Hebert, D.N. 2003. N-linked glycans direct the cotranslational folding pathway of influenza hemagglutinin. *Mol. Cell* 11(1): 79–90.

Kasaikina, M.V., Fomenko, D.E., Labunskyy, V.M., Lachke, S.A., Qiu, W., Moncaster, J.A. et al. 2011. Roles of the 15-kDa selenoprotein (Sep15) in redox homeostasis and cataract development revealed by the analysis of Sep15 knockout mice. *J. Biol. Chem.* 286(38): 33203–33212.

Korotkov, K.V., Kumaraswamy, E., Zhou, Y., Hatfield, D.L & Gladyshev, V.N. 2001. Association between the 15-kDa selenoprotein and UDP-glucose:glycoprotein glucosyltransferase in the endoplasmic reticulum of mammalian cells. *J. Biol. Chem.* 276(18): 15330–15336.

Labunskyy, V.M. 2005. A novel cysteine-rich domain of Sep15 mediates the interaction with UDP-glucose: glycoprotein glucosyltransferase. *J. Biol. Chem.* 280: 37839–37845.

Labunskyy, V.M., Yoo, M.H., Hatfield, D.L. & Gladyshev, V.N. 2009. Sep15, a thioredoxin-like selenoprotein, is involved in the unfolded protein response and differentially regulated by adaptive and acute ER stresses. *Biochemistry* 48(35): 8458–6845.

Shchedrina, V.A., Zhang, Y., Labunskyy, V.M., Hatfield, D.L. & Gladyshev, V.N. 2010. Structure-function relations, physiological roles, and evolution of mammalian ER-resident selenoproteins. *Antioxid. Redox. Signal.* 12(7): 839–849.

Tian, J., Liu, Q., Dong, S., Qiao X. & Ni, J. 2010. A New Method for Multi-Site-Directed Mutagenesis, *Anal. Biochem.* 406(1): 83–85.

Vyacheslav, M.L. 2005. A Novel cysteine-rich domain of Sep15 mediates the interaction with UDP-glucose: Glycoprote in Glucosyltransferase. *J. Biol. Chem.* 280(45): 37839–37845.

You, Z., Lei, K.*, Zhu, L., Xia, J. & Wang, B. 2013. Prediction of protein-protein interactions from amino acid sequences with ensemble extreme learning machines and principal component analysis. *BMC Bioinformatics* 14(8): S10.

Selenium in the Environment and Human Health – Bañuelos, Lin & Yin (Eds)
© 2014 Taylor & Francis Group, London, ISBN 978-1-138-00017-9

Overexpression of 15-kDa selenoprotein induces cell apoptosis in Hepg2 cell

J. Zheng
College of Life Science, Shenzhen Key Laboratory of Marine Bioresources and Ecology, Shenzhen, China

J. Tian
College of Life Science, Shenzhen Key Laboratory of Microbial Genetic Engineering, Shenzhen University, Shenzhen, China

1 INTRODUCTION

15-kDa selenoprotein (Sep15) was first described in 1998 in human T cells (Gladyshev *et al.*, 1998). The human gene of Sep15 is located at position p31 on human chromosome 1. This region is frequently associated with tumor progression, which suggests a potential association of Sep15 with cancer pathogenesis (Kumaraswamy *et al.*, 2000).

The relationship between Sep15 and cancers, however, is very complicated. On the one hand, Sep15 was a preventive for cancer. The levels of Sep15 expression in human and mouse liver, brain, testes and prostate were higher than that in hepatocarcinoma and prostate cancer cell lines (Gladyshev *et al.*, 1998; Behne *et al.*, 1997; Ganther, 1999; Apostolou *et al.*, 2004). On the other hand, recent research has demonstrated an opposite role for Sep15 in the promotion of growth of colon cancer cells (Méplan *et al.*, 2010; Irons *et al.*, 2010; Sutherland *et al.*, 2010). Mice injected with the Sep15-deficient cells developed fewer lung tumors and did not form pulmonary metastases compared to mice injected with control cells (Gresner *et al.*, 2008; Jablonska *et al.*, 2008).

The expression level of Sep 15 was very low in hepatocarcinoma than in para-carcinoma tissue, so the higher expression level of Sep 15 possibly prevented liver tumor. ROS (reactive oxygen species) was identified to be higher in tumor tissues than in normal tissues (Gotoh & Cooper, 1998; Friesen *et al.*, 1999). To study the role of Sep15 in hepatocarcinoma cell lines, HepG2, Sep15 was over expressed in HepG2. The cells were treated with or without exogenous oxidative stress, tert-bulty hydroperoxide (tBOOH), cell viability and apoptosis were detected by CCK-8 and AnnexinV-FITC/PI staining, respectively.

2 MATERIALS AND METHODS

HepG2 cells were cultured in Dulbecco's Modified Eagle Medium (Hyclone, Thermo), with 10% Fetal Bovine Serum (FBS, Gibco), 100U/ml penicillin and streptomycin in a CO_2 incubator at 37°C. The open reading frame of the human Sep15 was PCR-amplified from cDNA clone with the following primers: 5'-AGTCGTC-GACATGGTAGCGATGGCGGCTG-3' and 5'-GTCTCTAGATTAAGCGTAATCTGGAA-CATCGTATGGGTATATGCGTTCCAACTTT-TCAC -3'. The PCR product was digested with Sal I/Xba I and ligated into the pSelExpress1 vector, which was a highly efficient selenocysteine insertion vector (Novoselov *et al.*, 2007). HepG2 cells were transfected with the plasmid with Lipofectamine LTX & PLUS (Invitrogen) according to the manufacturer's instructions. The Sep15 overexpression vector was transfected to cells with 500 nm sodium selenite. 48 h after transfection, the Sep15 overexpression cells were treated with 50,100 and 150 μM tert-bulty hydroperoxide (tBOOH, sigma, USA) for 24 h and cell viabilities were performed by CCK-8 with microplate reader and Annexin V/PI with flow cytometer.

3 RESULTS AND DISCUSSIONS

We demonstrated the Sep15ORF-pSelExpression vector could develop the expression of Sep15, as was detected by western blotting. Overexpression of Sep15 in HepG2 cell results in significantly decreased viability of these cells with treatment of tBOOH for 24 h (Fig. 1). The results are in agreement with the conclusion provided by Ferguson et al. (2006). To further determine the cell apoptosis in response to tBOOH, Sep15 overexpression cells were incubated with AnnexinV-FITC. This binds to pohsphatidylserine exposed on the cell surface during apoptosis, and propidium iodide, which accumulates in late apoptotic cells (Cabreiro *et al.*, 2008). As shown in Figure 2, in the Annexin V-FITC/PI experiment, the sum of early apoptosis cells increased when Sep15 overexpression occurred in HepG2 cells. There was a decrease

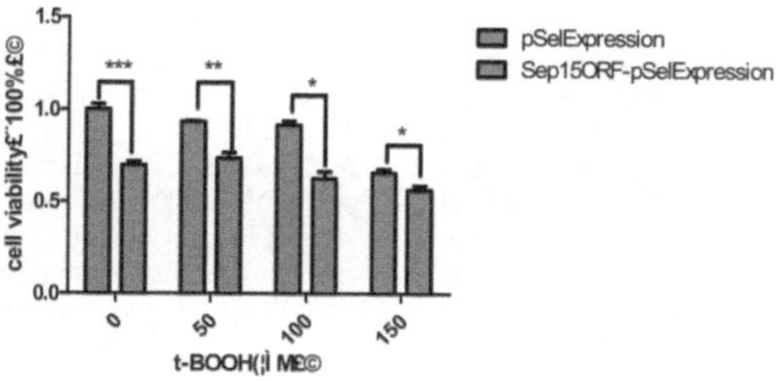

Figure 1. Cell viability of Sep15 overexpression cells and control cells treated with 50, 100 and 150 μM tBOOH for 24 h. T test (n = 3), * P < 0.05, ** P < 0.01, ***P < 0.001.

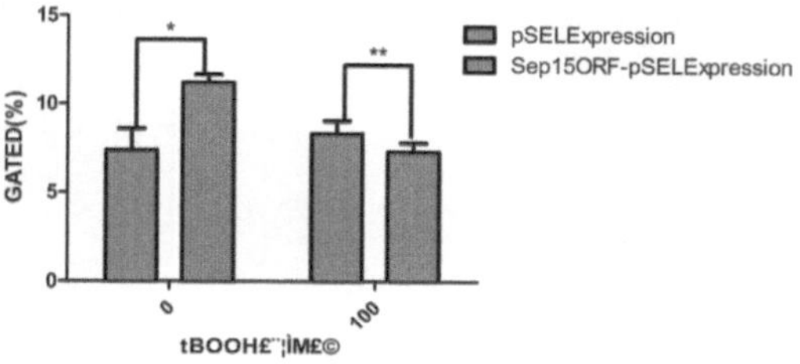

Figure 2. Effect of Sep15 overexpression cells treated with 100 μM tBOOH for 24 h. Apoptosis was assayed by flow cytometry using an AnnexinV-FITC/PI apoptosis detection kit. The sum of early apoptosis cells was shown. T test (n = 3), * P < 0.05 and ** P < 0.01.

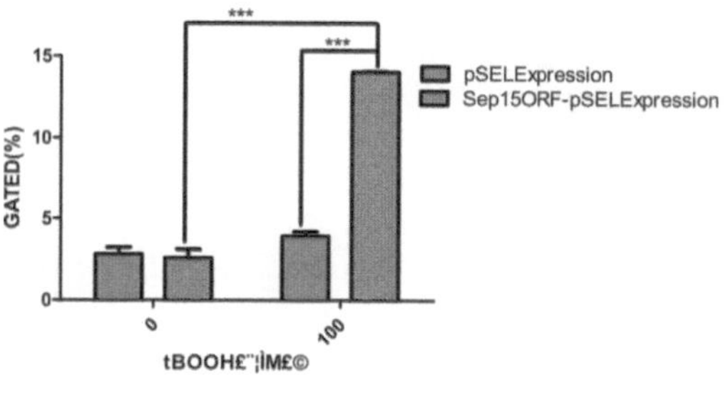

Figure 3. Effect of Sep15 overexpression cells treated with 100 μM tBOOH for 24 h. Apoptosis was assayed by flow cytometry using an AnnexinV-FITC/PI apoptosis detection kit. The sum of late apoptosis cells was shown. T test (n = 3), ** P < 0.01.

when overexpression Sep15 cells were treated with tBOOH. Interestingly, the sum of late apoptosis significantly increased in the Sep15 overexpression cells treated with tBOOH (Fig. 3). The tendency of total sum of early apoptosis and late apoptosis in Sep15 overexpression cells treated with tBOOH was corresponding to the cell viability result shown in Figure 1.

Sep15 played an important but varied role in different cancer pathogenic mechanisms (Sinoula *et al.*, 2004; Sergey *et al.*, 2007; Robert *et al.*, 2010). Our results showed that overexpression of Sep15 would weaken the antioxidative ability of HepG2 cells. To improve the expression level of Sep15 may be a good sign for killing hepatocacinoma cells. However, the role of Sep15 on the apoptosis and whether supplement of Sep15 can prevent liver carcinoma needs to be identified in future work.

4 CONCLUSIONS

In summary, increased Sep15 protein level can weaken the ability of HepG2 cells to adapt to tBOOH exposure.

REFERENCES

Apostolou, S., Klein, J.O., Mitsuuchi, Y., Shetler, J.N., Poulikakos, P.I., Jhanwar, S.C. *et al.* 2004. Growth inhibition and induction of apoptosis in mesothelioma cells by selenium and dependence on selenoprotein SEP15 genotype. *Oncogene.* 23: 5032–5040.

Behne, D., Kyriakopoulos, A., Kalcklösch, M., Weiss, N.C., Pfeifer, H., Gessner, H. et al. 1997. Two new selenoproteins found in the prostatic glandular epithelium and in the spermatid nuclei. *Europe pubMed Centra.* 10(23): 340–345.

Cabreiro, F., Picot, C.R., Perichon, M., Castel, J., Friguet, B. & Petropoulos, I. 2008. Overexpression of mitochondrial methionine sulfoxide reductase B2 protects leukemia cells from oxidative stress-induced cell apoptosis and protein damage. *The Journal of Biological Chemistry.* 283: 16673–166681.

Ferguson, A.D., Labunskyy, V.M., Fomenko, D.E., Araç, D., Chelliah, Y., Amezcua, C.A. *et al.* 2006. NMR structures of the selenoproteins sep15 and selM reveal redox activity of a new thioredoxin-like family. *The Journal of Biological Chemistry.* 281(6): 3536–3543.

Friesen, C., Fulda, S. & Debatin, K.M. 1999. Induction of CD95 ligand and apoptosis by doxorubicin is modulated by the redox state in chemosensitive and drug-resistant tumor cells. *Cell Death and Differentiation.* 6: 471–480.

Ganther, H.E. 1999. Selenium metabolism, selenoproteins and mechanisms of cancer prevention: complexities with thioredoxin reductase. *Carcinogenesis.* 20(9): 1657–1666.

Gladyshev, V.N., Jeang, K.T., Wootton, J.C. & Hatfield, D.L. 1998. A new human selenium-containing protein. *The Journal of Biological Chemistry.* 273(15): 8910–8915.

Gotoh, Y., & Cooper, J.A. 1998. Reactive oxygen species and dimerization- induced activation of apoptosis signal regulating kinase 1 in tumor necrosis factor-αsignal transduction. *The Journal of Biological Chemistry.* 273(28): 17477–17482.

Gresner, P., Gromadzinska, J., Jablonska, E., Kaczmarski, J. & Wasowicz, W. 2008. Expression of selenoprotein-coding genes SEPP1, SEP15 and hGPX1 in non-small cell lung cancer. *Lung Cancer.* 65: 34–40.

Irons, R., Tsuji, P.A., Carlson, B.A., Ouyang, P., Yoo, M.H., Xu, X.M. *et al.* 2010. Deficiency in the 15 kDa selenoprotein inhibits tumorigenicity and metastasis of colon cancer cells. *Cancer Prevention Research.* *3*(5): 630–639.

Jablonska, E., Gromadzinska, J., Sobala, W., Reszka, E. & Wasowicz, W. 2008. Lung cancer risk associated with selenium status is modified in smoking individuals by Sep15 polymorphism. *European Journal of Nutrition.* 47: 47–54.

Kumaraswamy, E., Malykh, A., Korotkov, K.V., Kozyavkin, S., Hu, Y., Kwon, S.Y. *et al.* 2000. Structure-expression relationships of the 15-kDa selenoprotein gene. *The Journal of Biological Chemistry.* 275(45): 35540–35547.

Méplan, C., Hughes, D.J., Pardini, B., Naccarati, A., Soucek, P., Vodickova, L. *et al.* 2010. Genetic variants in selenoprotein genes increase risk of colorectal cancer. *Carcinogenesis* 31(6): 1074–1079.

Novoselov, S.V., Lobanov, A.V., Hua, D., Kasaikina, M.V., Hatfield, D.L. & Gladyshev, V.N. 2007. A highly efficient form of the selenocysteine insertion sequence element in protozoan parasites and its use in mammalian cells. *Proceedings of the National Academy of Sciences.* 104(19): 7857–7862.

Sutherland, A., Kim, D.H., Relton, C., Ahn, Y.O. & Hesketh, J. 2010. Polymorphisms in the selenoprotein S and 15-kDa selenoprotein genes are associated with altered susceptibility to colorectal cancer. *Genes & Nutrition* 5: 215–223.

Selenium in the Environment and Human Health – Bañuelos, Lin & Yin (Eds)
 ISBN 978-1-138-00017-9

Selenium nucleic acids in chemical and structural biology

Zhen Huang
Department of Chemistry and Department of Biology, Georgia State University, Atlanta, GA, USA

1 INTRODUCTION

Although selenium (Se) has not yet been found in DNA, identification of the presence of Se in natural tRNAs has led to discovery of the naturally-occurring 2-selenouridine and 5-methylaminomethyl-2-selenouridine (mnm^{5}se^{2}U). Selenium can provide unique biochemical and biological functions and properties to macromolecules, including protein and RNA. The presence of a Se moiety in natural RNAs provides for greater exploration of function and structure space, which leads to better properties and increased adaptability. Thus, my research group has pioneered the incorporation of Se into DNAs and RNAs by atom-specific substitution of oxygen, discovering novel properties, materials and applications. We have found that the ^{2Se}T/A (or ^{2Se}U/A) base pair maintains virtually the same structure and recognition as the corresponding native T/A (or U/A) base pair (Hassan, 2010; Sun, 2012). We have also revealed that a bulky selenium atom at the thymidine 2-position in DNAs can largely increase the mismatch discrimination (including the wobble pairing discrimination). This atom-specific substitution and probing provide a novel tool to investigate specific recognition of base pairs, which is the basis for high fidelity during replication, transcription and translation. The Se-nucleobase-derivatized DNA and RNA have X-ray crystal structures virtually identical to the corresponding native DNA and RNA (Fig. 1).

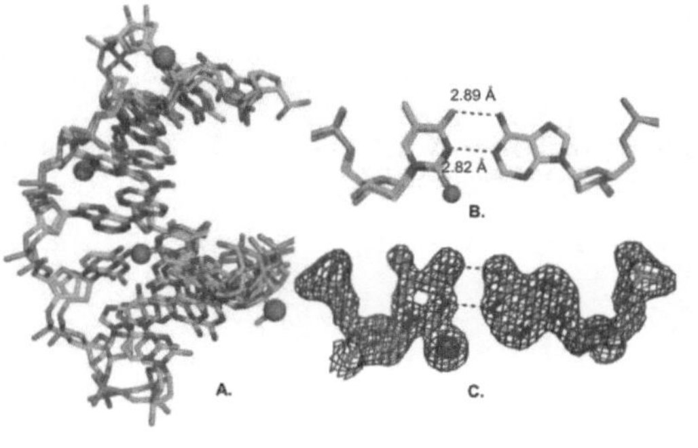

Figure 1. Global and local structures of the 2-Se-T-DNA [(5'-GdU$_{2'-Se}$-G-^{Se}T-ACAC-3')$_2$], with a resolution of 1.58 Å. (A) The superimposed comparison of the Se-DNA duplex (3HGD, in green) with its native counterpart (1D78, in cyan). The red balls represent the selenium atoms. (B) The superimposed comparison of the local ^{Se}T4/A5 (in green) and native T4/A5 (in cyan) base pairs. (C) The experimental electron density map of the ^{Se}T4/A5 base pair with $\sigma = 1.0$.

Furthermore, as the atomic size and electron delocalization ability of Se are larger than those of oxygen, we predicted a UV change when the nucleobase oxygen is replaced by Se. Surprisingly, the UV red-shift is so large that normally invisible DNA (or RNA) becomes visible (yellow color) when replacing the thymine 4-oxygen with Se (Lin *et al.*, 2011; Zhang *et al.*, 2013). We are the first to observe this exciting DNA and RNA discovery by only a single atom engineering. This breakthrough opens a novel avenue for nucleic acid-based pathogen and disease detection.

2 STRUCTURE DETERMINATION OF SELENIUM NUCLEIC ACIDS WITH HIGH BASE-PAIRING FIDELITY

Via chemogenetic and chemical synthetic strategies, my laboratory recently has pioneered and developed atom-specific substitution of nucleic acid oxygen with Se (Lin *et al.*, 2011; Sheng *et al.*, 2011, 2012), which can be used as an atomic probe for structure and function studies of nucleic acids and nucleic acid recognition (Fig. 1). The Se derivatization and the atom-specific substitution lead to a novel paradigm of nucleic acids and allow discoveries of new applications and materials, such as high-fidelity recognition, self-assembling nanostructures, and molecular electronic materials. Our nucleic acid chemogenetic strategy with Se has demonstrated a great potential as a general methodology for structure and function studies of nucleic acids, their protein complexes and their recognition principles. Furthermore, we find that the Se-derivatized nucleic acids have virtually identical structures to the corresponding natives, while the conventional Br-derivatization caused severe perturbations on the local backbone and hydration. Furthermore, we found that the Se-derivatization can facilitate crystallization and the diffraction quality is high.

We postulate that the incorporation of Se into DNAs and RNAs at a variety of positions (especially the nucleobases: C, G, T and U) allows nucleic acids to have useful functions and properties, thereby leading to a new paradigm for nucleic acids (Caton-Williams *et al.*, 2011; Salon *et al.*, 2011). Research in this exciting research area will largely deepen our fundamental understandings on nucleic acid as well. By following our novel research leads and recent breakthroughs in visible DNA (or colored DNA), we have explored the DNAs and RNAs derivatized with Se on nucleobases (C, G, T and U), and focused on their chemistry, physical chemistry, biochemistry and structure investigations. The discovery of visible DNAs opens a new research area in nucleic acid-based detection and DNA sequencing, which is important in post-genomic era. Due to the Se high phasing power, one Se atom per 30 nucleotides is typically introduced for X-ray crystallography.

3 CONCLUSIONS

Our biophysical studies on the ^{2Se}T-DNAs and ^{2Se}U-RNAs indicate that the perfectly-matched Se-nucleic acid duplexes have almost the same stability as the native ones. Interestingly, the 2-Se-substitution largely increases specificity of the base pair recognition by further discouraging the T/G (or U/G) wobble and T/C (or U/C) base-pairs, providing a unique chemical strategy to further enhance base-pairing fidelity. Consistently, our crystal structure study further reveals that the selenium-atomic substitution does not significantly alter the native T/A (or U/A) base pairing and overall duplex structure. Our experimental results indicate that this 2-Se-substitution facilitates higher specificity of the thymidine and uridine pairing with the natural nucleobases. Moreover, these 2-Se-derivatizations provide a useful tool in derivatization, crystallization, high-resolution, and phasing for X-ray crystal structure studies of nucleic acids and their protein complexes. The Se-atom-specific probing will open new research opportunities for further investigating base-pair recognition and high fidelity of DNA polymerase replication, RNA polymerase transcription, and mRNA translation.

ACKNOWLEDGEMENT

This work is supported by NIH (R01GM095881 and GM095086) and NSF (MCB-0824837 and CHE-0750235).

REFERENCES

Caton-Williams, J., Lin, L., Smith, M. & Huang, Z. 2011. Convenient Synthesis of Nucleoside 5'-Triphosphates for RNA Transcription. *Chemical Communications* 47: 8142–8144.

Hassan, A.E.A., Sheng, J., Zhang, W. & Huang, Z. 2010. High Fidelity of Base Paring by 2-Selenothymidine in DNA. *Journal of American Chemical Society* 132: 2120–2121.

Lin, L., Sheng, J. & Huang, Z. 2011. Nucleic Acid X-ray Crystallography via Direct Selenium Derivatization. *Chemical Society Reviews* 40: 4591–4602.

Salon, J., Sheng, J., Gan, J. & Huang, Z. 2011. Synthesis and crystal structure of 2'-Se-modified guanosine containing DNA. *J. Org. Chem.* 75: 637–641.

Sheng, J., Hassan, A.E.A., Zhang, W., Zhou, J., Xu, B., Soares, A. & Huang, Z. 2011. Synthesis, Structure and Imaging of Oligodeoxyribonucleotides with Tellurium-nucleobase Derivatization. *Nucleic Acids Res.* 39: 3962–3971.

Sheng, J., Zhang, W., Hassan A.E.A., Gan, J., Soares, A., Geng, S., Ren, Y. & Huang, Z. 2012. Hydrogen bond formation between the naturally modified nucleobase and phosphate backbone. *Nucleic Acids Res.* 40: 8111–8118.

Sun, H., Sheng, J., Hassan A.E.A., Jiang, S., Gan, J. & Huang, Z. 2012. Novel RNA base pair with higher specificity using single selenium atom. *Nucleic Acids Res.* 40: 5171–5179.

Zhang, W., Hassan A.E.A. & Huang, Z. 2013. Synthesis of novel di-Se-containing thymidine and Se-DNAs for structure and function studies. *Science China: Chemistry.* 56: 273–278.

Selenium in the Environment and Human Health – Bañuelos, Lin & Yin (Eds)
 ISBN 978-1-138-00017-9

Selenocysteine lyase: Delivering selenium in biosynthetic pathway

H. Mihara
College of Life Sciences, Ritsumeikan University, Kusatsu, Shiga, Japan

T. Kurihara & N. Esaki
Institute for Chemical Research, Kyoto University, Uji, Kyoto, Japan

1 INTRODUCTION

Selenium is mainly utilized for biosynthesis of selenocysteine residues in selenoproteins. Biosynthesis of selenocysteine is catalyzed by selenocysteine synthetase in the presence of the *O*-phosphoseryl-tRNA and selenophosphate, which is synthesized by selenophosphate synthetase (reviewed by Allmang *et al.*, 2009). Selenophosphate synthetase utilizes a selenide substrate, which is derived either from diet or degraded selenoproteins. Selenocysteine lyase (SCL) putatively salvages free selenocysteine liberated from selenoproteins and recycles selenium for the synthesis of selenoprotein (Kurokawa *et al.*, 2011).

SCL is a pydiroxal 5′-phosphate-dependent enzyme that catalyzes decomposition of L-selenocysteine to form L-alanine and an enzyme-bound selenylsulfide intermediate (Omi *et al.*, 2010). In the presence of a reducing agent, the bound selenium is converted into selenide *in vitro* (Esaki *et al.*, 1982). However, the detailed reaction mechanism and the physiological role of the enzyme are still largely unknown. In this paper, we provide a concise overview of recent developments in our understanding of the function and mechanism of SCL.

2 STRUCTURE AND FUNCTION OF SELENOCYSTEINE LYASE

2.1 *Catalytic properties*

SCL has been purified to homogeneity and characterized from pig liver (Esaki *et al.*, 1982) and from *Citrobacter freundii* (Chocat *et al.*, 1985). The bacterial enzyme is markedly different from the mammalian enzyme in physicochemical properties and amino acid composition. The mammalian enzyme is a homodimer with the subunit *M*r of 48,000, whereas the bacterial enzyme is monomeric with the M_r of 64000. Nevertheless, both enzymes are very similar in their catalytic properties, degrading L-selenocysteine into L-alanine and selenium but not acting on L-cysteine at all. By the action of this enzyme, L-selenocysteine is converted stoichiometrically to hydrogen selenide and L-alanine in the presence of an excess amount of DTT in the reaction mixture. Further experiments clearly show that selenocysteine lyase catalyzes the removal of elemental selenium from L-selenocysteine, and that the formation of hydrogen selenide is due to non-enzymatic reduction of the product, elemental selenium, with DTT (Esaki *et al.*, 1982).

2.2 *Sequence comparison*

The amino acid sequence of SCL was first provided by the cDNA cloning of mouse SCL (Mihara *et al.*, 2000). The enzyme is shown to be distantly related in primary structure to cysteine desulfurases, which catalyzes the desulfurization of L-cysteine to provide sulfur for iron-sulfur clusters and other sulfur-containing cofactors (Hidese *et al.*, 2011). A database search queried with SCL retrieves sequences of putative homologs from vertebrates like mammals, chickens, fishes, and frogs as well as invertebrates such as coelenterates and worms. Overall, sequence similarity (~30%) is found between SCLs and cysteine desulfurases. However, they are clearly classified into two distinct groups. In particular, the regions corresponding to Gln105-Gly121 and Asn205-Pro214 of mSCL are not found in mammalian cysteine desulfurases. The lysine residue that forms a Schiff's base with PLP is identified in rat SCL (Omi *et al.*, 2010) and bacterial cysteine desulfurases (Fujii *et al.*, 2000). Alignment of SCL members shows a conserved motif (T[IVT][VT]GHKFY[GA]P[RK]IGA) around the PLP-binding lysine. In contrast, the corresponding conserved motif in cysteine desulfurase members is [ST]X[SA][ASG]HK[FIL][YNE][GA]PKG[VI]GA.

2.3 *Reaction mechanism*

SCL has a conserved cysteine residue that corresponds to the persulfide-forming cysteine residue in cysteine desulfurases (Mihara & Esaki, 2002). A mutant SCL protein (C375A) in which Cys375 is replaced by an alanine residue exhibits no activity to L-selenocysteine, suggesting that Cys375 is a catalytically essential residue of the enzyme similar to the conserved catalytic cysteine residue of cysteine desulfurases (Omi *et al.*, 2010; Mihara & Esaki, 2002).

ESI-MS analysis revealed that one selenium atom is bound to Cys375 of SCL after cleavage of L-selenocysteine, indicating the formation of an *S*-selenyl-cysteine intermediate during the catalytic reaction (Omi *et al.*, 2010). Spectrum and X-ray crystal structure analyses showed that L-cysteine enters into the active site pocket but can not form an external aldimine complex with PLP. The formation of an external aldimine complex between SCL and L-selenocysteine requires the active-site nucleophile Cys375. An interaction between the protonated thiol group of Cys375 and the deprotonated selenol group of the substrate L-selenocysteine probably directs the formation of an aldimine linkage between the substrate and PLP (Fig. 1). Thus, Cys375 of SCL plays a central role in the discrimination between selenium and sulfur in a substrate and its analog.

It remains unclear whether selenium is directly released from Cys375–S–Se(H) or trapped by a selenium-transferring protein and subsequently released by a reductant in the reaction system. If the selenium in *S*-selenyl-cysteine, which is more sensitive to oxygen than persulfide sulfur, is released directly from an intermediate and diluted by the bulk solvent, then delivery of selenium to a specific acceptor molecule would be inefficient. Plausibly, selenium is shipped to the target protein through interaction with the selenium-transferring protein in a manner similar to the reaction of cysteine desulfurases (Hidese *et al.*, 2011).

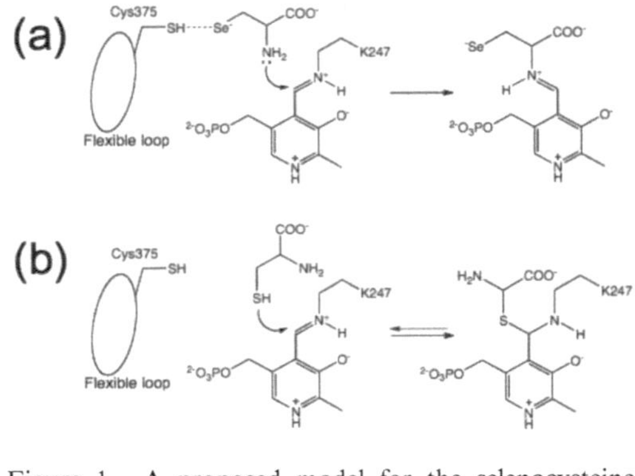

Figure 1. A proposed model for the selenocysteine/cysteine discrimination by SCL. (a) Cys375 interacts with the substrate selenocysteine to direct the normal formation of external aldimine complex. (b) Cys375 cannot direct the substrate analog cysteine, resulting in the formation of non-productive adduct.

REFERENCES

Allmang, C., Wurth, L. & Krol, A. 2009. The selenium to selenoprotein pathway in eukaryotes: more molecular partners than anticipated. *Biochim. Biophys. Acta* 1790: 1415–1423.

Chocat, P., Esaki, N., Tanizawa, K., Nakamura, K., Tanaka, H. & Soda, K. 1985. Purification and characterization of selenocysteine β-lyase from *Citrobacter freundii*. *J. Bacteriol.* 163: 669–676.

Esaki, N., Nakamura, T., Tanaka, H. & Soda, K. 1982. Selenocysteine lyase, a novel enzyme that specifically acts on selenocysteine. Mammalian distribution and purification and properties of pig liver enzyme. *J. Biol. Chem.* 257: 4386–4391.

Fujii, T., Maeda, M., Mihara, H., Kurihara, T., Esaki, N. & Hata, Y. 2000. Structure of a NifS homologue: X-ray structure analysis of CsdB, an *Escherichia coli* counterpart of mammalian selenocysteine lyase. *Biochemistry* 39: 1263–73.

Hidese, R., Mihara, H. & Esaki, N. 2011. Bacterial cysteine desulfurases: versatile key players in biosynthetic pathways of sulfur-containing biofactors. *Appl. Microbiol. Biotechnol.* 91: 47–61.

Kurokawa, S., Takehashi, M., Tanaka, H., Mihara, H., Kurihara, T., Tanaka, S. *et al.* 2011. Mammalian selenocysteine lyase is involved in selenoprotein biosynthesis. *J. Nutr. Sci. Vitaminol. (Tokyo)* 57: 298–305.

Mihara, H., Kurihara, T., Watanabe, T., Yoshimura, T. & Esaki, N. 2000. cDNA cloning, purification, and characterization of mouse liver selenocysteine lyase. Candidate for selenium delivery protein in selenoprotein synthesis. *J. Biol. Chem.* 275: 6195–6200.

Mihara, H. & Esaki, N. 2002. Bacterial cysteine desulfurases: their function and mechanisms. *Appl. Microbiol. Biotechnol.* 60: 12–23.

Omi, R., Kurokawa, S., Mihara, H., Hayashi, H., Goto, M., Miyahara, I. *et al.* 2010. Reaction mechanism and molecular basis for selenium/sulfur discrimination of selenocysteine lyase. *J. Biol. Chem.* 285: 12133–12139.

Selenium in the Environment and Human Health – Bañuelos, Lin & Yin (Eds)
© 2014 Taylor & Francis Group, London, ISBN 978-1-138-00017-9

Knock-down of SelR gene induced the apoptosis of human neuroblastoma SH-SY5Y cells

C. Wang & J.Z. Ni
Changchun Institute of Applied Chemistry, Chinese Academy of Sciences, Changchun, China
College of Life Sciences, Shenzhen University, Shenzhen, China
University of Chinese Academy of Sciences, Chinese Academy of Sciences, Beijing, China

Z.Y. Zeng & Q. Liu
College of Life Sciences, Shenzhen University, Shenzhen, China

R.L. Zhang
Shenzhen Center for Disease Control and Prevention, Shenzhen, China

1 INTRODUCTION

Selenium (Se) exerts its biological function mainly through selenoproteins in which Se is present in the form of selenocysteine (Sec) that is encoded by a traditional stop codon TGA in the open reading frame of the gene (Atkins & Gesteland, 2000).

Selenoprotein R (SelR, also named SelX or MsrB1), was first identified as a selenoprotein through bioinformatics methods (Kryukov *et al.*, 1999). SelR can be found in both the cytoplasm and nucleus of mammalian cells. It can stereospecifically catalyze the reduction of oxidized methionine residue in proteins (Kim & Gladyshev, 2005). Recent studies have shown that SelR is relevant to the lens cell survival, and silencing SelR can increase oxidative stress that causes lens cell death. This indicates that SelR plays a key role in conferring oxidative stress resistance and possibly preventing cataract formation (Marchetti *et al.*, 2005). Overexpression of Methionine sulfoxide reductase (Msr) leads to increased ability to resist oxidation and to a prolonged lifespan (Cabreiro *et al.*, 2006). Selenium has been proposed to play a role in preventing Alzheimer's Disease (AD) (Loef *et al.*, 2011). As it is highly expressed in mammalian brain (Zhang *et al.*, 2008), SelR may have anti-aging properties and neuronal protective functions.

In the present study, we specifically knocked down SelR in SH-SY5Y cells by packaged lentivirus, and detected by real time PCR the mRNA levels of Bax, Bad and Bcl-2 that are associated with cell apoptosis. The results showed that knock-down of SelR induced apoptosis in SH-SY5Y cells, indicating that SelR plays a key role in protecting SH-SY5Y cells.

2 MATERIALS AND METHODS

2.1 *Cell culture*

Human neuroblastoma SH-SY5Y cells were cultured in Dulbecco's Modified Eagle's Medium (DMEM) and 10% heat-inactivated fetal bovine serum supplemented with 100 IU/ml penicillin and 100 mg/ml streptomycin maintained at 37°C and 5% CO_2. 293FT cells were cultured in RPMI 1640 medium supplemented with 10% heat-inactivated fetal bovine serum, 100 U/ml penicillin G and 100 mg/ml streptomycin.

2.2 *Plasmid construction*

The vector lentiviral vector pLKO.1-puro was used to construct the pLKO.1-puro-SelR-1 plasmid. Based on the sequence of human SelR gene, two shRNAs were designed to knock-down SelR. The plasmid pLKO.1-puro-SelR-1 was constructed by inserting annealed oligonucleotides into the appropriate insertion sites. The plasmid for negative control was named pLKO.1-puro-nc. BLAST analysis showed these two plasmids have no homology with other human genes.

2.3 *Lentivirus packaging and cell infection*

The day before transfection, cells were seeded into a 10-mm dish. Plasmid pLKO.1-puro-SelR-1/ pLKO.1-puro-nc plasmid and packaging vector were co-transfected into 293FT cells using Lipofectamine 2000 according to the manufacturer's instruction. The resulting supernatant was collected after 48 h. The packaged Lentivirus were named Lenti-SelR-1/ Lenti-nc.

SH-SY5Y cells were infected with the appropriate Multiplicity Of Infection (MOI). Live cells were selected through puromycin two to three days after infection.

2.4 *Real-time PCR detection*

Total RNA was isolated from the SH-SY5Y cells using RNAiso Plus (TAKARA) according to the manufacturer's instruction.

Real-time PCR was performed using the SYBR Green PCR Master Mix kit (TAKARA) and followed the vendor's protocol. Primers for SelR and the internal control (β-actin) are shown in Table 1. The expression level of the target gene was normalized by the expression level of β-actin. Relative mRNA expression levels were calculated using the $2^{-\Delta\Delta Ct}$ rules.

2.5 *Statistical analysis*

Statistical analysis was performed using two-tailed Student's *t*-test, and a *P* value of <0.05 was considered to be statistically significant. Data were expressed as the mean ±SD of triplicate samples. All results were confirmed in at least three independent experiments.

3 RESULTS AND DISCUSSION

3.1 *Knock-down of SelR in SH-SY5Y cells*

To estimate the efficiency of SelR knock-down after shRNA transfection, the mRNA levels of SelR and β-actin were examined using real-time PCR. Treatment of the SH-SY5Y cells with the Lenti-SelR-1 was quite effective and resulted in significant knock-down of the cellular SelR in mRNA level (Fig. 1). The mRNA level of SelR was down-regulated by 50% compared with the negative control ($P < 0.05$). The negative control transfected with the random shRNA did not affect the mRNA and protein expression levels of SelR.

3.2 *Cell apoptosis caused by SelR knock-down*

To investigate the effect of SelR knock-down on cell apoptosis, the mRNA levels of Bax, Bad and Bcl-2 were determined by real time PCR. When SelR was knocked down, the mRNA levels of Bax (Fig. 2A) and Bad (Fig. 2B) in the cells were increased by 120% ($P < 0.05$) and 190% ($P < 0.001$), respectively, compared with the negative control. Meanwhile, the mRNA level of Bcl-2 (Fig. 2C) was decreased

Table 1. Primers for SelR and β-actin.

Gene name	Primer(5'-3')
SelR	F: ATGTCGTTCTGCAGCTTCTTC R: CACACTTGCCACAGGACAC
β-actin	F: TAAAGACCTCTATGCCAACACAG R: CACGATGGAGGGGCCGGACT-CATC

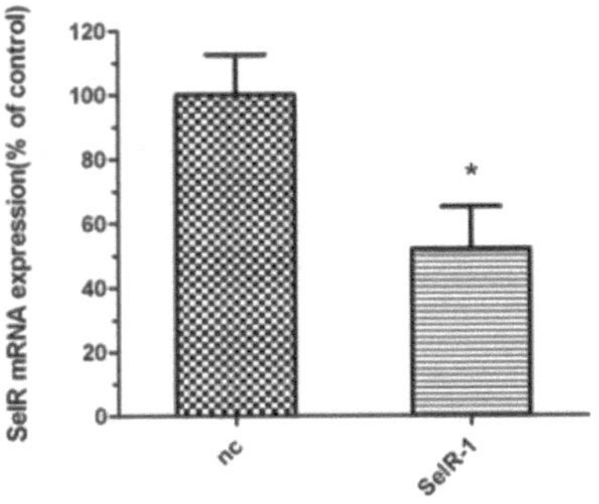

Figure 1. The efficiency of SelR knock-down in SH-SY5Y cells. *: Significance ($P < 0.05$).

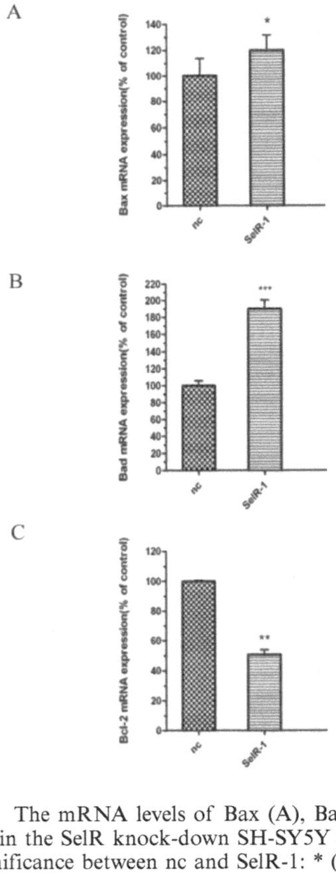

Figure 2. The mRNA levels of Bax (A), Bad (B) and Bcl-2 (C) in the SelR knock-down SH-SY5Y cells. Levels of significance between nc and SelR-1: * ($P < 0.05$), ** ($P < 0.01$) and *** ($P < 0.001$).

by 51% ($P < 0.01$). These results suggested that SelR knock-down led to cell apoptosis.

It is known that SelR plays an important role in protecting cell against oxidative damage and scavenging ROS (Li *et al.*, 2013). In this study, apoptosis of SH-SY5Y cells caused by SelR knock-down is probably due to the shortage of SelR in resisting oxidative damage. Therefore, it is reasonable to think that SelR in brain can protect neurocytes from oxidative damage.

4 CONCLUSIONS

SelR gene knocked down by lentivirus in the SH-SY5Y cells significantly resulted in cell apoptosis. Those results indicate that SelR may play an important role in protecting neurocytes from apoptosis.

ACKNOWLEDGEMENTS

This work was financially supported by the National Natural Science Foundation of China (No. 31070731 and 21271131), the Natural Science Foundation of Guangdong Province (No. 10151806001000023). The authors C. Wang and Z.Y. Zeng contributed equally to this paper.

REFERENCES

Atkins, J.F. & Gesteland, R.F. 2000. The twenty-first amino acid. *Nature* 407: 463–465.

Cabreiro, F., Picot, C.R., Friguet, B. & Petropoulos, I. 2006. Methionine sulfoxide reductases: relevance to aging and protection against oxidative stress. *Ann. N. Y. Acad. Sci.* 1067: 37–44.

Li, Y., Jia, Y., Zhou, J. & Huang, K.X. 2013. Effect of methionine sulfoxide reductase B1 silencing on high-glucose-induced apoptosis of human lens epithelial cells. *Life Sciences* 92: 193–201.

Loef, M., Schrauzer, G.N. & Walach, H. 2011. Selenium and Alzheimer's disease: a systematic review. *J. Alzheimers Dis.* 26: 81–104.

Kim, H.Y. & Gladyshev, V.N. 2005. Role of structural and functional elements of mouse methionine-S-sulfoxide reductase in its subcellular distribution. *Biochemistry* 44: 8059–8067.

Kryukov, G.V., Kryukov, V.M. & Gladyshev, V.N. 1999. New mammalian selenocysteine-containing proteins identified with an algorithm that searches for selenocysteine insertion sequence element. *J. Biol. Chem.* 274: 33888–33897.

Marchetti, M.A., Pizarro, G.O., Sagher, D., DeAmicis, C., Brot, N., Hejtmancik, J.F. *et al.* 2005. Methionine sulfoxide reductases B1, B2, and B3 are present in the human lens and confer oxidative stress resistance. *Invest. Ophthalmol. Vis. Sci.* 46: 2007–2112.

Zhang, Y., Zhou, Y., Schweizer, U., Savaskan, N.E., Hua, D., Kipnis, J. *et al.* 2008. Comparative analysis of selenocysteine machinery and selenoproteome gene expression in mouse brain identifies neurons as key functional sites of selenium in mammals. *J. Biol. Chem.* 283: 2427–2438.

Selenium in the Environment and Human Health – Bañuelos, Lin & Yin (Eds)
© 2014 Taylor & Francis Group, London, ISBN 978-1-138-00017-9

Nano-Se° and selenite ions impact on DNA double-strand breaks induction, repair and signaling: Structural, chemical and biological features

L. Charlet, C. Bochot & F. Bardelli
ISTerre, University Joseph Fourier, CNRS, BP 53, Grenoble Cedex 1, France

N. Foray, C. Abou Faycal, M. Ferlazzo, C. Devic, A. Granzotto & L. Bodgi
INSERM, Radiobiology Group, UMR1052, Cancer Research Centre, Lyon, France

1 INTRODUCTION

Selenium (Se) is an essential trace mineral for mammals and is widely distributed in inorganic forms in soils and in organic forms in food. Excessive and deficient amounts of Se occur all over the world, with both conditions leading to health-related diseases. Many crops including wheat, rice, and pasture are poor in Se and it is estimated that 100 million people worldwide suffer from Se deficiency. In humans, Se has one of the narrowest ranges between dietary deficiency (<40 μg per day) and toxic levels (>400 μg per day). International agencies have therefore set dietary reference values in the range of 30–55 μg per day for Se intake. Insufficient Se supply has been associated with growth retardation and impaired bone metabolism, and is thought to cause abnormalities in thyroid function. In central China and South East Siberia, very low Se content of locally-produced food have been related to geographically widespread endemic diseases, such as Keshan disease (a cardiomyiopathy) that was first recognized to be related to selenium deficiency in 1935, or the chronic, degenerative osteoarthrosis Kashin Beckdisease (Winkel *et al.*, 2013).

Selenium dietary supplementation is usually achieved either by direct ingestion of selenocysteine or selenomethionine nutritional supplements, or via Se fortified food products, such as mustard, tea, broccoli or carrots. In Finland and in UK, experimental wheat fields are fertilized with 20 g/ha Se(VI), which was enough to provide the Se daily requirements via the consumption of 3 slices of bread half (S. McGrath, pers. comm.). In China and USA, fortified plant products are produced in fields that are either naturally rich in Se (such as the Great Plains or the California Central Valley) or amended with grounded Se-rich shales, where Se is present as selenide minerals or Se° (Fernandez-Martinez & Charlet, 2009). While the solubility of metallic Se in the absence of oxygen has been claimed to be very low (10^{-8} to 10^{-9} mol/L; Bruggenwert *et al.*, 2005), a variety of abiotic and microbiological processes may enhance this solubility in presence of an oxidant such as oxygen (Tomei *et al.*, 1995; Stolz & Oremland, 1999; Oremland & Stolz, 2000; Oremland *et al.*, 2004; Lee *et al.*, 2007) or of a reductant, such as hydrazine (Ida *et al.*, 2012). The resulting solubility of Se drastically differs from that of bulk metallic Se.

The aim of the present study was to assess the effect of aqueous vs. nano-particulate Se onto DNA double-strand breaks induction, repair and signaling.

2 MATERIALS AND METHODS

2.1 *Synthesis, structure and morphology of Se° nanoparticles*

Se° nanoparticles (Se°NP) were produced in two different modes, i.e., either biologically (Buchs *et al.*, 2013), or by the inorganic synthesis route (Zang *et al.*, 2001). Briefly, in the latter case, particles were synthesized with or without Bovine Serum Albumin (BSA) by inorganic reduction of Se(IV), while, in the former case, they were produced by *Bacillus selenatarsenatis*, a dissimilatory Se reducer, harvested in late stationary phase. The microstructure (*e.g.*, structure, aggregation, shape, size and composition) of Se°NP were all characterized by Tecnai G2 Biotwin (FEI) Transmission Electron Microscope (TEM) with an accelerated voltage of 120 kV.

High-energy X-ray total scattering data of Se°NP loaded into 0.8 mm diameter polymide capillaries and sealed with wax under anoxic conditions ($[O_2]$ < 2 ppm) were acquired from 0 to 23.2 $Å^{-1}$ at beamline ID15B of the European Synchrotron Radiation Facility, Grenoble, France. Pair Distribution Functions (PDF) were obtained using the PDFGetX2 software (Qiu *et al.*, 2004). A crystalline standard (LaB_6) was measured and used to estimate the instrumental resolution (Toby & Egami, 1992). Fits of the PDF were performed using the PDFGui software (Farrow *et al.*, 2007).

Zeta potentials and size distribution of Se°NP were measured with a Malvern Nanozetasizer. Dissolution rates were measured in Se°NP suspended either in the cell growth medium (see below) or in 10^{-3} M NaCl at various fixed pH values either in the glovebox, ($P_{O2} < 2$ ppm) or under a 1 atm O_2 atmosphere. The dispersion was then separated into particles and solution by ultracentrifugation. The amount of Se in the filtrate (soluble Se) was detected by ICP-Mass Spectroscopy,

2.2 *Cell culture and exposure*

The human skin fibroblast (161 BR) cells, provided from a dermal biopsy, were cultured as monolayers with EMEM (Sigma-Aldrich), and supplemented with 12% fetal calf serum, penicillin and streptomycin. More than 90% of cells were maintained in plateau phase. Exposure to Se was performed by adding sodium selenite directly or Se°NP at various times and loads, followed by X-ray irradiation by an Olympus X-ray clinical irradiator at a 2 Gy/min dose-rate. Culture medium was changed immediately before irradiation of Se-treated cells, to inhibit the influence of Se compounds that have not entered into cells.

2.3 *Immunofluorescence pH_2AX, pATM and MRE11 assays*

The immunofluorescence detailed protocol was described elsewhere (Foray *et al.*, 2003). Briefly, cells were fixed in 3% paraformaldehyde, 2% sucrose PBS for 15 min at room temperature and permeabilized in 20 mM HEPES, pH 7.4, 50 mM NaCl, 3 mM $MgCl_2$, 300 mM sucrose 0.5% Triton X-100 for 3 min. Thereafter, coverslips were washed in PBS prior to immunostaining. Primary antibody incubations were performed for 40 min at 37°C in PBS supplemented with 2% bovine serum fraction V albumin (BSA) and followed by PBS washing. For anti-pH_2AXser139 (1:800), primary antibodies were used at 1:100. Anti-pH_2AXser139 and Anti-pATMser1981 antibodies were provided by Millipore, while Anti-MRE11 antibodies were purchased from Abcys. Incubations with anti-mouse FITC or with anti-rabbit TRITC secondary antibodies (1:100; Sigma-Aldrich) were performed at 37°C in 2% BSA for 20 min. Coverslips were mounted in Vectashield containing 4,6-Diamidino-2-phenylindole (DAPI) to counterstain the nuclei. Coverslips were examined with an Olympus fluorescence microscope. The micronuclei in binucleated DAPI-stained cells were scored during immunofluorescence experiments. DAPI-staining permitted also to indirectly evaluate yield of G1 cells (nuclei with homogeneous DAPI staining), S cells (nuclei showing numerous pH2AX foci), G2 cells (nuclei with heterogeneous DAPI staining) and metaphase (visible chromosomes).

3 RESULTS AND DISCUSSION

3.1 *Morphology and structure of biotic and abiotic Se°NP*

Se° NP diameter was estimated to be 20 nm from TEM image, while 80 nm is the hydrodynamic diameter measured by Dynamic Light Scattering (DLS). PDF analysis of high-energy X-ray total scattering data retrieved a value of 0.9 nm, but this value represent size of the coherent scattering domain and not the physical diameter, which is always smaller than the size observed by TEM or DLS. Se°NP are probably constituted by the aggregation/agglomeration of smaller particles. PDF analysis indicates that the crystallites of Se°NP are compatible with the α, β, or γ, Se8-monoclinic structures, while Se1-trigonal and Se6-rhomboedric structures can be ruled out.

In presence of oxygen, Se°NP solubility, as SeO_3^{2-}, increases regularly with pH above the Point of Zero Charge, *i.e.*, above pH 4, the PZC of both for abiotic (this study) and biotic (Buchs *et al.*, 2013) Se°NP. In contrast, in presence of a reductant such hydrazine monohydrate, hydrogen selenide (HSe^-) concentration increased linearly with decreasing pH below pH 8 (Ida *et al.*, 2013). Depending on wether the dissolution will occur outside (oxic cell growth medium) or inside (anoxic cytoplasm), either one of the above dissolution mechanism will occur.

3.2 *DNA double-strand break and repair*

The DNA Double-Strand Break (DSB) was assessed by pH_2AX recorded after exposing fibroblast cells to various concentrations of aqueous (SeO_3^{2-}) or particulate (Se°) concentration. In absence of irradiation, no DNA break is observed up to 100 μM Se(IV) exposure concentration, and DSB yield increased with Se at higher concentration. Micronuclei production, quantified as cell % showing micronuclei, increases with Se(IV) exposure concentration up to 30 μM. At higher Na_2SeO_3 concentrations, the percentage of cells showing micronuclei was found to be lower. The number of pH_2AX foci per cell (measured by fluorescence), decrease with time after irradiation, however, this decrease was slower in presence of 20 μM Se(IV), indicating a DSB repair inhibition at such low doses. MRE11 foci, involved in non homologous recombination or misrepair, indicate a significant impact at 100 μM or more. At 1000 μM, cell cycle is blocked. Similar studies are being performed with Se°NP and results will be shown and discussed in Heifei.

4 CONCLUSIONS

The exposure of human skin cells (fibroblasts) to selenite (Se(IV)) ions has a different effect on DSB, depending on the selenite dose. At very low concentration (20 μM) the exposure does not induce DSB, but inhibits their repair. At higher concentrations (100 μM), the exposure induces DNA DSB, and activates DSB misrepair. At even larger concentration (1000 μM) micronuclei block cell division is observed.

Se°NP given as supplementation undergo oxidative dissolution above their point of zero charge, releasing Se(IV) in the growth culture medium. However if the Se°NP were internalized, the reductive dissolution would lead to HSe^- release in the cytoplasm or nucleus. Both inorganic and biologically produced Se°NP were determined to have the same size, PZC, and structure.

ACKNOWLEDGEMENTS

We thank Markus Lenz for providing the bio Nano Se°NP sample. We acknowledge partial funding from the AVIESAN "Hemibreaks" programme, and from the Institut Universitaire de France.

REFERENCES

Aurelio, G., Fernandez-Mart´ınez, A., Cuello, G.J., Rom´an-Ross, G., Alliot, I. & Charlet, L. 2010. Structural study of selenium(IV) substitutions in calcite *Chem. Geol.* 270: 249–256.

Buchs, B., Evangelou, M.W.H., Winkel, L.H.E. & Lenz, M. 2013. Colloidal properties of nanoparticular biogenic selenium govern environmental fate and bioremediation effectiveness. *Environ. Sci. Technol.* 47: 2401–2407.

Charlet, L., Mingliang, K., Bardelli, F., Kirsch, R., Géhin, A., Grenèche, J.M. *et al.* 2012. Nanocomposite pyrite-greigite reactivity towards Se(IV)/Se(VI). *Environ. Sci. Technol* 46: 4869–4876.

Farrow, C.L., Juhás, P., Liu, J.W., Bryndin, D., Božin, S.D., Bloch, J. *et al.* 2007. PDFfit2 and PDFgui: computer programs for studying nanostructure in crystals, *J. Phys.: Condens. Matter* 19: 335219–335226.

Fernandez-Martinez, A. & Charlet, L. 2009. Selenium environmental cycling and bioavailability: a structural chemist point of view. *Rev. Environ. Sci. Biotechnol.* 8: 81–110.

Foray, N., Marot, D., Gabriel, A., Randrianarison, V., Carr, A.M., Perricaudet, M. *et al.* 2003. A subset of ATM- and ATR-dependent phosphorylation events requires the BRCA1 protein. *EMBO J.* 22: 2860–2871.

Ida, Y., Yamaguchi, T., Tanaka, T. & Nakayama, S. 2010. Solubility of selenium at high ionic strength under anoxic conditions *J. Nuclear Sci. and Technol.* 47: 431–438.

Kang, M., Chen, F., Wu, S., Yang, Y., Bruggeman, C. & Charlet, L. 2011. Effect of pH on aqueous Se(IV) reduction by Pyrite *Environ. Sci. Technol.* 45: 2704–2710.

Qiu, X., Thompson, J.W. & Billinge, S.J.L. 2004. PDFgetX2: A GUI driven program to obtain the pair distribution function from X-ray powder diffraction data. *J. Appl. Cryst.* 37: 678.

Winkel, L., Johnson, A., Lenz, M., Grundl, T., Leupin, O., Amini, M. *et al.* 2012. Environmental selenium research—from microscopic processes to global understanding. *Environ. Sci. Technol* 46: 571–579.

Zhang, J.S., Gao, X.Y., Zhang, L.D. & Bao, Y.P. 2001. Biological effects of a nano red elemental selenium. *Biofactors* 15: 27–38.

Selenium in the Environment and Human Health – Bañuelos, Lin & Yin (Eds)
© 2014 Taylor & Francis Group, London, ISBN 978-1-138-00017-9

Effect of kappa-selenocarrageenan combined with Epirubicin on proliferation and cell cycle of HepG2 cells

N. Ling, Y.B. Ji, X.M. Cao, N. Cheng, Q.Y. Sun & X. Zhong
Research Center on Life Sciences and Environmental Sciences, Harbin University of Commerce, Harbin, Heilongjiang, China

1 INTRODUCTION

Epirubicin (EPI), an anthracycline anti-tumor antibiotic which is structurally related to doxorubicin, is among the most active single agents. Kappa-selenocarrageenan (KSC) is a new-type Se-containing polysaccharide. KSC has demonstrated its high anti-tumor activities and possible effects on other molecular mechanisms through anti-oxidation, induction of tumor cell apoptosis, blockade of cell cycle and enhancement of immunity (Zhang *et al.*, 2005). In addition, KSC can better improve anti-tumor efficacy, enhance immune functions and decrease nephrotoxicity and bone marrow suppression of cisplatin-contained chemotherapy (Hu *et al.*, 1997). In this study, we investigated the anti-tumor effects of KSC in combination with EPI on HepG2 cells and the possible mechanisms, which may provide an experimental basis for clinical application.

2 METHODS

2.1 *Inhibition of cell proliferation*

KSC and EPI were added to HepG2 cells and cultured for 48h in RPMI1640 containing 10% FBS. MTT assay were performed to measure cell proliferation. The drug concentrations inhibiting cell growth by 50% (IC_{50}) were obtained by graphical analysis. The drug interaction was evaluated by Q value, which was calculated as follows:

$$Q = \frac{E(AB)}{EA + (1 - EA) \times EB}$$

where $E(AB)$ = tumor inhibition rate of drug combination; EA = inhibition rate of drug A; $E(B)$ = inhibition rate of drug B.

When Q value was equal to 0.85–1.15, or less than 0.85 or more than 1.15, additive or antagonistic or synergistic interaction was thought to occur (Wang, 2001).

2.2 *Flow cytometry cell cycle analysis*

The cells were harvested after drug exposure, fixed in 70% ethanol, stained with propidium iodide and analyzed by flow cytometry. The data were analyzed using CellQuest™ software to determine the percent cells in each phase of the cell cycle.

2.3 *Detection of Cyclin A, Cdk2 and Cdc25A protein*

Cells were treated with KSC and EPI for 48h in RPMI1640 with 10% FBS. After treatment, cells were lysed and 20 μg of soluble protein were separated by electrophoresis on a SDS-PAGE gel. Protein levels were detected by western blot analysis.

3 RESULTS

3.1 *Inhibition of cell proliferation*

As shown in Table 1, either KSC or EPI alone inhibited HepG2 activity in a dose-dependent

Table 1. Inhibition of cell proliferation and Q value.

Group	Dose (mg/L)	IR (%)	IC50 (mg/L)	Q values
CK	–	–	–	–
KSC	15	22.83		
	30	38.72	53.295	–
	60	50.88		
	120	68.69		
EPI	0.5	30.09		
	1	38.39		
	2	43.54		
	4	52.45	3.612	–
	8	58.76		
	16	63.52		
	32	70.15		
KSC + EPI	30 + 0.5	45.86		0.802
	30 + 1	53.66		0.862
	30 + 2	58.90		0.901
	30 + 4	60.48	0.807	0.853
	30 + 8	64.85		0.868
	30 + 16	73.51		0.947
	30 + 32	78.67		0.963

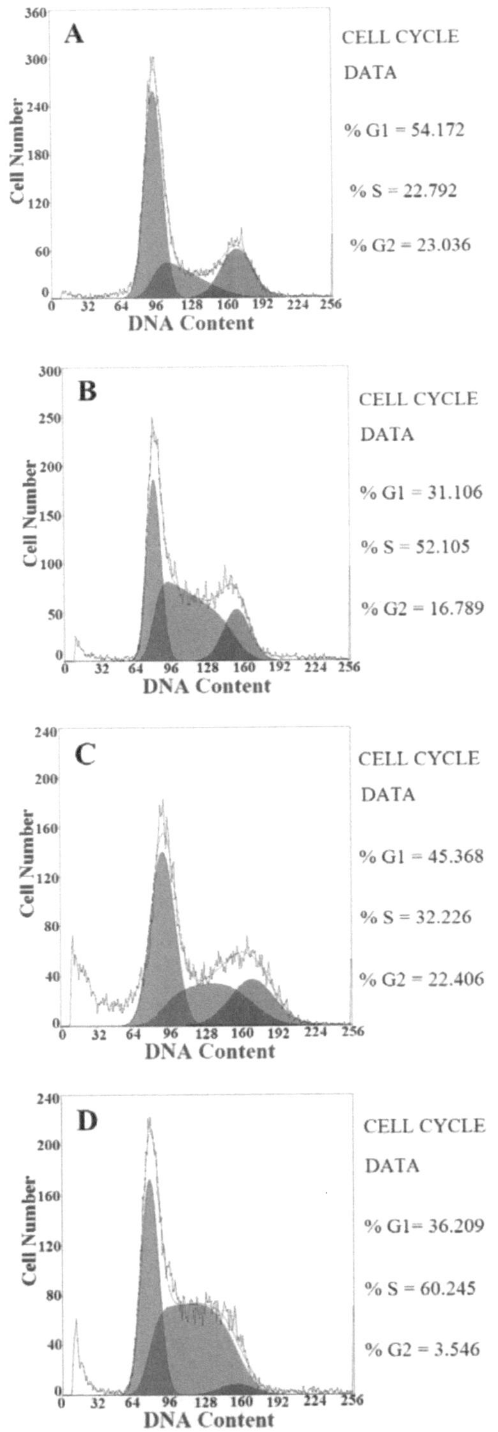

Figure 1. KSC and EPI on Cell cycle of HepG2 cells. A. CK, B. 30 mg/L KSC, C. 2 mg/L EPI, D. KSC + EPI.

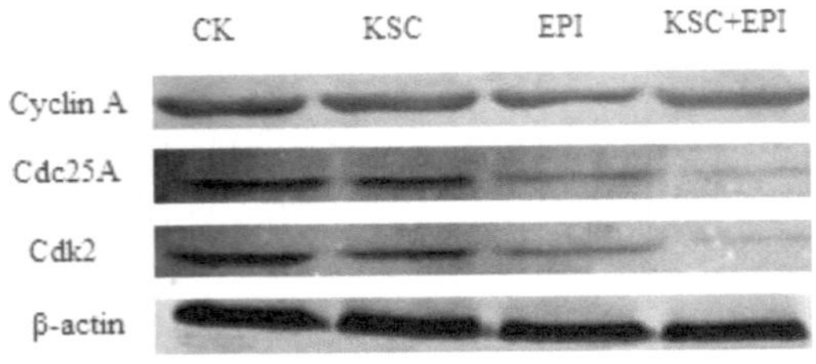

Figure 2. Expressions of Cyclin A, Cdc25A and Cdk2. KSC (30 mg/L), EPI (2 mg/L), KSC (30 mg/L) + EPI (2 mg/L).

manner, but the effect was greater when the two agents were combined. IC_{50} value of EPI alone was 3.612 mg/L, and decreased to 0.807 mg/L when combined with KSC. The studies demonstrated that interaction between KSC and EPI in HepG2 cells was additive ($0.85 < q < 1.15$).

3.2 *Effect of KSC and EPI on cell cycle and apoptosis rate*

FCM revealed that KSC or EPI could make HepG2 cells G_1 phase decrease and S phase increase, which indicated S phase arrest. The rate of S phase in combination group increased significantly. In addition, the apoptosis rate of combination group showed an upward tendency as compared with KSC or EPI alone (Fig. 1).

3.3 *Detection of Cyclin A, Cdk2 and Cdc25A protein*

As shown in Figure 2, treatment of HepG2 cells for 48h with KSC and EPI resulted in a prominent decrease in the expressions of Cyclin A, Cdc25A and Cdk2, as compared with the untreated control. Meanwhile, expression of the three proteins decreased more obviously in combination group.

4 DISCUSSION

Epirubicin is accepted as the most effective treatment for breast cancer and other metastatic disease. The major adverse effects of epirubicin are haematological and cardiac toxicity, mucositis, nausea and vomiting (Coukell & Faulds, 1997). Reducing drug side-effects would be a most important approach to improving the success of anticancer chemotherapy. In this study, KSC in combination with EPI showed an additive effect on the anti-proliferation of HepG2 cells and make IC_{50} value of EPI decrease, which indicated KSC could improve efficacy and decrease its toxicity. The main mechanism of anti-tumor may be due to induce the S-phase block and influence expression of cell-cycle regulatory proteins.

5 CONCLUSIONS

KSC in combination with epirubicin can markedly improve the inhibitory effect on HepG2 cells and induce S phase arrest, which may be associated with down-regulating the expression of S-phase related proteins.

ACKNOWLEDGEMENTS

This work was supported by grants from the Education Department of Heilongjiang Province Science and Technology Research (NO. 12531151).

REFERENCES

Coukell, A.J. & Faulds, D. 1997. Epirubicin. *Drugs* 53(3): 453–482.

Hu, Y.J., Chen, Y., Zhang, Y.Q., et al. 1997. The Protective Role of Selenium on the Toxicity of Cisplatin-Contained Chemotherapy Regimen in Cancer Patients. *Biol. Trace Elem. Res.* 56: 331–341.

Wang, W., Qin, S.K., Chen, B.A. & Chen, H.Y. 2001. Experimental study on antitumor effect of arsenic trioxide in combination with cisplatin or doxorubicin on hepatocellular carcinoma. *World J. Gastroenterol.* 7(5): 702–705.

Zhang, Z.W., Wei, H.L. & Su, H.X. 2005. Immunoloregulation and anti-tumor effects of kappa-selenocarrageenan. *J. Lanzhou Univ. (Med. Sci.)* 31(2): 88–91.

6 *Selenium pollution and control*

Selenium in the Environment and Human Health – Bañuelos, Lin & Yin (Eds)
 ISBN 978-1-138-00017-9

Increasing environmental emission of selenium and its countermeasures in China

Lin-Hua Jiang, Chun-Lei Zhu, Yu-Cheng Wen, Ge Zhang & Ning Duan
Technology Center for Heavy Metal Cleaner Production Engineering, Chinese Research Academy of Environmental Sciences, Beijing, China

Ying Liu
Suzhou Institute for Advanced Study, University of Science and Technology of China, Suzhou, Jiangsu, China

Fa-Hua Zhu
State Power Environmental Protection Research Institute, Nanjing, Jiangsu, China

1 INTRODUCTION

Selenium (Se) is an important nutritional essential element, and insufficient intake of Se can lead to some diseases. However, the excessive amount and time span of Se intake will lead to acute poisoning or chronic poisoning. Moreover, high concentrations of Se in organisms can also cause remarkable genetic toxicity (Luchli, 1993; Díaz, 1996; Lemly, 1997). So far, a large number of accidents caused by excessive Se intake have been reported in foreign countries. Accordingly, those countries (such as France, Germany, Japan and the former Soviet Union) have passed comprehensive laws, regulations and environmental standards to limit Se emission. China is generally low in Se supplies or as a country with Se deficiency even although selenosis accidents have been reported in Enshi (Hubei) and Ziyang (Shanxi) and a few other places. However, since Se played more and more important roles in hydrometallurgy industry, China has become a net importer of Se, and the amount of Se emissions to the environment is also growing in recent years (Fig. 1). It is necessary and urgent to prospectively study and analyze the likelihood of ecological accidents caused by growing Se in Chinese environment.

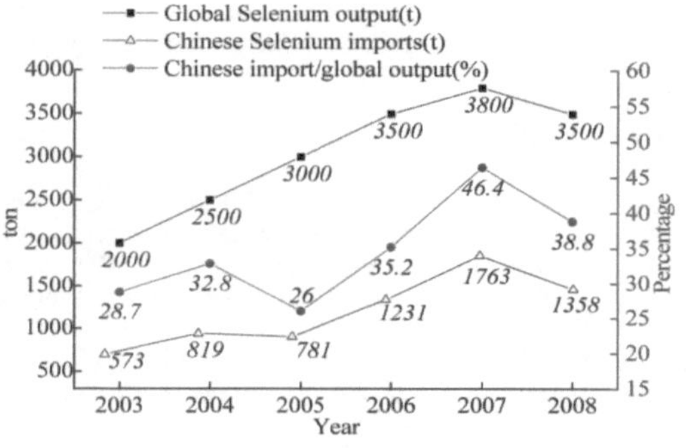

Figure 1. Global selenium annual production and the amount of Chinese annual selenium import.

2 MATERIALS AND METHODS

On the basis of the mass fraction of Se in coal, China could be divided into three zones: high Se areas (>5 mg/kg), including Chongqing, Anhui, Jiangsu and Hubei (except in Enshi); low Se areas (2–5 mg/kg), including Jilin, Guizhou, Yunnan, Sichuan, Shandong, Shanxi, and Hunan. If the investigation was based on the industrial manufacture process, the electrolytic manganese enterprises exhaust lots of Se, mainly in Chongqing, Hunan, Guangxi, Ningxia, and Hubei. Additionally, firms extracting Se from copper anode mud and making special glass were also investigated. Based on above investigations, the counter measures were proposed in this study.

3 RESULTS

Concentrations of Se in Chinese coals range from 4 to 5 mg/kg, which is much higher than the average of the world (3 mg/kg). Chemical analysis and chemical equilibrium calculation indicate that the ratios of Se in slag, filter fly ash, fine fly ash, and flue gas are 11.2%, 71.2%, 1.1% and 16.5%, respectively. Selenium discharged from coal power generation will eventually enter into the environment. In China, the coal combustion has increased continually from 1978 to 2008, resulted in a similar national increase in Se emission (Fig. 2).

Considering different manufacture processes, such as oil refinery, copper refining, sulfuric acid manufacture, pigment, dyestuff, and special glass industry,

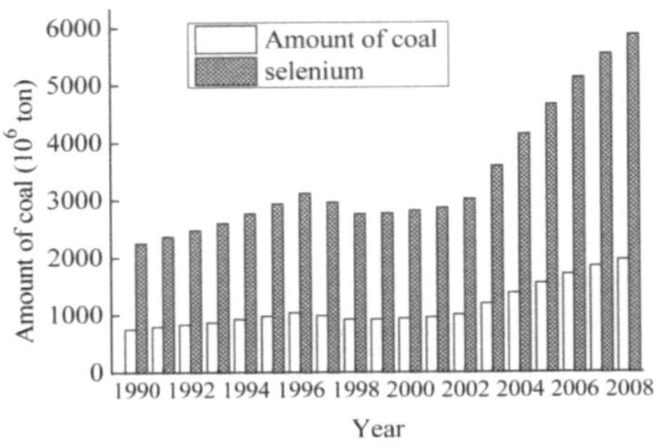

Figure 2. Total annual coal combustion in China and the amount of selenium emission from coal combustion.

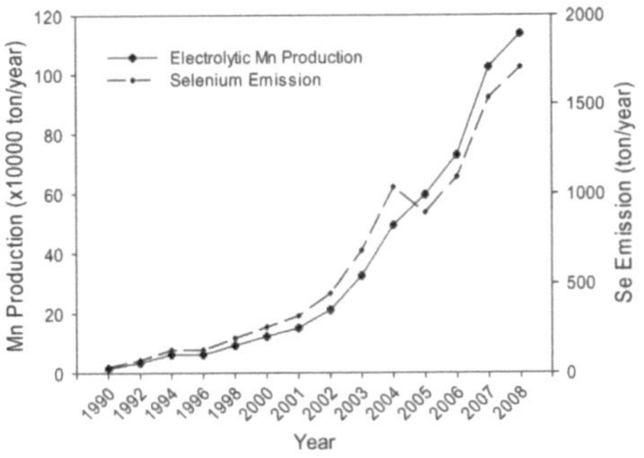

Figure 3. Chinese annual production of electrolytic manganese and the total annual selenium emission.

are the primary discharge sources of Se sewage and Se pollutants. For example, Se in the electrolytic copper will eventually be released into the environment. Selenium dioxide has been extensively used as electrolytic additives in the electrolytic manganese. The increase of Chinese electrolytic manganese quantity from 1990 to 2008, as well as Se emission amount, in which included Se entering downstream industries with manganese products (Fig. 3).

4 DISCUSSIONS

4.1 *Periodically systematic investigations on Se emissions and Se concentrations*

Focusing on the studied industries and regions in this paper, situations of Se emission in the environment are regularly tracked to clarify origin, stock, transport, and distribution. Jointly epidemiological investigations will also be conducted to evaluate Se environmental risks in China.

4.2 *Establishing Se environmental standards*

In order to provide a reliable basis for environmental standards, comprehensive studies should be performed on Se cycling, Se enrichment, migration mechanisms, Se bioavailability, and toxicity. On the basis of environmental baseline, a new environmental standard should be established and current standards should be modified, added to improve Se emission limits.

4.3 *Strengthening environmental monitoring and controlling Se environmental risks*

The modern monitoring tools have been used to establish monitoring network in key zones (zones with high Se), key enterprises and other regions with high Se emissions. Sewage containing Se, secondary pollution, surface water, groundwater, soil and air will be monitored to ensure exact and reliable on data. With early warning and emergency monitoring, rapid identification of Se environmental risks is achieved, and Se pollution accidents can be effectively prevented to cause harm on human beings.

4.4 *Developing new technology to reduce Se emissions*

Clean production and advanced technologies should be promoted to reduce Se emissions during coal combustion. Se supplement substitutes should be vigorously developed, and the use of Se should be reduced and cancelled for electrolytic manganese, glass manufacturing and other industries. Efficient separation technology should be accelerated to improve Se extraction from electrolytic copper anode mud, and the solid waste containing Se should be properly disposed to prevent causing secondary pollution.

5 CONCLUSIONS

With the fast development of Chinese economy, the consumption of Se in the economic system increases rapidly. China has become a net importer of Se, and the amount of Se emissions to the environment has also largely increased. It is necessary and urgent to prospectively study and analyze the likelihood of ecological impacts caused by the increasing amount of Se in Chinese environment.

REFERENCES

Diaz, J.P., Navarro, M., Lopez, H. & Lopez, M.C. 1996. Selenium (IV) and (VI) levels in potable, irrigation and waste waters from an industrial zone in southeastern Spain. *Science of the Total Environment* 186(3): 231–236.

Lemly, A.D. 1997. Environmental implications of excessive selenium: a review. *Biomedical and Environmental Sciences* 10(4): 415–435.

Luchli, A. 1993. Selenium in plants: uptake, functions and environmental toxicity. *Botanica Acta* 106(6): 455–468.

Selenium in the Environment and Human Health – Bañuelos, Lin & Yin (Eds)
© 2014 Taylor & Francis Group, London, ISBN 978-1-138-00017-9

Concentrations of selenium and other heavy metals in workers at a metal refining factory in Hunan, China

Lin-Hua Jiang, Zhi Zhang, Yu-Cheng Wen, Fu-Yuan Xu, Chao Zhou & Ning Duan
Technology Center for Heavy Metal Cleaner Production Engineerings, Chinese Research Academy of Environmental Sciences, Beijing, China

Ying Liu
Suzhou Institute for Advanced Study, University of Science and Technology of China, Suzhou, Jiangsu, China

1 INTRODUCTION

Selenium (Se) is a critical trace element to support life in nature. The lack of Se will lead to some diseases, including Keshan Disease (a degenerative heart disease erupting in Keshan, Heilongjiang, China) and Kaschin-Beck Disease (an osteoarthropathy which causes deformity of the affected joints) (Zhang *et al.*, 2010). However, excessive Se intake will cause Se toxicological symptoms, including hair and nail loss and induce nervous system disorders. Thus, the World Health Organization (WHO) recommended the required dietary intake of Se to be 50–200 μg/day for adults (WHO, 1987). It is reported that workers engaging in the processing of Se rectifiers for long term experience decreased hemoglobin type anemia, decrease of white blood cells and impaired fingernails (Yoshizawa *et al.*, 1998; Beck, 1999). One hydrometallurgy industry exploits large quantities of Se dioxides as an electrolytic additive in China. The corresponding consumption of Se is above 2000 tons in the hydrometallurgy industry of China, was about 1/3 of Se totally consumed in China. In the present study, an investigation was conducted on the metal refining workers to evaluate their Se levels and related health conditions.

2 RESEARCH METHODS

The metal refining factory in Hunan, China, was selected for this study. Hair, fingernail, and urine samples were collected from the workers named as stripping workers who stripped the finished metal from electrolytic cathode plates, including 26 hair samples, 23 fingernail samples and 26 urine samples. In addition, 73 fingernail samples and 80 urine samples were also collected from those named as surface workers who worked around the surfaces of electro-baths. Samples were acid-digested in mixed HNO_3-HCl-HF, and the total Se and other heavy metal concentrations were determined by ICP-MS (Thermo, USA). The other heavy metal elements mainly include As, Cd, Pb and Cr. As, Cd and Pb come from raw materials. Cr is from the electrolytic additive in the electrolytic process.

3 RESULTS AND DISCUSSION

3.1 *Se levels in hair samples*

Selenium content of hair samples from the stripping workers are shown in Figure 1. The concentrations show that the stripping workers had extremely high Se in their hair (15 mg Se/kg), which is about 70 times higher than that of common people (about 0.2 mg Se/kg) (Chen *et al.*, 2007). Regarding other heavy metals in the hair, the content of Cr was 8 mg/kg, which was about 26 times higher than the normal level (0.3 mg /kg) (Chen *et al.*, 2007). The content of Pb was about 6 mg/kg, which was about 2 times higher than the normal level (3 mg/kg) (Chen *et al.*, 2007). Arsenic and Cd were detected at relatively low levels, but

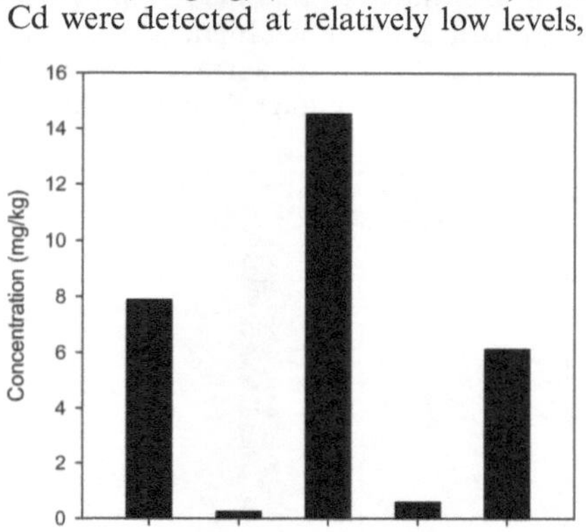

Figure 1. Concentrations of Se and Cr, As, Cd, and Pb in hair samples of the stripping workers.

the Cd content was about 11 times higher than the normal level (0.05 mg/kg) (Chen *et al.*, 2007).

3.2 *Selenium levels in fingernails*

The fingernail Se concentrations of the stripping and surface workers are shown in Figure 2. The surface workers had about 398 mg Se/kg in their fingernails, which was about 349 times higher than the normal level, *i.e.*, 1.14 mg Se/kg (Gu, 1982). It was much higher than that of the stripping workers (14 mg Se/kg), indicating the surface workers likely exposed under much higher Se environment. Similar with hair, As and Cd were quite low in fingernails, but they were about 2.6 and 2 times higher than the normal levels, *i.e.*, 0.2 mg As/kg and 0.08 mg Cd/kg, respectively (Gu, 1982). Moreover, concentrations of Cr and Pb in fingernails of the surface workers were obviously higher than those of the stripping workers. The excessive Cr and Pb reached about 3.6 and 1.2 times of the normal levels, *i.e.*, 6.2 mg Cr/kg and 13.8 mg Pb/kg, respectively (Gu, 1982).

3.3 *Selenium levels in urines*

Concentrations of Se and other metals in urine samples are displayed in Figure 3. The stripping workers had about 130μg Se/L, 35μg Cr/L, 80μg As/L and 5 μg Cd/L in urine, which is 2 times greater than those in the surface workers. However, the surface workers had about 83 μg Pb/kg in urine, which was much higher than that of the stripping workers. The levels of Se, Cr, and As were about 27, 59, 5 and 8 times higher than the normal levels, *i.e.*, 4.8 μg Se/L, 0.6 μg Cr/L, 15 μg As/L and 0.64 μg Cd/L, respectively (Gu, 1982). The excessive Pb in the surface workers was about 6 times of those in the stripping workers and reached 13 times of the normal levels (6.3 μg /L) (Gu, 1982). It indicated that the stripping workers likely exposed under much higher Se, Cr, As and Cd environment, but the surface workers likely exposed under much higher Pb environment.

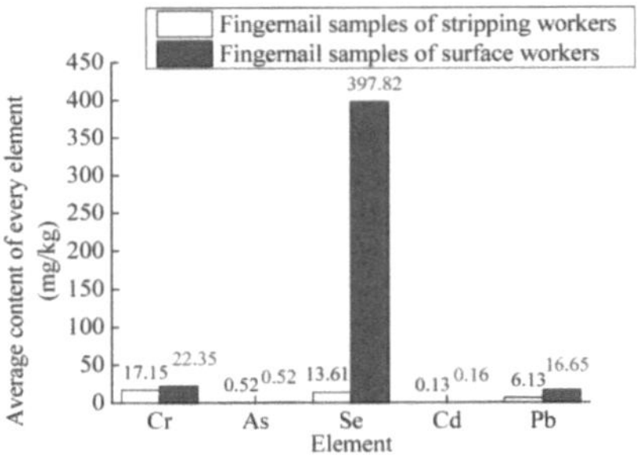

Figure 2. Concentrations of Se and Cr, As, Cd, and Pb in fingernails of the stripping workers and the surface workers.

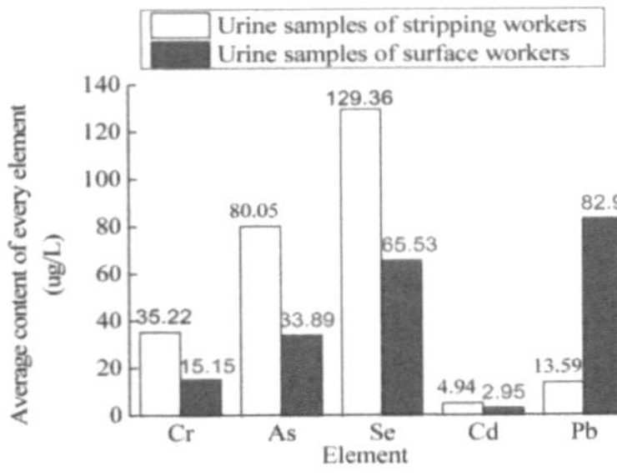

Figure 3. Concentrations of Se and Cr, As, Cd, and Pb in urine samples of the stripping workers and the surface workers.

4 CONCLUSIONS

Selenium is one of the essential trace elements in human needs. However, a human's long-term exposure to environments rich of Se could eventually lead to Se poisoning in human bodies. By investigating the hair, fingernails and urine of workers who operate in the environment rich of Se, we found that the levels of Se measured in workers' hair, fingernails and urine are about 70 times, 12 to 349 times and 14 to 27 times higher than the corresponding normal levels, respectively. The maximum levels of other heavy metals, including Cr, As, Cd, and Pb, inside workers' bodies approximately reach 59 times, 5 times, 11 times, and 13 times higher than the corresponding normal levels, respectively. This study indicates that human health could be largely jeopardized by industrial Se and other heavy metals pollution.

REFERENCES

Beck, M.A. 1999. Selenium and host defence towards viruses. *Proceedings of the Nutrition Society* 58(3): 707–711.

Chen, X.Y., Zhang, H.Y., Qin, J.F., Liang, D.D., Wang, G.Y. & Li, Z.X. 2007. The standards of Trace Elements Science Association of China, *World Elemental Medicine* 14(3): 39–46.

Gu, Z.W. 1982. The normal values of elements in body blood, urine, hair and fingernail for foreign people, *Industrial Health and Occupational Diseases*, (6): 380–383.

Yoshizawa, K., Willett, W.C., Morris, S.J., Stampfer, M.J., Spiegelman, D., Rimm, E.B. *et al.* 1998. Study of prediagnostic selenium level in toenails and the risk of advanced prostate cancer. *Journal of the National Cancer Institute* 90(16): 1219–1224.

Zhang, A., Cao, J.L., Yang, B., Chen, J.H., Zhang, Z.T., Li, S.Y. *et al.* 2010. Effects of moniliformin and selenium on human articular cartilage metabolism and their potential relationships to the pathogenesis of Kashin-Beck disease. *Journal of Zhejiang University-Science* B 11(3): 200–208.

Selenium in the Environment and Human Health – Bañuelos, Lin & Yin (Eds)
© 2014 Taylor & Francis Group, London, ISBN 978-1-138-00017-9

Chemical characterization on selenium in residues of fourteen factories in Hunan, China

Lin-Hua Jiang, Xiao-Juan Gao, Zhi Zhang, Zhi-Gang Dan, Wen-Lin Guo & Ning Duan
Technology Center for Heavy Metal Cleaner Production Engineerings, Chinese Research Academy of Environmental Sciences, Beijing, China

Ying Liu
Suzhou Institute for Advanced Study, University of Science and Technology of China, Suzhou, Jiangsu, China

1 INTRODUCTION

In 2008, Chinese annual output of Electrolytic Manganese Metal (EMM) reached 1.88 million tons, and accounted for 97.4% of the total world annual production and 98.58% of the total world production capacity (Tan, 2009; Duan, 2010). However, to produce one ton of EMM, about 1 kg selenium (Se) as SeO_2 was used, including 60.7% of Se in the products, 22.3% in anode mud, and 17% in waste residues (Duan *et al.*, 2010). Indeed, the toxicity of Se in the environment must be considered if a large quantity of Se wastes is released into the environment without being properly treated (Karen, 2009).

China is one of the three major industrial countries using Se in the world industrial purposes. Therefore, environmental impacts of Se pollution need to be assessed and managed. In the present study, the total Se contents in residues at fourteen EMM manufactures and their distributions in different chemical fractions were determined, which provides the essential information and understanding for the evaluation of Se release from industrial residue dams and their potential impacts on the environment.

2 MATERIALS AND METHODS

A total of 14 factories in Hunan, China, were selected to be studied, including: 14 samples from new residue dumps in recent year, 14 from old residue dumps 5 years ago, and 14 background soil samples. All samples were digested in mixed HNO_3-HCl-HF and total Se contents were determined by Inductively-Coupled Plasma-Mass Spectrometry (ICP-MS). In addition, water-soluble Se, exchangeable Se, acid-soluble Se, organic-bound Se, simple substitute Se, and residue Se were separated, respectively, by sequential chemical-extraction technique according to Zhu *et al.* (2008).

3 RESULTS AND DISCUSSION

The results of total Se contents are shown in Figure 1. Overall, total Se contents in most samples were in the range of 20 to 50 mg/kg, except two values in new residue were as high as 70–80 mg/kg, and three old residue samples contained very low concentrations of Se. Selenium contents in old residue were slightly higher than those of new ones. In contrast, background soils from study site contained no Se. Reduction of Se contents in new residue were likely related with improved technology to increase efficiency on Se addition (Duan *et al.*, 2010).

The distributions of water-soluble Se, exchangeable Se, acid-soluble Se, organic-bound Se, simple-substitute Se and residue Se are displayed in Figure 2. In new residue samples, the Se fractionation was as follows within total Se: 13.03% as organic-bound Se, 16.33% as exchangeable Se, 19.94% as simple substitute Se, 15.28% as water-soluble Se and 9.58% Se in residues. Generally,

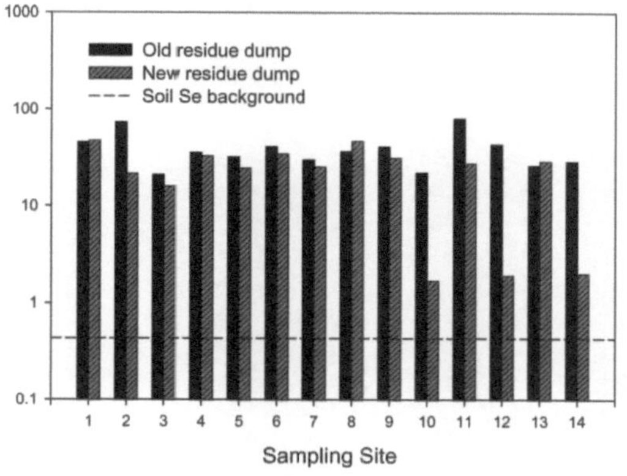

Figure 1. Concentrations of total selenium (mg/kg) in new residue and old residue dumps at 14 industrial factories in Hunan, China.

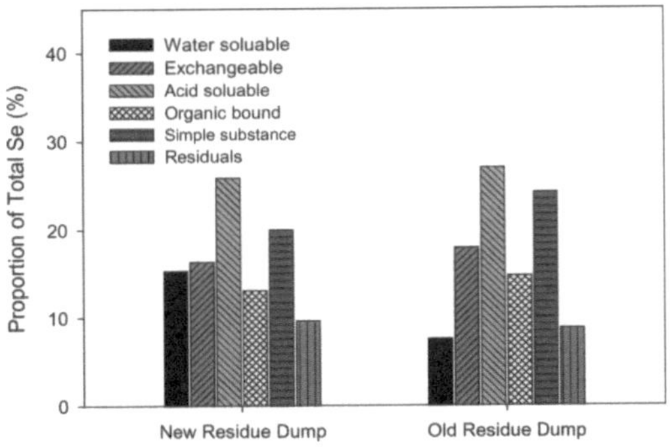

Figure 2. Fractionation of Se in the new and old residue dumps at the stacking area of different factories.

the water-soluble Se and the exchangeable Se were regarded as bioavailable Se, and the acid-soluble Se and organic-bound Se were considered as transferable Se (Yin & Yuan, 2012). Thus, about 31.61% and 38.85% Se in new residue could be bioavailable and transferable, respectively. In other words, about 70.46% Se could be re-released into environment to cause potential toxicity. Interestingly, compared with new residue, the proportions of water-soluble Se, organic-bound Se and simple substitute Se, slightly increased along with obviously decreasing in residue Se in old residue. This last observation indicates that residue Se could be transformed to release during stacking.

4 CONCLUSIONS

In conclusion, the residue Se could be transformed into bioavailable Se during stacking. Moreover, about 70.46% Se could be released into the environment and the Se contents in industrial residues from Hunan fourteen factories can reach as high as 20–50 mg/kg exceeding 81 times the soil Se background on average, which may cause potential ecological problems. Therefore, it is urgent for industrial field to develop and research new technologies on reducing Se and Se substitutes.

REFERENCES

Duan, N., Wang, F., Zhou, C.B., Zhu, C.L. & Yu HB. 2010. Analysis of pollution materials generated from electrolytic manganese industries in China. *Resources, Conservation and Recycling* 54: 506–511.

Karen, H. 2009. Globally sustainable manganese metal production and use. *J. Environ. Manage.* 90(12): 3736–3740.

Tan, Z.Z. 2009. Retrospect of 2008. Chinese electrolytic manganese industry circumstances and responding measures in 2009. *International Forum on Electrolytic Mn Product.*

Yin, X.B. & Yuan, L.X. (eds.) 2012. *Phytoremediation and Biofortification: Two Sides of One Coin.* Germany: Springer Press.

Zhu, J.M., Qin, H.B., Li, L., Feng, Z.G. & Su, H.C. 2008. Fractionation of Se in high-Se soils from Yutangba, Enshi, Hubei. *Acta Scientiae Circumstantiae* 28(4): 772–777.

Selenium in the Environment and Human Health – Bañuelos, Lin & Yin (Eds)
© 2014 Taylor & Francis Group, London, ISBN 978-1-138-00017-9

Interaction of selenium and mercury in soil-plant system in mercury mining areas and their combined exposure and risk

Hua Zhang
Institute of Geochemistry, Chinese Academy of Sciences, Guiyang, China
Norwegian Institute for Water Research, Oslo, Norway

Xin-Bin Feng
Institute of Geochemistry, Chinese Academy of Sciences, Guiyang, China

Thorjørn Lassen
Norwegian Institute for Water Research, Oslo, Norway

1 INTRODUCTION

Selenium (Se) has recently been proposed to be an emerging global metalloid contaminant with potential ecological risk second only to mercury (Hg) (Hu *et al.*, 2009). However, Se is also an essential trace element and nutrient that is of vital importance to human health, and many biological functions. With its dual role as an essential micronutrient at physiologically appropriate levels and a toxic substance at higher levels, Se has attracted a great deal of attention from both the scientific community and the public in recent decades (Fordyce, 2013; Rayman, 2012).

Mercury (Hg) is an exogenous, toxic, and ubiquitous trace element that is nonessential to humans and animals. Its organic form, such as methyl-Hg (MeHg), is of great concern due to the high toxicity and potentially severe effects on humans. MeHg can bioaccumulate and be biomagnified in aquatic food webs and even some terrestrial plants (*e.g.*, rice (Zhang *et al.*, 2010b)), eventually posing a serious threat to humans through the consumption of fish and/or rice (Zhang *et al.*, 2010a).

The antagonism by Se of the toxicity of Hg in organisms has been known since Pařízek and Ošťádalová (1967) showed that Se effectively prevented mortality caused by high doses of inorganic Hg (IHg) in rats. Ganther *et al.* (1972) further demonstrated that Se decreased the toxicity of MeHg in fish. Since then, a large number of subsequent studies have confirmed that Hg-Se antagonism is a widespread phenomenon in microorganisms, fish, poultry, humans and other mammals. Most of this abundant literature has been well summarized by several review articles (Zhang *et al.*, 2012), suggesting that a universal Se-Hg antagonism may exists among various species.

The Wanshan mercury-mining area, known as the "mercury capital" of China, was the largest mercury industrial base in China. The historic large-scale mining and retorting activities over more than 2,000 years, have produced approximately 22,000 tons of Hg, 6,000 tons of cinnabar, and large quantities of mine wastes in Wanshan. Cinnabar and mercury ores in Hg mining areas usually contain a considerable amount of Se. In the Wanshan area, the Se content in common cinnabar was reported to be 0.02–0.87% (wt) and attained 2.7% (wt) in metacinnabar, as the Sulfur (S) in the cinnabar and ores can be replaced by Se to form an isomorphous series of HgS-HgSe (Zhang *et al.*, 2012). Mining and retorting activities of this cinnabar and ore can result in the release of large quantities of Se into the surrounding environment. For example, Se in local paddy soils and rice grains was observed to be 0.16 to 36.6 mg/kg and 0.02 to 0.67 mg/kg, respectively (Zhang *et al.*, 2012). Furthermore, during the retorting processes used to mobilize and separate the desired Hg, Se is readily volatilized and can be emitted into the air. This Se cools and can adhere to atmospheric dust particles, subsequently reaching distant terrestrial and aquatic systems by either dry or wet deposition.

This study was designed (1) to test the hypothesis that levels of Se in the rhizosphere may affect the uptake and translocation of IHg and MeHg in the rice plant (*Oryza sativa* L.) in Wanshan Hg mining aera, China; and (2) to evaluate mercury exposure risk that considered Se-Hg interactions.

2 METHODS AND MATERIALS

The total Se in all samples was determined by Hydride Generation-Atomic Fluorescence Spectrometry (HG-AFS). Using Method 1631, Hg_T was measured in the soil samples using Cold Vapor Atomic Absorption Spectrometry (CVAAS), and Hg_T in the rice samples was determined using the dual-stage gold amalgamation method and Cold Vapor Atomic Fluorescence Spectrometry (CVAFS) detection. Using Method 1630, MeHg was determined using aqueous ethylation, purge, trap, and GC CVAFS detection. Exposure and risk assessment were evaluated using the Se-Health Benefit Value (Se-HBV), which is calculated as Se-HBV = Se(Se/Hg) – Hg(Hg/Se) (Kaneko & Ralston, 2007). This equation includes both the absolute molar concentrations and the relative molar ratios of Se and Hg. The Se-HBV indicates the health benefits (if positive) or health risks (if negative) of Se in terms of Hg exposure.

3 RESULTS AND DISCUSSION

The results showed that, in general, the Se levels were inversely related to the levels of both IHg and MeHg in the rice grains. In addition, a consistent reduction in translocation of both Hg and MeHg in the aerial shoots (*i.e.*, the stem, leaf, husk, and grain) with increasing Se levels in the soils was observed. Furthermore, the Se levels were positively correlated with the IHg levels in the soils and the roots. These results suggest that Se may play an important role in limiting the bioaccessibility, absorption, and translocation/bioaccumulation of both IHg and MeHg in the aerial rice plant, which may be related to the formation of an Hg-Se insoluble complex in the rhizosphere and/or root surface.

We found that all the sites showed health benefits rather than health risks when assessed using criteria that considered the protective interactions between Se and MeHg based on their molar concentrations, which indicated that the health risks of MeHg exposure were offset by Se intake. The reverse was also true: the health risks of excessive Se intake were neutralized by moderate MeHg exposure. Results indicate from previous study (Zhang *et al.*, 2010a) may have overestimated the level of risk for the local residents (34% of sites shown MeHg exposure exceeded 0.1 μg/kg/day established by the USEPA) because we only considered Hg in the environment and in foods in this area. This conclusion may be ubiquitous for previous Hg exposure assessment as a molar ratio of Se:Hg > 1:1 are commonly observed in marine fish or rice (Zhang *et al.*, 2012).

However, we should mention that the Se-HBV criterion and the traditional Se:Hg molar ratio both have a serious limitation: in certain extreme cases, although the safety requirement (Se:Hg molar ratio > 1 or Se-HBV > 0) is met, the Se intake may be either below the level required for normal selenoenzyme activity and synthesis (deficiency) or above the safe range (poisoning). Although the Se-HBV and Se:Hg molar ratio may both appear ideal, these ratios are associated with hidden risks. Therefore, an assessment based on either criterion may be misleading.

We therefore propose a new criterion for concurrently evaluate Se and Hg co-exposure: $BRV = PDI_{Se} - \Delta_{Se} - PDI_{Hg}$ (Zhang *et al.*, unpubl.), where BRV represents the benefit-risk value, which indicates either health benefits (if $0 < BRV < \nabla_{Se}$) or health risks (if $BRV < 0$ or $BRV > \nabla_{Se}$); Δ_{Se} represents the minimal Se amount required for normal biological function when Hg exposure is zero; ∇_{Se} represents a threshold value for Se poisoning which considered the protective effects against Se toxicity from Hg exposure; PDI represents the probable daily intake of Se or Hg. All of the calculations are based on units of molar concentrations.

4 CONCLUSIONS

Selenium may inhibit the uptake and translocation of Hg from rhizosphere soils to the aerial tissues of rice plant, which may be related to the formation of an inert Hg-Se insoluble complex in the rhizosphere or root surface. A novel criterion for concurrently evaluating the combined benefits and risks of co-exposure to Hg and Se was proposed.

REFERENCES

Fordyce, F.M. 2013. Selenium deficiency and toxicity in the environment. In O. Selinus, B. Alloway, J.A. Centeno, R.B. Finkelman, R. Fuge, U. Lindh, *et al.* (eds.), *Essentials of Medical Geology*: Revised Edition, 375–416, New York: Springer.

Hu, X., Wang, F. & Hanson, M.L. 2009. Selenium concentration, speciation and behavior in surface waters of the Canadian prairies. *Sci. Total Environ.* 407: 5869–5876.

Kaneko, J.J. & Ralston, N.C. 2007. Selenium and mercury in pelagic fish in the central north pacific near Hawaii. *Biol. Trace Elem. Res.* 119: 242–254.

Rayman, M.P. 2012. Selenium and human health. *Lancet* 379: 1256–1268.

Zhang, H., Feng, X., Larssen, T., Qiu, G. & Vogt, R.D. 2010a. In inland China, rice, rather than fish, is the major pathway for methylmercury exposure. *Environ. Health Perspect.* 118: 1183–1188.

Zhang, H., Feng, X., Larssen, T., Shang, L. & Li, P. 2010b. Bioaccumulation of methylmercury versus inorganic mercury in rice (*Oryza sativa* L.) grain. *Environ. Sci. Technol.* 44: 4499–4504.

Zhang, H., Feng, X., Zhu, J., Sapkota, A., Meng, B., Yao, H. *et al.* 2012. Selenium in soil inhibits mercury uptake and translocation in rice (*Oryza sativa* L.). *Environ. Sci. Technol.* 46: 10040–10046.

Selenium in the Environment and Human Health – Bañuelos, Lin & Yin (Eds)
© 2014 Taylor & Francis Group, London, ISBN 978-1-138-00017-9

Selenium status and intake influences mercury exposure risk assessments

N.V.C. Ralston & L.J. Raymond
Energy and Environmental Research Center, University of North Dakota, Grand Forks, ND, USA

1 INTRODUCTION

Mercury (Hg) occurs in a variety of molecular forms that differ in source, exposure, tissue distribution, and toxicity. Elevated Hg exposures are hazardous, but low-level exposures are ubiquitous and without biological consequences. Elemental Hg (Hg^0) a chemically unreactive but volatile liquid at room temperature, readily disperses into air and can be easily absorbed by the body following inhalation. In the body, it can be oxidized to inorganic Hg^+ or Hg^{2+} forms that are chemically reactive with chalcogens such as sulfur or selenium (Se) in tissues.

Anthropogenic release of Hg^0 joins the continually cycling global atmospheric pool, until oxidation results in deposition of $Hg^{+/2+}$ forms that can be biologically absorbed. Anaerobic bacteria methylate $Hg^{+/2+}$, forming MeHg. Since Hg is electrophilic, it reacts with nucleophilic thiols of cysteine (Cys) to form MeHg-Cys. Since the MeHg-Cys molecule is similar in size and structure to methionine (Met), its transport into cells is competitively inhibited by Met, and consequently its distribution in reflects that of Met. All aquatic ecosystems possess at least trace quantities of MeHg-Cys which is retained and possibly even used in place of Met in tissue proteins. Predators absorb MeHg-Cys, from proteins of prey animals they have consumed, and it subsequently redistributes throughout their tissue proteins.

1.1 *Selenium physiology*

Because of distinctions in geological parent material, drainage characteristics and/or pH, the availability of Se in soils and waters can vary dramatically, even in adjacent locales. Sulfur and Se are not biochemically differentiated by prokaryotes and plants, thus their respective incorporation into Met or the Se analogue selenomethionine (SeMet) occurs in quantities that reflect their relative concentrations in the environment. However, animal cells degrade SeMet into selenide, the precursor required for *de novo* synthesis of selenocysteine (Sec). Although Cys molecules are repeatedly recycled during protein synthesis, Sec must be degraded to inorganic selenide before its Se can be reincorporated into a newly formed Sec.

Selenide is the precursor that is required for Sec synthesis during its co-translational insertion into nascent selenoproteins. This even occurs in coordination with an in-frame UGA linked to stem-loop structures within the 3'-untranslated regions of selenoprotein transcripts that bind unique accessory factors (Hatfield & Gladyshev, 2002). The unique redox range of the ionized Sec enables it to catalyze reactions that are beyond the capacity of Cys. Since the pKa of Cys is 8.3, while that of Sec is 5.5, Cys is predominantly protonated (uncharged) at physiological pH, while the Se of Sec is in the ionized selenoate form.

2 MERCURY SELENIUM INTERACTIONS

As a soft electrophile, Hg forms strong covalent bonds with soft nucleophiles such as the sulfur of thiols, but even stronger bonds with Se (Dyrssen & Wedborg, 1991). Mercury is notable for its ability to exchange binding partners and form covalent bonds with other molecules, but transitions to new partners exclusively involve the formation of associations that are of equal or greater thermodynamic stability. The MeHg bound to the Cys of glutathione, thioredoxin, lipoic acid, etc. forms a "suicide substrate". When these substrates are acted upon by selenoenzymes, MeHg will irreversibly transfer from thiol to the Se of Sec: the most powerful intracellular nucleophile, and the preferred target of soft electrophiles.

The inhibition of selenoenzymes by MeHg has been repeatedly shown (*e.g.*, GPx, see Prohaska & Ganther, 1978; TRx, see Carvalho *et al.*, 2008) in both animal and *in vitro* models. Once the Hg-Se bond has formed, it is extremely stable. Degradation of these complexes in the lysosome results in the accumulation of inorganic mercury selenide (HgSe) precipitates that can be analyzed by X-ray Absorption Fine-Structure (XAFS) and related spectroscopic methods (Huggins *et al.*, 2009) years after formation (Korbas *et al.*, 2010).

2.1 *Selenium Health Benefit Values (HBV_{Se})*

Because it was formerly thought that Hg's molecular mechanism of toxicity involved Hg binding to thiols, pseudo-first order reaction kinetics were assumed to be appropriate. However, now that it is known that Hg irreversibly inhibits selenoenzymes at cell concentrations that approach and especially those that exceed equimolar stoichiometries (approximately 1–2 μM), bimolecular reaction rate equations need to be applied to properly evaluate Hg exposure risks.

To simplify these assessments for regulatory agencies and consumers, equations have been developed to reflect relative benefits and risks associated with foods that contain MeHg. The initial Se-HBV equation that was applied provided a reliable index of risk from Hg exposures but it did not reliably reflect beneficial effects of dietary Se intakes.

Thus, the equation has been modified to directly indicate benefits of dietary Se as well as Hg risks. To differentiate this updated version from the earlier iteration of the equation, the present version has been designated HBV_{Se}. As with the previous version of this calculation, Se and Hg are exclusively in molar concentration units. It is calculated:

$$HBV_{Se} = (Se\text{-}Hg)/Se \bullet (Se + Hg)$$

Seafoods with a negative HBV_{Se} indicate their consumption increases risks of adverse effects on child neurodevelopment. However, a positive value indicates expectations of beneficial outcomes. The magnitude of the positive or negative value indicates the relative benefits or risk.

2.2 *HBV_{Se} of ocean fish*

The HBV_{Se} of ocean fish vary among species, and tend to diminish with increasing size within species, but almost all species are uniformly positive. Certain varieties of shark, and some of the larger specimens of swordfish and halibut appear to be the only exceptions (see Table 1). Maternal consumption of seafood varieties with a negative HBV_{Se} may pose risks to fetal development consistent with those that have been observed in association with MeHg exposure from eating shark meat in New Zealand (Crump *et al.*, 1998). However, since most ocean fish are Se-rich, their consumption would protect against the adverse effects of high MeHg exposures, as has been observed in the Faroes Study (Budtz-Jorgenson *et al.*, 2005). Populations exposed to MeHg from ocean fish consumption that are even higher on average than those observed in the Faroes find substantial health benefits accompany increasing ocean fish consumption (Davidson *et al.*, 2011). Child development outcomes that reflect the nutritional benefits of ocean fish consumption have been seen in numerous studies (*e.g.*, Hibbeln *et al.*, 2007).

2.3 *HBV_{Se} of marine mammals*

In the Faroes Study, ~85% of the study populations' total MeHg exposure arose from eating pilot whale meats. Increased MeHg exposures were associated with adverse effects in the children that is consistent with the negative HBV_{Se} of that seafood (see Table 1). Since the HBV_{Se} of beluga whale meat is negative, it should also be avoided by pregnant women.

2.4 *HBV_{Se} of freshwater fish*

It is not surprising that the presence of Se in the environment influences MeHg bioaccumulation in fish, since high MeHg exposures have a direct impact on Se metabolism. In the presence of increased Se, more HgSe would form and retire Hg

Table 1. Hg, Se, and HBV_{Se} of selected seafoods.

Seafood	n†	μM Hg	μM Se	HBV_{Se}
Skipjack tuna	10	1.68 ± 0.52	19.83 ± 11.71	19.6 ± 11.8
Bigeye tuna	50	3.00 ± 1.23	12.38 ± 3.48	11.5 ± 3.8
Albacore tuna	20	2.49 ± 1.18	11.11 ± 2.4	10.4 ± 2.7
Chum salmon	289	0.21 ± 0.07	3.29 ± 0.70	3.3 ± 0.7
Halibut	1821	1.44 ± 1.37	4.90 ± 1.92	3.9 ± 3.2
Swordfish	50	5.32 ± 2.98	5.43 ± 1.48	0.8 ± 9.2
Thresher shark	10	4.86 ± 1.60	6.55 ± 1.51	2.7 ± 2.0
Mako shark	10	9.01 ± 1.99	4.07 ± 0.48	16.4 ± 8.6
Beluga whale	10	31.80 ± 13.33	20.19 ± 2.15	38.3 ± 44.4
Pilot whale	15	16.45 ± 8.47	3.17 ± 1.39	82.3 ± 39.3

† number of samples assessed and values indicate means ± standard deviations of the indicated data.

from further cycling in biota. However, the effects of environmental Se go beyond Se-dependent Hg-retirement.

Recent studies have found that improved Se status increased the rate of MeHg-excretion from fish (Bjerregaard *et al.*, 2011). Similar effects have been observed in human studies, which have found that improved Se intake increased urinary excretion of Hg and diminished oxidative stress (Li *et al.*, 2012). Therefore, improved environmental Se availability diminishes MeHg bioaccumulation and toxicity.

3 CONCLUSIONS

Previous MeHg risk assessment approaches were inadequately informed regarding important biological and geochemical interactions of Hg and Se. Since MeHg irreversibly inhibits selenoenzyme activities, it is clear that MeHg and Se must both be assessed to solve biomolecular reaction rate equations required to describe their interactions. A simplified assessment of MeHg exposure risk is obtained using HBV_{Se} as a guide. The HBV_{Se} criterion provides an index that predicts the risks associated with MeHg exposures with reliable accuracy. Adverse effects of MeHg exposures from consumption of pilot whale meat indicates the risks that are associated with eating that seafood, but it is profoundly inappropriate to base advice regarding ocean fish consumption on the effects of eating whale meat.

Populations with poor dietary Se intakes are more vulnerable to MeHg exposures. This vulnerability may accentuate risks of MeHg exposures from freshwater fish, and will require increased focus on assessing the amounts of MeHg and Se that are present in the fish being consumed, as well as in the population's background diets. Increased monitoring of water bodies will be needed to identify regions where diminished environmental Se availability may exacerbate MeHg exposure risks, particularly in areas where increased Hg pollution originates from point sources.

REFERENCES

Alexander, J., Thomassen, Y. & Aaseth, J. 1983. Increased urinary excretion of selenium among workers exposed to elemental mercury vapor. *J. Appl. Toxicol.* 3: 143–145.

Berntssen, M.H.G., Hylland, K., Lundebye, A.K. & Julshamn, K. 2004. Higher faecal excretion and lower tissue accumulation of mercury in Wistar rats from contaminated fish than from methylmercury chloride added to fish. *Food Chem. Toxicol.* 42(8): 1359–1366.

Bjerregaard, P., Fjordside, S., Hansen, M.G. & Petrova, M.B. 2011. Dietary selenium reduces retention of methylmercury in freshwater fish. *Environ. Sci. Technol.* 45: 9793–9798.

Bjoernberg, A.A. 1989. Decontamination of mercury from Swedish "Black Listed" lakes by addition of selenium. In S.C. Carapella, Jr. (ed.), *Proceedings of the 4th International Symposium of Uses of Selenium and Tellurium*, 357–360. Banff, Canada: Selenium Tellurium Dev. Assoc. Darien, CT.

Budtz-Jørgensen, E., Grandjean, P. & Weihe, P. 2007. Separation of risks and benefits of seafood intake *Environ. Health Perspect.* 115: 323–327.

Carvalho, C.M.L., Chew, E.H., Hashemy, S.I., Lu, J. & Holmgren, A. 2008. Inhibition of the human thioredoxin system: A molecular mechanism of mercury toxicity. *J. Biol. Chem.* 283: 11913–11923.

Crump, K.S., Kjellstrom, T., Shipp, A.M., Silvers, A. & Stewart, A. 1998. Influence of prenatal mercury exposure upon scholastic and psychological test performance: Benchmark analysis of a New Zealand cohort. *Risk Anal.* 18: 701–713.

Davidson, P.W., Cory-Slechta, D., Thurston, S.W., Huang, L.S., Shamlaye, C.F., Gunzler, D. *et. al.* 2011. Fish consumption and prenatal methylmercury exposure: Cognitive and behavioral outcomes in the main cohort at 17 years from the Seychelles child development study. *Neurotoxicol.* 32: 711–717.

Dyrssen, D. & Wedborg, M. 1991. The Sulfur-mercury (II) system in natural waters. *Water, Air, Soil Pollut.* 56: 507–519.

Hatfield, D.L. & Gladyshev, V.N. 2002. How selenium has altered our understanding of the genetic code. *Mol. Cell. Biol.* 22: 3565–3576.

Hibbeln, J.R., Davis, J.M., Steer, C., Emmett, P., Rogers, I., Williams, C. *et al.* 2007. Maternal seafood consumption in pregnancy and neurodevelopmental outcomes in childhood (ALSPAC study): an observational cohort study. *Lancet.* 369: 578–585.

Hill, K.E., Zhou, J., McMahan, W.J., Motley, A.K., Atkins, J.F., Gesteland, R.F. *et al.* 2003. Deletion of Selenoprotein P alters distribution of selenium in the mouse, *J. Biol. Chem.* 278: 13640–13646.

Huggins, F., Raverty, S.A., Nielsen, O.S., Sharp, N., Robertson, J.D. & Ralston, N.V.C. 2009. An XAFS investigation of mercury and selenium in beluga whale tissues. *Env. Bioindicators* 4: 291–302.

Korbas, M., O'Donoghue, J.L., Watson, G.E., Pickering, I.J., Singh, S.P., Myers, G.J. *et al.* 2010. The chemical nature of mercury in human brain following poisoning or environmental exposure. *ACS Chem. Neurosci.* 1: 810–818.

Li, Y.-F., Dong, Z., Chen, C., Li, B., Gao, Y., Qu, L. *et al.* 2012. Organic selenium sup-plementation increases mercury excretion and decreases oxidative damage in long-term mercury-exposed residents from Wanshan, China. *Environ. Sci. Technol.* 46: 11313–11318.

Prohaska, J.R. & Ganther, H.E. 1977. Interactions between selenium and methylmercury in rat brain. *Chemico-Biological Interactions* 16: 155–167.

Selenium in the Environment and Human Health – Bañuelos, Lin & Yin (Eds)
© 2014 Taylor & Francis Group, London, ISBN 978-1-138-00017-9

Synchrotron studies of selenium interactions with arsenic

I.J. Pickering, O. Ponomarenko & G.N. George
University of Saskatchewan, Saskatoon, Saskatchewan, Canada

P.F. La Porte & K. Strait
Emory University, Atlanta, Georgia, USA

J. Gailer
University of Calgary, Calgary, Alberta, Canada

E.M. Leslie
University of Alberta, Edmonton, Alberta, Canada

J. Spallholz
Texas Tech University, Lubbock, Texas, USA

1 INTRODUCTION

1.1 *Arsenic is a major world health problem*

More than 100 million people world-wide are thought to consume drinking water potentially containing harmful levels of arsenic. In Bangladesh alone approximately 57 million people are affected in what has been called the World's Worst Mass Poisoning; in the worst affected areas up to one-quarter of all deaths are attributable to arsenic exposure. Chronic arsenic exposure causes what is known as arsenicosis, a syndrome featuring characteristic skin lesions (hypermelanosis and hyperkeratosis), vascular and endocrine pathologies, several types of cancer and increased mortality.

A number of measures have been proposed to combat the problem in Bangladesh. Deeper wells into less contaminated aquifers have proved successful in some localities, although these wells have become contaminated with time. Home-based water filter systems employing iron chips have been used to remove arsenic from drinking water but their effectiveness requires that filter materials be changed on a regular basis, with maintenance proving to be problematic in the rural Bangladeshi environment. Moreover, unfiltered water used for agricultural irrigation results in arsenic's uptake into food plants, especially rice, and this food-borne arsenic contributes to the human dose. At the time of writing, despite considerable efforts by a number of groups, there is no adequate solution to the crisis in place.

1.2 *A selenium solution?*

The antagonistic interaction of arsenic and selenium has been known since the pioneering work of Moxon in the 1930's (Moxon, 1938). This work and subsequent work by others established that a toxic dose of arsenic could be cancelled by an equal and also otherwise toxic dose of selenium. More recently we have discovered the molecular basis for this surprising antagonism; the formation of the seleno-*bis*-S-glutathionyl arsinium ion or $[(GS)_2AsSe]^-$ (Fig. 1, Gailer *et al.*, 2000). This novel molecule contains selenide covalently bound to trivalent arsenic and two glutathione thiolates complete the coordination of the latter. $[(GS)_2AsSe]^-$ is rapidly formed on co-administering arsenite and selenite to mammals, and is excreted in the bile. Informed by this chemical foundation, several clinical trials have explored a possible palliative role for selenium compounds in arsenicosis, reporting some limited success.

Recently, progress has been made in understanding the transport and metabolism of $[(GS)_2AsSe]^-$ at the molecular level. Carew and Leslie (2010) have identified that Multidrug Resistance

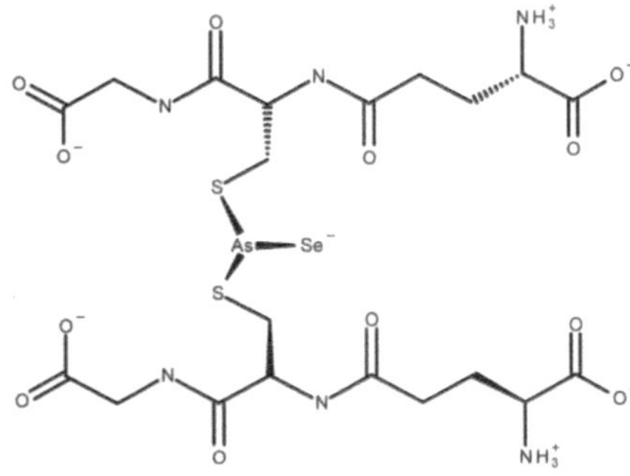

Figure 1. Schematic structure for the seleno-*bis*-S-glutathionyl arsinium ion.

protein 2 (MRP2) transports $[(GS)_2AsSe]^-$ in MRP2-enriched membrane vesicles, suggesting a role for this transporter in the excretion of the compound. Despite this progress, much remains to be elucidated about the formation, transport and excretion of $[(GS)_2AsSe]^-$ in mammals. We have used synchrotron X-ray methods, and in particular X-ray absorption spectroscopy and X-ray fluorescence imaging, to investigate the molecular mechanisms underlying arsenic-selenium antagonism.

2 SYNCHROTRON METHODS

2.1 *X-ray absorption spectroscopy*

Synchrotron X-ray Absorption Spectroscopy (XAS) uses an X-ray beam of tunable energy to excite a core level electron. For incident energies just above the absorption edge, the near-edge spectrum probes bound states, showing sensitivity to chemical environment or speciation, including oxidation state, nature of ligands and their geometry. The near-edge can determine chemical types for unknowns by giving a fingerprint match with standard compounds. The EXAFS arises from quantum photoelectron interference by surrounding atoms and can give accurate bond-length and neighbor atom information.

2.2 *X-ray fluorescence imaging*

Monitoring characteristic X-ray fluorescence lines while scanning a micro-beam across a sample can yield elemental maps. This technique of X-ray Fluorescence Imaging (XFI) can be applied to sections of samples or even to whole small animals. Figure 2 shows a schematic experimental setup and an example of XFI data for a section of a rodent.

The advantages of XFI and XAS are that they can give localization and speciation information on samples without complex extraction, fixing or staining procedures, which are likely to modify chemical form or distribution of the different species present.

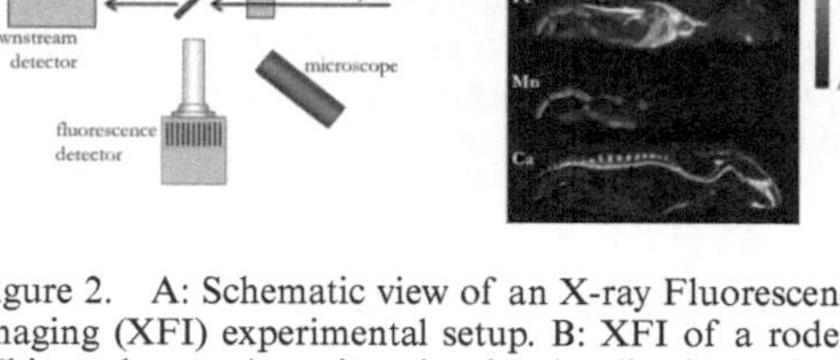

Figure 2. A: Schematic view of an X-ray Fluorescence Imaging (XFI) experimental setup. B: XFI of a rodent (Chinese hamster) section showing localizations of different elements.

3 PROGRESS

We have used XFI and XAS, together with conventional analysis methods, to follow arsenic and selenium transformation, localization, transport and excretion in a model mammalian system. XFI is used to determine the distributions of arsenic, selenium and other elements, in particular whether arsenic and selenium co-localize and concentrate in certain organs. XAS of specific tissues elucidates chemical form, including whether arsenic and selenium are bound to each other. These measurements allow further elucidation of the biochemical relationship between arsenic and selenium in mammals, providing greater clarity on a possible palliative role for selenium in arsenicosis.

ACKNOWLEDGEMENTS

This work was funded by the Natural Sciences and Engineering Research Council of Canada (NSERC), and by the Canadian Institutes of Health Research (CIHR). IJP and GNG are Canada Research Chairs; OP is a CIHR-THRUST Associate. Portions of this research were carried out at the Stanford Synchrotron Radiation Lightsource, a Directorate of SLAC National Accelerator Laboratory and an Office of Science User Facility operated for the U.S. Department of Energy Office of Science by Stanford University. The SSRL Structural Molecular Biology Program is supported by the DOE Office of Biological and Environmental Research, and by the National Institutes of Health, National Institute of General Medical Sciences (including P41GM103393). The contents of this publication are solely the responsibility of the authors and do not necessarily represent the official views of NIGMS, NCRR or NIH.

REFERENCES

Carew, M.W. & Leslie, E.M. 2010. Selenium-dependent and independent transport of arsenic by the human multidrug resistance protein 2 (MRP2/ABCC2): implications for the mutual detoxification of arsenic and selenium. *Carcinogenesis* 31: 1450–1455.

Gailer, J., George, G.N., Pickering, I.J., Prince, R.C., Ringwald, S.C., Pemberton, J.C. *et al.* 2000. A metabolic link between arsenite and selenite: The seleno-bis(S-glutathionyl) arsinium ion. *J. Am. Chem. Soc.* 122: 4637–4639.

Moxon, A.L. 1938. The effect of arsenic on the toxicity of seleniferous grains. *Science* 88: 81.

Selenium in the Environment and Human Health – Bañuelos, Lin & Yin (Eds)
© 2014 Taylor & Francis Group, London, ISBN 978-1-138-00017-9

Protective role of selenium on arsenic trioxide (As_2O_3)-induced cytotoxicity in HUECVs

Zhang Man, Lin Xinying, Jiang Yanyan & Zou Xiaoyan
College of Public Health, Shandong University, Jinan, Shandong, China

Bian Jianchao
Shandong Institute for Endemic Disease Control and Research, Jinan, Shandong, China

1 INTRODUCTION

Endemic arsenic (As) poisoning is widely distributed around the world, and is especially popular and diverse in China. It is reported that long-term exposure to As is closely related to hypertension, cardio-cerebral vascular disease etc, and the relationship was dose-dependent. Countries such as Japan, Poland, Mexico and Argentina also reported that high-arsenic water can lead to peripheral vascular diseases. Selenium (Se) has been found to boost immunity and scavenge free radicals as an antioxidant. The aim of this study was to observe the damage induced by As on cell growth and activities of superoxide dismutase (SOD), glutathione peroxidase (GSH-PX), endothelial Nitric Oxide Synthase (eNOS), inducible Nitric Oxide Synthase (iNOS), and the content of malondialdehyde (MDA), as well as to observe the protective effect of Se in Human Umbilical Vein Endothelial Cells (HUVECs) in vitro. By exploring the protective mechanism of selenium on arsenic-induced toxicity, we hope to provide more experimental evidence for preventive application of Se in high As areas.

2 MATERIALS AND METHODS

The HUVECs were grown in cell culture flasks at 30°C with 5% CO_2 in RPMl-l640 culture medium containing 100 IU penicillin/ml and 100 IU streptomycin/ml supplemented with 5% fetal bovine serum. Cells were treated in four groups: control group (RPMI-l640 culture medium), Se group (Se^{4+}, 0.05, 0.1, 0.5, 1, 2 μmol/L), As group (As^{3+}, 0.05, 0.1, 0.5, 1.5, 3, 6, 12 μmol/L), arsenic + selenium group (As^{3+}, 0.05, 0.1, 0.5, 1.5, 3, 6, 12 μmol/L; Se^{4+}, 0.1 μmol/L). Culture medium and cells were collected after 48h. The viability and capacity of proliferation of HUVECs were assessed by MTT colorimetric assay. Morphology change was observed through Wright Giemsa staining. The activities of SOD, GLU-PX, iNOS, eNOS and the content of MDA in cell culture solution were determined with the colorimetric method. The expressions of eNOS mRNA and iNOS mRNA were tested by Reverse Transcription-Polymerase Chain Reaction (RT-PCR).

3 RESULTS

3.1 *Determination of the intervention dosage of Se*

The dosage of Se was determined through observing the activity of cells cultured with varied concentrations of Se. Selenium can promote cells' growth without adverse effects at 0.1 μmol/L.

3.2 *The effect of Se on As_2O_3 induced cellular activity decrease*

Arsenic dose dependently inhibited cell proliferation. Viable cell numbers were further reduced at higher exposure of As. Compared with the As treatments at 0.1 and 0.5 μmol/L, Se protected cellular activity significantly ($P < 0.05$) in the As + Se treatment.

3.3 *The effect of Se on As_2O_3 induced antioxidant function decrease*

The activity of GSH-PX in As + Se group were greater ($P < 0.05$) compared to it was in As group at As concentrations of 0.05, 0.1 μmol/L. Compared with As group, the activity of SOD in As Se group were higher ($P < 0.05$) at the As concentrations of 0.05, 0.5μmol/L. The content of MDA were lower ($P < 0.05$) compared to it was in As group when the concentrations of As were at 0.05 and 0.1 μmol/L.

3.4 *The effect of Se on As_2O_3 caused eNOS and iNOS activitiy changes*

The activity of eNOS in As + Se group was strengthened significantly ($P < 0.05$) in contrast with As group when the As concentrations were 0.05, 0.1, 0.5 μmol/L, respectively. The iNOS activity of As + Se group was significantly ($P < 0.05$) reduced compared with As group when the As concentrations were 0.05 and 0.1 μmol/L. The outcome by RT-PCR revealed that at the As concentrations of 0.05, 0.5, 1.5 μmol/L, respectively, the content of eNOS mRNA increased significantly ($P < 0.05$), while the level of iNOS mRNA decreased significantly ($P < 0.05$) in As + Se group compared with As group at concentrations of 0.05, 0.1, and 1.5 μmol/L.

4 DISCUSSION

Superoxide dismutase can scavenge superoxide anion radicals and protect cells. As one of the most important antioxidant enzymes, GSH-PX plays an essential role in the protection of cell membrane against oxidative damage. MDA can reflect the level of lipid peroxidation in organism. Our research showed that activities of SOD and GSH-PX increased and the content of MDA decreased when Se was added to As. This observation suggests that Se protected the cell from As damage by antagonizing oxidation.

Endothelial cells express two types of NOS, iNOS and eNOS. NO derived from oxidation of L-arginine catalyzed by NOS plays dual roles of protection and cytotoxicity. Recent research by Deng et al. found excessive NO can induce cell apoptosis through lipid peroxidation and reduced membrane fluidity. Our research found Se can increase eNOS mRNA level, and decrease iNOS mRNA thus to normalize their transcription which helps to explain the role of Se in antagonizing arsenic's damage.

Another recently published study (Shweta & Judit, 2012.) had confirmed selenium's protective role in reducing arsenic's toxicity in rats. In our study, arsenic reduced cellular proliferation rate, cell's antioxidant capability, and increased peroxide content. Arsenic also increased iNOS transcription and reduced eNOS transcription. These could be mechanical pathways to hypertension and cardiovascular disease. Selenium has strong antioxidative activity and significant scavenging activity to free radicals. In our study, Se antagonized the damage caused by As through different approaches. In this respect, we inferred that Se plays an important protective role in cardiovascular disease.

Selenium protected at a lower As level in our study, which diminished at higher As concentrations. This could be due to the intervention dose of Se for varied concentrations of As in our study was only a pre-fixed, suitable and lower concentration (0.1 μmol/L). Whether a higher Se level can still help to antagonize the arsenic's damage needs to be further researched.

REFERENCES

Deng, Y., Huang, K.X. & Xu, H.B. 2002. The inhibition of selenium on the damage of endothelial cell induced by NO. *Huazhong Univ. of Sci. & Tech. (Nature Science Edition).* 30(5): 101–104.

Engel, R.R., Hopenhayn-Rich, C., Receveur, O. & Smith, A.H. 1994. Vascular effects of chronic arsenic exposure: a review. *Epidemiol. Rev.* 16: 184–209.

Fagan, K.A., Morrissey, B., Fouty, B.W., Sato, K., Harrall, J.W., Morris, K.G. *et al.* 2001. Upregulation of nitric oxide synthase in mice with severe hypoxia-induced pulmonary hypertension. *Respir. Res.* 2: 306–313.

Shweta, S. & Judit, S. 2012. Dietary selenium fortification: a potential solution to chronic arsenic toxicity. *Toxicol Environ Chem.* 94(7): 1453–1465.

Uğuz, A.C., Nazıroğlu, M., Espino, J., Bejarano, I., González, D., Rodríguez, A.B. *et al.* 2009. Selenium modulates oxidative stress-induced cell apoptosis in human myeloid HL-60 cells through regulation of calcium release and caspase-3 and -9 activities. *J. Membr. Biol.* 232(1–3): 15–23.

Yang, W.H. & Xia, Y.J. 2011. Arsenic exposure on cardiovascular system. *Environ. Health* 28(8): 746–747.

Selenium in the Environment and Human Health – Bañuelos, Lin & Yin (Eds)
© 2014 Taylor & Francis Group, London, ISBN 978-1-138-00017-9

Removal of selenite from wastewater using microbial fuel cells inoculated with *Shewanella oneidensis* MR-1

Yu Zhang, Zhilin Wu, Qian Wang, Limin Yang & Miao Li
Key Laboratory of Agri-Food Safety of Anhui Province, School of Resources and Environment and Plant Protection, Anhui Agriculture University, Hefei, Anhui, China

Z.-Q. Lin
Department of Biological Sciences and Environmental Sciences Program, Southern Illinois University Edwardsville, Edwardsville, Illinois, USA

Gary Bañuelos
USDA-ARS, Parlier, California, USA

1 INTRODUCTION

Selenium (Se) is important trace element with key physiological roles in oxidative stress and immune function. However, the release of large amounts of Se into the environment from industrial and agricultural activities can lead to a bioaccumulation of Se into the food chain (Bañuelos and Lin, 2005). High concentrations of Se are known to be toxic to many forms of living organisms. Selenite and selenate are the two major species of inorganic Se. Selenite is taken-up faster by aquatic plants and is more toxic than selenate to aquatic organisms. Hence, remediation of Se is important and similar to other approaches that have been proposed to deal with Se contamination in soil and water (Bañuelos and Lin, 2005).

Microbial Fuel Cells (MFCs) can utilize microbes to oxidize organic and inorganic substrates and generate electricity (Logan and Rabaey, 2012). There has been much research into the applications of MFCs, including recent investigations that suggest that the cathode of MFC can be used for bioremediation and the extraction of valuable metals such as Ag (Choi and Cui, 2012). *Shewanella oneidensis* MR-1 is a facultative anaerobic bacterium known for its versatility in using a variety of soluble and insoluble electron acceptors, which makes it an interesting candidate for use in the bioremediation of toxic heavy metals (Lee *et al.*, 2007).

In the present study, both simultaneous Se removal and electricity generation were evaluated in Microbial Fuel Cells (MFCs) utilizing *Shewanella oneidensis* MR-1. The capability of those MFCs, the effect of selenite concentration on electricity generation, and Se of Se-containing wastewaters will be investigated and discussed.

2 MATERIALS AND METHODS

2.1 *Microbial fuel cell assembly*

MFCs utilizing *S. oneidensis* MR-1 were constructed as described previously (Catal *et al.*, 2009; Lee *et al.*, 2007). Non-wet proofed and wet-proofed carbon cloths were used as anode and cathode, respectively. The air-facing side of the cathode was prepared according to a published procedure (Catal *et al.*, 2009; Lee *et al.*, 2007). The liquid-facing side of the cathode was coated with platinum using Nafion as a binder. The electrodes were placed 1.7 cm away from each other.

2.2 *Operation of microbial fuel cell*

The medium solution of MFCs utilizing *S. oneidensis* MR-1 was prepared as reported previously (Catal *et al.*, 2009; Lee *et al.*, 2007). Glucose (1.2 g/L) was used as carbon source. Se-containing solutions were prepared by dissolving sodium selenite in the medium solution containing glucose. The MFCs were inoculated with *S. oneidensis* MR-1 and operated with the Se-free medium solution. When a stable voltage output was obtained, medium solution was refreshed with Se-containing medium solution. Selenite concentration was gradually increased from 0 to 500 mg/L Se for the glucose-fed MFCs utilizing *S. Oneidensis* MR-1 to investigate the effect of Se concentration on MFC performance. The medium solution was collected at the end of each batch operation and residual selenite was analyzed after filtering the solution through a 0.2-μm cellulose acetate Millipore filter.

2.3 *Chemical analyses*

Selenite concentration was determined according to a procedure reported previously (Catal *et al.*, 2009), including the calculation of power density and Coulombic Efficiency (CE).

3 RESULTS AND DISCUSSION

3.1 *Effect of selenite concentration on electricity generation*

In the MFCs utilizing *S. oneidensis* MR-1 fed with glucose, CE increased from 22% to 33% with the increase of selenite from 0 to 100 mg/L (Table 1). However, further increasing the Se concentration resulted in a decrease in CE. Though selenite has a potential to accept electrons from the substrate and thus reduce the CE, the slightly increased CE at low/moderate selenite concentrations indicates that the loss of electrons to selenite did not significantly affect the CE. It is possible that some of the nonelectrogens in the culture were inhibited, resulting in an increase in CE. It is also possible that the deposition of elemental Se on the cathode surface blocked O2 diffusion from air, thus increasing the CE. At higher selenite concentrations, anodic bacterial activities were inhibited, resulting in lower CE.

3.2 *Selenite removal*

In MFCs utilizing *S. oneidensis* MR-1 fed with glucose, over 90% of the selenite was removed from the solution with initial selenite concentration up to 200 mg/L (Table 2). Such a removal is comparable to other efficient biological Se removal processes (Table 3).

Table 1. Effect of Se concentration on Coulombic Efficiency (CE) of MFCs utilizing *S. oneidensis* MR-1 fed with glucose.

Se concentration (mg/L)	Coulombic efficiency (%)
0	22.4 ± 1.6
100	33.6 ± 2.1
200	28.9 ± 1.8
300	11.5 ± 0.8
400	0.8 ± 0.2
500	0 ± 0

Table 2. Selenite removal from wastewater using MFCs utilizing *S. oneidensis* MR-1 fed with glucose.

Se concentration (mg/L)	Se removal (%)
0	100 ± 0
100	92.6 ± 2.3
200	90.1 ± 2.1
300	86.5 ± 1.7
400	80.4 ± 1.4
500	71.8 ± 1.1

Table 3. Microbes of inoculum and removal efficiency for selenite/selenate.

Inoculum	Se (mg/L)	Removal (%)	Time (h)	References
S. oneidensis	100	92	100	Current study
Pseudomonas sp.	50	100	100	Morita *et al.* (2007)
Mixed culture	200	99	72	Catal *et al.* (2009)

4 CONCLUSIONS

In conclusion, the results indicated that power output was not affected by selenite up to <200 mg/L with glucose as substrate. Coulombic efficiencies of MFCs with glucose increased from 22% to 33% at <200 mg Se/L. About 92% of 100 mg/L as selenite was removed in 100 h for MFCs fed with glucose. The results suggest that the potential of using MFC technology for Se bioremediation. These results suggest that *Shewanella oneidensis* MR-1 enriched for electricity generation in the MFC system could use selenite as electron acceptor. By using these electrogens, the simultaneous generation of electricity and Se removal can be achieved, indicating the potential of using MFCs utilizing *Shewanella oneidensis* MR-1 for Se wastewater treatment.

REFERENCES

Bañuelos, G.S. & Lin, Z.Q. 2005. Phytoremediation management of selenium-laden drainage sediments in the San Luis Drain: a greenhouse feasibility study. *Ecotoxicol. Environ. Safety* 62(3): 309–316.

Choi, C. & Cui, Y.F. 2012. Recovery silver from wastwater coupled with power generation using a microbial fuel cell. *Bioresour Technol.* 107: 522–525.

Catal, T., Bermek, H. & Liu, H. 2009. Removal of selenite from wastewater using microbial fuel cells. *Biotechnol Lett.* 31: 1211–1216.

Lee, J.H., Han, J., Choi, H. & Hur, H.G. 2007. Effects of temperature and dissolved oxygen on Se(IV) removal and Se(0) precipitation by *Shewanella* sp. HN-41. *Chemosphere* 68(10): 1898–1905.

Logan, B.E. & Rabaey, K. 2012. Conversion of wastes into bioelectricity and chemicals using microbial electrochemical technologies. *Science* 337: 686–690.

Morita, M., Uemoto, H. & Watanabe, A. 2007. Reduction of selenium oxyanions in wastewater using two bacterial strains. *Eng. Life Sci.* 7(3): 235–240.

Selenium in the Environment and Human Health – Bañuelos, Lin & Yin (Eds)
ISBN 978-1-138-00017-9

Simultaneous removal of selenite and electricity production from Se-laden wastewater by constructed wetland coupled with microbial fuel cells

Limin Yang, Zhilin Wu, Jie Wu, Yu Zhang & Miao Li
Key Laboratory of Agri-Food Safety of Anhui Province, School of Resources and Environment and Plant Protection, Anhui Agriculture University, Hefei, Anhui, China

Z.-Q. Lin
Department of Environmental Sciences, Southern Illinois University, Edwardsville, Illinois, USA

Gary Bañuelos
ARS-USDA, Parlier, California, USA

1 INTRODUCTION

Constructed wetlands (CWs) are effective and reliable wastewater treatment systems, which are relatively inexpensive to construct and operate, easy to maintain, and one of the most productive ecosystems in terms of biomass (Lin *et al.*, 2010). CWs could be used in the remediation of waters contaminated with heavy metals and other toxic trace elements. They are an attractive biological technology for the treatment of wastewater from industry and from agriculture (Lin *et al.*, 2010).

Microbial Fuel Cells (MFCs) have been intensively studied during the last decade as a promising technology for both wastewater purification (Catal *et al.*, 2009) and for simultaneous electricity generation. MFC technology not only demonstrated a great potential in power production from common organic matter in wastewater, but also has great potential for remediating toxic organics and heavy metals (Catal *et al.*, 2009).

Selenium (Se) is an essential element for animals and humans. However, high concentrations of Se can cause both acute and chronic toxicity in living organisms. Therefore, it is necessary to develop effective and cost efficient processes for Se removal in a Se-laden environment. Some approaches have been proposed to manage Se contamination in soil and water (Pilon-Smits *et al.*, 2010). In this regard, recent investigations have shown that utilizing CWs or MFC for the removal of Se compounds from the soil and water can be an efficient alternative option (Lin *et al.*, 2010; Catal *et al.*, 2009).

In this study, the performance of Constructed Wetland-Microbial Fuel Cell (CW-MFC) was investigated for removal of Se in conjunction with production of electricity from wastewater, and the effect of Se on CW-MFC performance was also discussed.

2 MATERIALS AND METHODS

2.1 *Preparation of Se containing agricultural wastewater*

An agricultural wastewater from Agricultural Park of Anhui Agriculture University containing glucose as a carbon source was used in the study. Sodium selenite (purchased from Sigma Ltd.) was used to enrich the agricultural wastewater with the following Se concentrations: 100, 200, 300, 400 and 500 mg Se/L, which refer to Catal *et al.* (2009).

2.2 *Configuration of CW-MFC*

A constructed wetland-microbial fuel cell (CW-MFC) was constructed as previously described (Yadav *et al.*, 2012). In the CW-MFC assembly, a vertical flow constructed wetland microcosm was constructed in the laboratory with a PVC pipe with an internal diameter of 10 cm and length of 60 cm. It was filled with gravel (2–4 mm) and planted with cattail (*Typha latifolia* L.). *T. latifolia* is an aquatic plant species and is used in many constructed wetlands studies (Lin et al., 2010). The upper end of microcosm was left open, while the lower end was sealed. In this microcosm, MFCs utilizing *Shewanella oneidensis* MR-1 was made inside the constructed wetland microcosm as previously reported (Lee *et al.*, 2007; Yadav *et al.*, 2012), A cathode electrode was placed near to the rhizospheric zone and the anode was placed near the bottom of microcosm. Insulated copper wires were used to connect both the electrodes (Fig. 1).

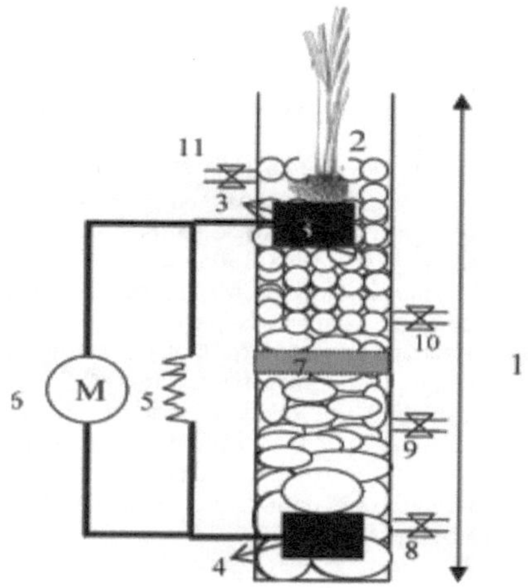

Figure 1. Schematic diagram of constructed wetland coupled with microbial fuel cell microcosm. (1) gravel filled constructed wetland-microbial fuel cell microcosm planted with cattail (*Typha latifolia* L.); (2) rhizospheric zone; (3) cathode, (4) anode; (5) resistance; (6) multimeter; (7) layer of glass wool; (8, 9, 10, 11) sampling points.

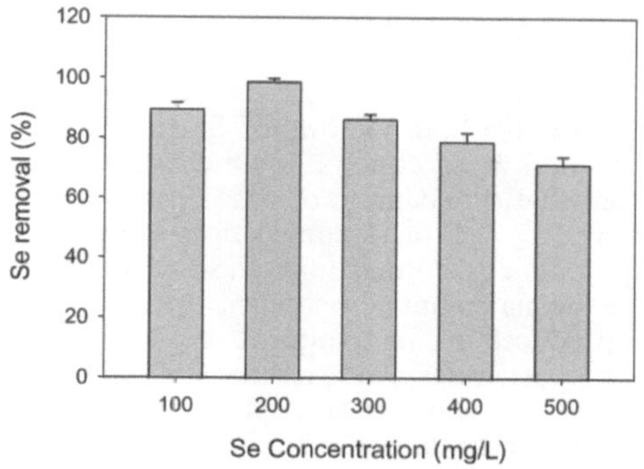

Figure 2. Selenium removal in a CW-MFC microcosm for a 72-hour time period.

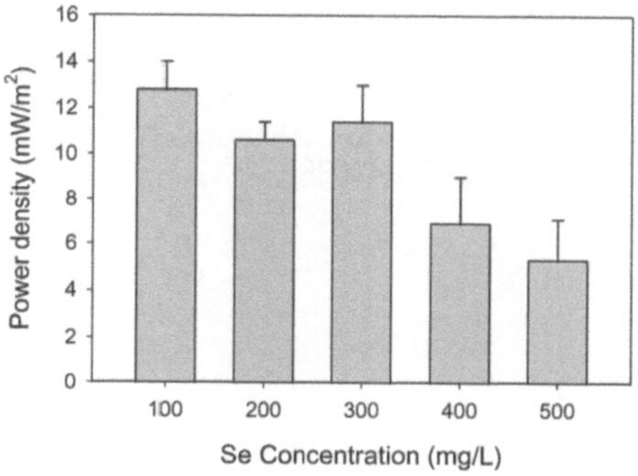

Figure 3. Power density in a CW-MFC microcosm for a 72-hour time period.

2.3 *Operation of CW-MFC*

The CW-MFC was inoculated with *Shewanella oneidensis* MR-1 and fed with agricultural wastewater containing glucose (10 mg/L) through a peristaltic pump. *S. oneidensis* MR-1 has considerable potential of generated electricity and removing pollutants (Lee *et al.*, 2007). In the beginning of the experiments, only agricultural wastewater with inoculum was fed to the microcosm and the systems are operated for more than two weeks to obtain a stable performance before evaluating the ability of the system to remove Se and produce electricity.

2.4 *Analysis of Se*

The analysis of Se concentration was carried out according to a procedure previously reported (Lin *et al.*, 2010). Sampling of water was performed both at the start and at the end of the experiments and analyzed for Se, as discussed by Lin *et al.* (2010). The power density was calculated according to the methods of Catal *et al.* (2009).

3 RESULTS AND DISCUSSION

3.1 *Selenium removal*

Selenium removal from wastewater at different Se concentration was found to be in the range of 71–98% within <72 h of the treatment time for all concentrations (Figure 2). We observed that 98% of Se was removed in 72 h at a Se concentration of 200 mg/L in the current system. The possible removal mechanism of Se in the system maybe due to the volatilization and adsorption of Se on the surface of gravels, accumulation and biotransformations by plants and microbes in the CW-MFC microcosm. This observation indicates that Se removal was achieved primarily by abiotic mechanisms.

3.2 *Generation of bioelectricity*

The power density produced at the different Se concentrations is presented in Figure 3. We observed that the maximum power density was found to be 12.8 mW/m^2 after 48 h of contact time during the treatment of wastewater containing 100 mg/L Se. In addition, the power density values decreased with an increase in the Se concentration in wastewater. This decrease in power density maybe due to increase in the level of Se toxicity on microbes and plants.

4 CONCLUSIONS

In this study, we evaluated the performance of CW-MFC for both removal of Se and electricity production from wastewater. Our results demonstrated that a maximum of 98% Se removal was achieved at 72 h of treatment from the wastewater containing 200 mg/L initial Se concentration, while the maximum power density of 12.8 mW/m^2 was produced during treatment of 100 mg/L containing wastewater. These results indicated that the CW-MFC system can be used as a cost-effective and environmentally-friendly Se wastewater treatment with simultaneous generation of electricity.

ACKNOWLEDGEMENTS

The authors wish to thank Dr. Xuebin Yin for his constructive suggestions. This work was partially supported by the National Natural Science Foundation of China (50949038).

REFERENCES

Catal, T., Bermek, H. & Liu, H. 2009. Removal of selenite from wastewater using microbial fuel cells. Biotechnol. Lett. 31: 1211–1216.

Lee, J.H., Han, J., Choi, H. & Hur, H.G. 2007. Effects of temperature and dissolved oxygen on Se(IV) removal and Se(0) precipitation by *Shewanella* sp. HN-41. *Chemosphere* 68(10): 1898–1905.

Lin, Z.-Q., Terry, N., Gao, S., Mohamed, S. & Ye, Z.H. 2010. Vegetation changes and partitioning of selenium in 4-year old constructed wetlands treating agricultural drainage. *Int. J. Phytoremediation* 12: 255–267.

Pilon-Smits, E.A.H., Valdez Barillas, J.R., Van Hoewyk, D. & Lin, Z.-Q. 2010. Phytoremediation of selenium. In: G. Plaza (ed.), Trends in Bioremediation and Phytoremediation, 355–372, Kerala, India: Research Signpost.

Yadav, A.K., Dash, P., Mohanty, A., Abbassi, R. & Mishra, B.K. 2012. Performance assessment of innovative constructed wetland-microbial fuel cell for electricity production and dye removal. *Eco. Eng.* 47: 126–131.

Selenium in the Environment and Human Health – Bañuelos, Lin & Yin (Eds)
© 2014 Taylor & Francis Group, London, ISBN 978-1-138-00017-9

Biological treatment systems for removing selenium from water

Norman Terry
University of California, Berkeley, California, USA

1 INTRODUCTION

Agricultural and industrial wastewaters contaminated with selenium (Se) represent a serious eco-toxic risk to humans and wildlife. The US EPA requires that levels of Se in surface waters should not exceed 5 μg/L. Of all the technologies available for removing Se from water, biological treatment systems, particularly those involving constructed wetlands, have the lowest construction and operation costs. Constructed wetlands have been shown to remove approximately 90% of the inflow Se from oil refinery wastewater (Hansen *et al.*, 1998) and from 65 to 85% of the Se in agricultural irrigation drainage water (Lin & Terry, 2003; Gao *et al.*, 2003). Because these wetlands were unable to lower Se concentrations in the outflow to below 5 μg Se/L, we began a series of mesocosm studies at UC Berkeley to improve their design. Using cattail-planted mesocosms we succeeded in lowering Se concentrations in the water column from 15 μg Se/L to 0.1 μg Se/L within 72 hours (Huang *et al.*, 2012).

Selenium removal by constructed wetlands occurs within sediments, principally through the microbially-mediated dissimilatory anaerobic reduction of oxyanions to water insoluble elemental Se (Se^0) (Masscheleyn & Patrick, 1993). However, there is always the risk that the buildup of Se in treatment wetlands may lead to Se eco-toxicity (Saiki & Lowe, 1987; Ohlendorf & Heinz, 2011). One way of mitigating this risk is to enhance Se removal by biological volatilization to the atmosphere (Lin *et al.*, 2002). Based on the fact that algae have a considerable propensity to take up and volatilize Se (Neumann *et al.*, 2003), we investigated the idea of incorporating an algal treatment unit within the water treatment system.

Our current research project was prompted by the need to develop a clean water supply for the Species Conservation Habitat (SCH) to be built by the California Department of Water Resources at the Salton Sea in southern California. The specific objectives were to screen algal species/strains to determine those best suited for the removal and volatilization of Se, and to determine the chemical forms of Se that are likely to accumulate within a combined algal-cattail wetland water treatment system.

2 MATERIALS AND METHODS

Six freshwater algae were tested including 2 different strains of *Chlorella vulgaris* (designated Cv), 2 isolates of *Chlorella* sp. collected from agricultural drains, *C. ellipsoidea*, and *Scenedesmus obliquus*. Using microcosms, measurements were made of Se uptake and volatilization (Lin & Terry, 2003). Total Se was determined using inductively-coupled plasma-dynamic reaction cell-mass spectrometry. Chemical forms of Se were determined by X-ray Absorption Near-Edge Structure (XANES) and Extended X-ray Absorption Fine Structure (EXAFS) at SSRL.

3 RESULTS AND DISCUSSION

3.1 *Algal screening*

Comparison of the six different algae with respect to their ability to remove Se from the water column revealed that an algal strain of *Chlorella vulgaris* designated Cv was the most effective. Cv removed 96% of the added Se (1580 μg Se/microcosm supplied as selenate) in 72 hours. This removal was accomplished by volatilization (61%) and by accumulation of Se in the algal biomass (27%).

3.2 *Selenium speciation*

X-ray Absorption Spectroscopy (XAS) speciation of Se in the algal biomass of Cv revealed that 42% was present in the form of selenate, 34% as selenomethionine (SeMet), a precursor of volatile Se, and 24% as elemental Se (Se^0). Algae, along with higher plants and bacteria, almost certainly metabolize selenate-Se through the sulfur-assimilation pathway generating selenocysteine and SeMet, a precursor of volatile Se. The presence of Se^0 in algae was confirmed using EXAFS.

3.3 *Algal-cattail treatment systems*

A conceptual design for an algal-cattail wetland treatment system is presented in Figure 1 (Huang *et al.*, submitted). Selenium-contaminated water is first pumped into a sedimentation pond, then into an algal treatment unit followed by a cattail wetland.

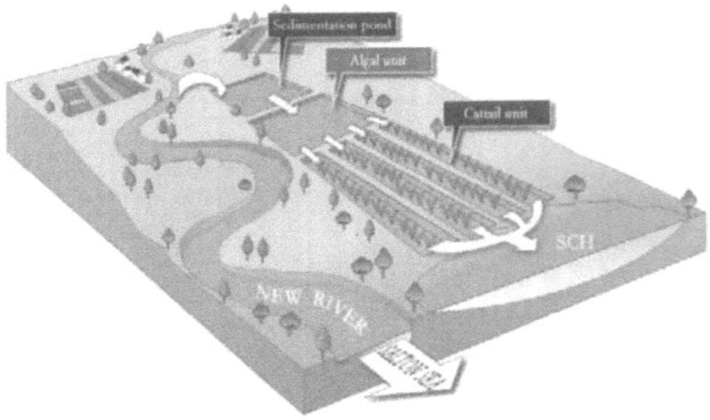

Figure 1. Conceptual design of a constructed wetland water treatment system for the removal of Se from river water flowing into the Salton Sea in California.

Our research suggests that the inclusion of an algal pretreatment unit containing appropriate algal strains should greatly enhance the efficiency of Se removal. Because much of the inflow Se will be volatilized by algae, cattails, and by microbes during decomposition of algae flowing into the wetland treatment unit, the risk of the buildup of potentially eco-toxic forms of Se within the wetland ecosystem will be attenuated.

4 CONCLUSIONS

The inclusion of an algal pretreatment unit in the conceptual design for an algal-cattail wetland water treatment system should (1) greatly enhance the efficiency of Se removal from Se-contaminated wastewater, and (2) minimize potential eco-toxic risks of Se buildup. The biological volatilization of Se should greatly reduce the risk of the buildup of potentially eco-toxic forms of Se within the wetland ecosystem.

REFERENCES

Gao, S., Tanji, K.K., Lin, Z.Q., Terry, N. & Peters, D.W. 2003. Selenium removal and mass balance in a constructed flow-through wetland system. *J. Environ. Quality* 32 (4): 1557–1570.

Hansen, D., Duda, P.J., Zayed, A. & Terry, N. 1998. Selenium removal by constructed wetlands: role of biological volatilization. *Environ. Sci. Technol.* 32 (3): 591–597.

Huang, J.C., Passeport, E. & Terry, N. 2012. Development of a constructed wetland water treatment system for selenium removal: use of mesocosms to evaluate design parameters. *Environ. Sci. Technol.* 46 (21): 12021–12029.

Huang, J.C., Suárez, M.C., Yang, S.I., Lin, Z.Q. & Terry, N. 2013. Development of a constructed wetland water treatment system for selenium removal: incorporation of an algal treatment component. *Environ. Sci. Technol.* (submitted).

Lin, Z.-Q., Cervinka, V., Pickering, I.J., Zayed, A. & Terry, N. 2002. Managing selenium-contaminated agricultural drainage water by the integrated on-farm drainage management system: role of selenium volatilization. *Water Research* 36 (12): 3150–3160.

Lin, Z.-Q. & Terry, N. 2003. Selenium removal by constructed wetlands: quantitative importance of biological volatilization in the treatment of selenium-laden agricultural drainage water. *Environ. Sci. Technol.* 37 (3): 606–615.

Masscheleyn, P.H. & Patrick, W.H., Jr. 1993. Biogeochemical processes affecting selenium cycling in wetlands. *Environ. Toxicol. Chem.* 12 (12): 2235–2243.

Neumann, P.M., De Souza, M.P., Pickering, I.J. & Terry, N. 2003. Rapid microalgal metabolism of selenate to volatile dimethylselenide. *Plant Cell Environ.* 26 (6): 897–905.

Ohlendorf, H.M. & Heinz, G.H. 2011. Selenium in birds. In W.N. Beyer, & J.P. Meador (eds), *Environmental Contaminants in Biota Interpreting Tissue Concentrations:* 669–701. Boca Raton, FL: CRC Press.

Saiki, M.K. & Lowe, T.E. 1987. Selenium in aquatic organisms from subsurface agricultural drainage water, San Joaquin Valley, California. *Arch. Environ. Contam. Toxicol.* 16 (6): 657–670.

Selenium in the Environment and Human Health – Bañuelos, Lin & Yin (Eds)
© 2014 Taylor & Francis Group, London, ISBN 978-1-138-00017-9

Engineering plant selenium accumulation: Potential uses and ecological impacts

E.A.H. Pilon-Smits
Colorado State University, Biology Department, Fort Collins, CO, USA

1 INTRODUCTION

Vegetation on most soils contains less than 1 mg selenium (Se) per kg dry weight. On seleniferous soils, most plant species contain <10 mg Se/kg, but so-called Se hyperaccumulator plants (*e.g.*, from the genera *Stanleya* and *Astragalus*) can accumulate 1,000–15,000 mg/kg (or 0.1–1.5% Se). Selenium is chemically similar to sulfur and metabolized in plants by the same mechanisms. Plants take up selenate via sulfate transporters and assimilate it into selenocysteine (SeCys) and selenomethionine (SeMet). Selenium hyperaccumulators appear to have several unique mechanisms: selenate-specific transporters, SeCys methyl transferase activity, and Se sequestration in specific organs and tissues.

While Se has not been shown to be an essential element for higher plants, there have been reports of beneficial effects of Se on plant growth, particularly in Se hyperaccumulator plants. Thus, Se appears to offer plants physiological benefit, perhaps via enhanced antioxidant activity. There is also mounting evidence that Se can protect plants from biotic stresses such as herbivory and microbial infections. Thus, fertilization with Se may promote plant growth and increase resistance to pests and other stresses. Since Se is essential at low levels for humans and other mammals, fertilization of crops with Se may have the additional benefit of enhancing its nutritional value.

The window between deficiency and toxicity is very narrow for Se. Hence, biofortification strategies with Se should always include careful monitoring of the Se levels in the final food products to avoid toxicity to the consumer. Different forms of Se in plants have different nutritional value and different toxicity levels for the consumer, and therefore it is also important to monitor the forms of Se present in crops. Finally, since Se in plants can affect ecological relationships, it is important to assess potential ecological implications of growing Se containing plants in large quantities, either for phytoremediation or biofortification.

In order to improve the efficiency and safety of Se phytoremediation or biofortification, better knowledge is needed on the biochemical mechanisms underlying plant Se uptake, metabolism and tolerance, as well as of the ecological effects of plant Se accumulation. Here, an overview is provided on our research in these areas of interest.

2 MATERIALS AND METHODS

These studies were carried out using Se hyperaccumulator *Stanleya pinnata* and related non-hyperaccumulator *Brassica juncea* (Brassicaceae). *B. juncea* (Indian mustard) is a popular species for phytoremediation and biofortification. A variety of experimental methods were used for conducting these multidisciplinary studies. Elemental analysis included acid digestion followed by Inductively-Coupled Plasma Atomic Emission Spectrometry (ICP-AES), while chemical speciation was analyzed using Liquid Chromatography Mass Spectrometry (LC-MS) or X-ray microprobe analysis at the Advanced Light Source (ALS) in Berkeley, California (USA).

3 RESULTS AND DISCUSSION

3.1 *Genetic and genomic approaches to study and manipulate plant Se metabolism*

Through manipulation of key genes involved in S/Se assimilation and volatilization, it has been possible to enhance Se accumulation, tolerance, and volatilization by plants (for a review see Pilon-Smits & LeDuc, 2009). Overexpression in *B. juncea* of ATP sulfurylase (APS), an enzyme involved in selenate-to-selenite conversion, led to enhanced selenate tolerance, reduction, and Se accumulation. Overexpression in *B. juncea* of Cystathionine gamma Synthase (CgS), the first enzyme in the conversion of SeCys to SeMet, enhanced volatilization rates. The CgS transgenics accumulated 40% less Se, probably because of their higher volatilization rates. The CgS transgenics were also more Se tolerant than wildtype plants, presumably owing to their lower tissue Se levels. Overexpression of the enzyme Selenocysteine Lyase (SL), which breaks down SeCys into alanine and elemental Se in *B. juncea* or *Arabidopsis thaliana,* resulted in reduced Se incorporation into proteins and up to 2-fold enhanced Se accumulation compared to wildtype plants.

When grown on naturally seleniferous soil in a greenhouse pot experiment the APS transgenics accumulated Se up to 3-fold higher levels than wildtype *B. juncea*, and the CgS transgenics contained 40% lower Se levels than wild type; all in agreement with the laboratory results. Furthermore, in two field experiments carried out by Gary Bañuelos and coworkers on Se-contaminated sediment in the San Joaquin Valley (California, USA), the APS and SL transgenics accumulated Se up to 4-fold and 2-fold higher levels than wildtype *B. juncea*, respectively; in agreement with the earlier laboratory experiments. Thus, in the field as well as the lab the various transgenics showed enhanced Se accumulation, volatilization and/or tolerance. These are promising traits for use in phytoremediation or in producing Se-fortified foods.

Selenium hyperaccumulators appear to have several unique mechanisms that facilitate their extreme capacity to accumulate and tolerate Se. Several genomic approaches (macroarrays, RNA sequencing) were used to elucidate key genes responsible for hyperaccumulation in *Stanleya pinnata*. The S assimilation pathway appears to be generally upregulated in the hyperaccumulator, when compared to non-hyperaccumualtor relatives *S. albescens* (Freeman *et al.*, 2010) or *S. elata* (unpublished). This may be related to elevated levels of certain plant growth regulators.

3.2 *Ecological implications of plant Se accumulation*

High plant Se levels reduced herbivory by a wide variety of invertebrate and vertebrate herbivores, due to both deterrence and toxicity (for a review see El Mehdawi & Pilon-Smits, 2012). Plant Se also reduced fungal pathogen infection. There is also evidence that high-Se litter deposition can inhibit Se-sensitive neighboring plants. There is no evidence that the hyperaccumulation of Se has a negative effect on reproductive functions or pollinator visitation. Further studies are needed to test the effects of high-Se plants on pollinators, since honey bees were not deterred by high-Se plants, and Se levels in nectar and pollen are high. While they are toxic to Se-sensitive ecological partners, there is mounting evidence that high-Se plants offer a niche for Se-tolerant partners including herbivores, pollinators, microbes, and neighboring plants (El Mehdawi & Pilon-Smits, 2012). Hyperaccumulators may even facilitate Se-tolerant partners through Se enrichment: neighboring plants with elevated Se levels enjoy enhanced growth and reduced herbivory. Selenium hyperaccumulators harbor many bacterial and fungal endophytes, which are often highly Se-tolerant. Some of these endophytes have plant growth promoting properties, and may be interesting to further explore for their capacity to enhance Se phytoremediation.

4 CONCLUSIONS

Genetic engineering of the S/Se assimilation pathway has been used successfully to enhance plant Se tolerance, accumulation and volatilization. Identification of key genes in Se hyperaccumulators may lead to future genetic engineering approaches.

Plant Se accumulation may affect ecological partners either negatively or positively, depending on their Se tolerance. This includes herbivores, pollinators, microbial symbionts and neighboring plants. These findings are of significance since they give insight into the ecological reverberations of Se (hyper) accumulation. This information will help the management of seleniferous areas and the agricultural production of Se-rich crops for phytoremediation or biofortification.

REFERENCES

El Mehdawi, A.F. & Pilon-Smits, E.A.H. 2012. Ecological Aspects of Plant Selenium Hyperaccumulation. *Plant Biol.* 14: 1–10.

Freeman, J.L., Tamaoki, M., Stushnoff, C., Quinn, C.F., Cappa, J.J., Devonshire, J. *et al.* 2010. Molecular mechanisms of selenium tolerance and hyperaccumulation in *Stanleya pinnata*. *Plant Physiol.* 153: 1630–1652.

Pilon-Smits, E.A.H. & LeDuc, D. 2009. Phytoremediation of selenium using transgenic plants. *Curr. Opin. Biotechnol.* 20: 207–212.

Selenium in the Environment and Human Health – Bañuelos, Lin & Yin (Eds)
© 2014 Taylor & Francis Group, London, ISBN 978-1-138-00017-9

The applications of selenium stable isotopes in environmental geochemistry

J.-M. Zhu
State Key Lab of Environmental Geochemistry, Institute of Geochemistry, Chinese Academy of Sciences, Guiyang, Guizhou, China
Department of Geology, University of Illinois at Urbana-Champaign, Urbana, Illinois, USA

T.M. Johnson
Department of Geology, University of Illinois at Urbana-Champaign, Urbana, Illinois, USA

1 INSTRUCTION

Selenium (Se) is an essential micronutrient for human beings and animals but can be toxic at high concentrations. The uneven distribution of Se on the earth surface formed both Se-excessive and Se-deficient geo-ecosystems, where human and animal health problems were well documented. Selenium has six stable isotopes in natural environment, ^{74}Se, ^{76}Se, ^{77}Se, ^{78}Se, ^{80}Se, and ^{82}Se, with abundance 0.89%, 9.37%, 7.64%, 23.77%, 49.61%, and 8.73%, respectively. Selenium occurs in four major oxidation states in the environment: -II, 0, +IV, and +VI. The different Se oxidation states have different geochemical behaviors that constrain the mobility, bioavailability, and toxicity of Se in surface environment. Geochemical reactions affecting Se speciation, whether abiotic or biotic, can alter Se isotopic compositions in reactants and products. To date, selenium isotope systematics have been preliminarily summarized by Johnson (2004), and Se isotope ratios are being used as the tracers of pollution sources, biogeochemical reactions, and geological processes.

2 SELENIUM ISTOPES MEASUREMENT AND NOTATION

Recently developed analytical methods have exploited on Hydride Generation (HG)-multicollector inductively coupled plasma- mass spectrometry (MC-ICP-MS) to attain accurate and precise measurements of Se isotope ratios using as little as 20 ng Se (Rouxel *et al.*, 2002; Zhu *et al.*, 2008a). Samples were introduced into the plasma as SeH_2 using a continuous flow HG system. External reproducibilities of ±0.1‰ (2σ) for reference materials NIST SRM 3149 and MH 495 and less than ±0.2‰ (±0.34‰ (2σ)) for natural samples have been obtained using ^{74}Se^{+77}Se (^{74}Se^{+82}Se) double-spike technique. These analytical techniques have greatly improved the analytical efficiency of measuring Se isotopes, and are promoting Se isotopic applications in environmental sciences and other fields.

At present, Se isotope ratios are generally reported relative to the NIST SRM 3149 reference solution. Selenium isotopic compositions are expressed in per mil (‰) and in the standard delta (δ) notation:

$$\delta^{82/76}\mathrm{Se} = \left[\frac{\left(\frac{^{82}\mathrm{Se}}{^{76}\mathrm{Se}}\right)_{\mathrm{sample}}}{\left(\frac{^{82}\mathrm{Se}}{^{76}\mathrm{Se}}\right)_{\mathrm{std}}} - 1 \right] \times 1000 \qquad (1)$$

3 APPLICATIONS IN ENVIRONMENTAL GEOCHEMISTRY

3.1 *The range of Se isotopes in natural samples*

$\delta^{82/76}$Se values of meteorites and basalts are near zero relative to the NIST SRM 3149, with mean values of 0.01 ± 0.32‰ and 0.38 ± 0.71‰, respectively. Altered basalts have a negative average composition (−1.24‰ for $\delta^{82/76}$Se). Mn nodule has a 0.41‰ of $\delta^{82/76}$Se value. Sulfides from modern hydrothermal fields yield negative average values of −1.51 ± 1.41‰. Sediments from deep sea and estuarine settings have a range of −1.09 to 3.39‰, and unweathered sedimentary rocks from different geologic periods fall within the range of −1.64 to 2.28‰. $\delta^{82/76}$Se values of Chinese coal ranged from −4.01 to 4.75‰. Higher plants have a range of −3.56 to 4.20‰, and $\delta^{82/76}$Se values of surface water from Colorado range from 0.6 to 4.51‰. Striking variation in $\delta^{82/76}$Se was found in an outcrop of weathered organic-rich shales at Yutangba (YTB), China, with spanning from −14.20 to 13.37‰. This variation suggested that redox cycling near the earth's surface can lead to significant isotopic fractionation during the weathering processes.

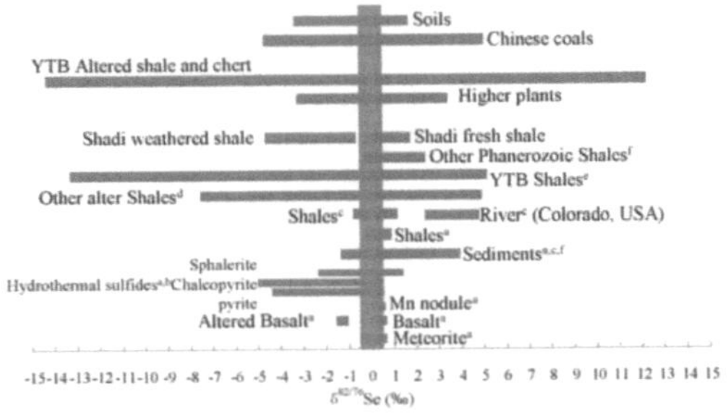

Figure 1. Summary of Se isotopic composition in natural samples. a: Rouxel *et al.* (2002); b: Rouxel *et al.* (2004); c: Clark and Johnson (2010); d: Johnson (2004); e: Wen and Carignan (2011); f: Mitchell *et al.* (2012).

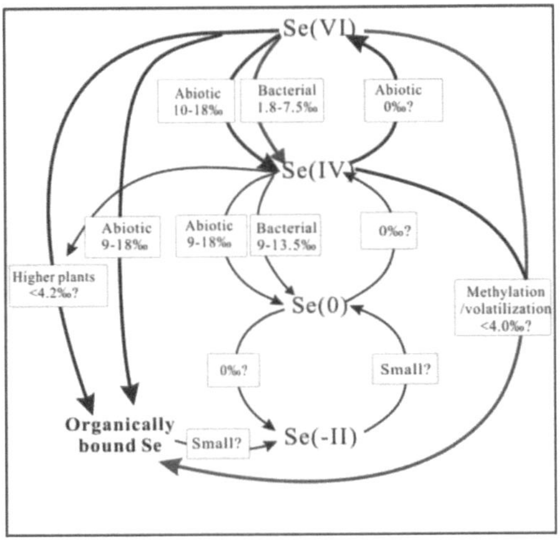

Figure 2. Summary of Se isotope systematics, modified from Johnson (2004).

3.2 *Fractionation processes*

Generally, the fractionation of Se isotopes in natural system is mass-dependent. As shown in Figure 2, reduction of Se oxyanions (VI and IV) by abiotic and biotic are dominant fractionation processes. Other processes included adsorption, precipitation, methylation/volatilization, and assimilation by biota produce a small or moderate fractionation. Isotopic fractionation of Se during oxidation of reduced phases is not well studied, but recent result also shows fractionation is small.

3.3 *Selenium isotopic applications*

Johnson *et al.* (2000) firstly explored the use of Se isotopic ratios to detect the transport pathways and related geochemical processes of Se-rich refinery wastewater into San Francisco Bay, California, although the isotopic differences between refinery and riverine inputs were not observed. Herbel et al. (2002) successfully use Se isotopes to identify biogeochemical processes of removing Se from an artificial wetland. An investigation of Se isotopic fractionation was carried out in sediments from Sweitzer Lake in Colorado, USA (Clark & Johnson, 2010), revealing a phenomenon that "reservoir effect" can decrease the isotopic shift produced by reactions in surface water-sediment systems, and make the associated effective fractionation to be small. Recent studies found that Se redox reactions occurring in subsurface systems, such as vadose zone, can cause large isotopic fractionation, particular in the weathering system of Se-rich rocks and mine waste. Additionally, selenium isotope ratios are becoming an important paleo-environmental proxy, revealing past redox changes in earth systems. Meanwhile, it is also worthwhile to explore trace uptake and translocation of Se in plants. In future work, Se isotope ratios is likely to be a powerful means to detect and possible quantify Se immobilization by reduction or other similar processes.

4 CONCLUSIONS

The fractionation systematics of Se isotopes in natural system has been preliminarily constructed. Selenium isotopes are becoming an indicator to trace pollutant source, geochemical processes and transport pathways of Se in environment, and its applications are exploring in other field such as agricultural, life sciences, and plant physiology.

REFERENCES

Clark, S.K. & Johnson, T.M. 2010. Selenium Stable Isotope Investigation into Selenium Biogeochemical Cycling in a Lacustrine Environment: Sweitzer Lake, Colorado. *J Environ Qual* 39: 2200–2210.

Herbel, M.J., Johnson, T.M., Tanji, K.K., Gao, S.D. & Bullen, T.D. 2002. Selenium stable isotope ratios in California agricultural drainage water management systems. *J Environ Qual* 31: 1146–1156.

Johnson, T.M., Bullen, T.D. & Zawislanski, P.T. 2000. Selenium stable isotope ratios as indicators of sources and cycling of selenium: Results from the northern reach of San Francisco Bay. *Environ. Sci. Technol.* 434: 2075–2079.

Johnson, T.M. 2004. A review of mass-dependent fractionation of selenium isotopes and implications for other heavy stable isotopes. *Chem. Geol.* 204: 201–214.

Mitchell, K., Mason, P.R.D., Van Cappellen, P., Johnson, T.M., Gill, B.C., Owens, J.D. *et al.* 2012. Selenium as paleo-oceanographic proxy: A first assessment. *Geochim. Cosmochim. Acta* 89: 302–317.

Rouxel, O., Ludden, J., Carignan, J., Marin, L. & Fouquet, Y. 2002. Natural variations of Se isotopic composition determined by hydride generation multiple collector inductively coupled plasma mass spectrometry. *Geochim. Cosmochim. Acta* 66: 3191–3199.

Rouxel, O., Fouquet, Y. & Ludden, J.N. 2004. Subsurface processes at the Lucky Strike hydrothermal field, Mid-Atlantic Ridge: Evidence from sulfur, selenium, and iron isotopes. *Geochim. Cosmochim. Acta* 68: 2295–2311.

Wen, H.J. & Carignan, J. 2011. Selenium isotopes trace the source and redox processes in the black shale-hosted Se-rich deposits in China. *Geochim. Cosmochim. Acta* 75: 1411–1427.

Zhu, J.M., Johnson, T.M., Clark, S.K. & Zhu, X.K. 2008. High precision measurement of selenium isotopic composition by hydride generation multiple collector inductively coupled plasma mass spectrometry with a Se-74-Se-77 double spike. *Chinese J. Anal. Chem.* 36: 1385–1390.

Author index